Matrices, Moments and Quadrature with Applications

Matrices, Moments and Quadrature with Applications

Gene H. Golub and Gérard Meurant

PRINCETON UNIVERSITY PRESS

PRINCETON AND OXFORD

Published by Princeton University Press, 41 William Street, Princeton, New Jersey 08540
In the United Kingdom: Princeton University Press, 6 Oxford Street, Woodstock, Oxfordshire OX20 1TW

ISBN: 978-0-691-14341-5

Library of Congress Control Number: 2009931397

British Library Cataloging-in-Publication Data is available

The publisher would like to acknowledge the author of this volume for providing the camera-ready copy from which this book was printed
This book has been composed in Times

Printed on acid-free paper ∞

press.princeton.edu

Printed in the United States of America

10 9 8 7 6 5 4 3 2 1

C'est une chose étrange à la fin que le monde
Un jour je m'en irais sans en avoir tout dit
Ces moments de bonheur ces midis d'incendie
La nuit immense et noire aux déchirures blondes

Rien n'est si précieux peut-être qu'on ne le croit
D'autres viennent Ils ont le cœur que j'ai moi-même
Ils savent toucher l'herbe et dire je vous aime
Et rêver dans le soir où s'éteignent des voix...

Il y aura toujours un couple frémissant
Pour qui ce matin-là sera l'aube première
Il y aura toujours l'eau le vent la lumière
Rien ne passe après tout si ce n'est le passant

C'est une chose au fond que je ne puis comprendre
Cette peur de mourir que les gens ont en eux
Comme si ce n'était pas assez merveilleux
Que le ciel un moment nous ait paru si tendre...

Malgré tout je vous dis que cette vie fut telle
Qu'à qui voudra m'entendre à qui je parle ici
N'ayant plus sur la lèvre un seul mot que merci
Je dirai malgré tout que cette vie fut belle

Louis Aragon (1897–1982), *Les Yeux et la Mémoire (1954)*

Contents

Preface

The project of this book was initiated by Gene Golub in mid-2005 during a visit to Cerfacs in Toulouse, France. Gene had been working on the relations between matrix computations, quadrature rules and orthogonal polynomials for more than thirty years. We collaborated on these topics during the 1990s and he envisioned the writing of this book as being a summary of our findings and of his achievements with many other collaborators. We met many times during recent years, especially during his sabbatical in Oxford in 2007, to work on this project. Gene was always very enthusiastic about it because this topic and the many applications that can be done were some of his favorites.

Unfortunately Gene Golub passed away on November 16, 2007. So he was not able to see the end of this work. This book can be considered as a tribute to the work he had done during more than three decades. For a short biography of Gene see [240] and for a more complete one see [61]. Even though the writing of the main parts of the book was almost finished when Gene passed away, all the remaining mistakes are mine.

Gérard Meurant

Acknowledgments

G. Meurant thanks C. Brezinski for his interest in this book and his encouragements. Thanks to G. Monegato and W. Gander for their readings of the manuscript and their suggestions. Many thanks to S. Elhay and J. Kautsky for their valuable comments and for providing codes implementing their numerical methods and to P. Tichý for his help. Finally, G.M. thanks Z. Strakoš for his readings of the manuscript, his wise comments and his lasting friendship.

Most of the Matlab codes used to produce the numerical experiments of this book are available at *http://pagesperso-orange.fr/gerard.meurant/*
Matlab is a trademark of the MathWorks company.

PART 1
Theory

Chapter One

Introduction

The aim of this book is to describe and explain the beautiful mathematical relationships between matrices, moments, orthogonal polynomials, quadrature rules and the Lanczos and conjugate gradient algorithms. Even though we recall the mathematical basis of the algorithms, this book is computationally oriented. The main goal is to obtain efficient numerical methods to estimate or in some cases to bound quantities like $I[f] = u^T f(A)v$ where u and v are given vectors, A is a symmetric nonsingular matrix and f is a smooth function. The main idea developed in this book is to write $I[f]$ as a Riemann–Stieltjes integral and then to apply Gauss quadrature rules to compute estimates or bounds of the integral. The nodes and weights of these quadrature rules are given by the eigenvalues and eigenvectors of tridiagonal matrices whose nonzero coefficients describe the three-term recurrences satisfied by the orthogonal polynomials associated with the measure of the Riemann–Stieltjes integral. Beautifully, these orthogonal polynomials can be generated by the Lanczos algorithm when $u = v$ or by its variants otherwise. All these topics have a long and rich history starting in the nineteenth century. Our aim is to bring together results and algorithms from different areas. Results about orthogonal polynomials and quadrature rules may not be so well known in the matrix computation community, and conversely the applications in matrix computations that can be done with orthogonal polynomials and quadrature rules may be not too familiar to the community of researchers working on these topics. We will see that it can be very fruitful to mix techniques coming from different areas.

There are many instances in which one would like to compute bilinear forms like $u^T f(A)v$. A first obvious application is the computation of some elements of the matrix $f(A)$ when it is not desired or feasible to compute all of $f(A)$. Computation of quadratic forms $r^T A^{-i}r$ for $i = 1, 2$ is interesting to obtain estimates of error norms when one has an approximate solution $\tilde{x}$ of a linear system $Ax = b$ and r is the residual vector $b - A\tilde{x}$. Bilinear or quadratic forms also arise naturally for the computation of parameters in some numerical methods for solving least squares or total least squares problems and also in Tikhonov regularization for solving ill-posed problems.

The first part of the book provides the necessary mathematical background and explains the theory while the second part describes applications of these results, gives implementation details and studies improvements of some of the algorithms reviewed in the first part. Let us briefly describe the contents of the next chapters.

The second chapter is devoted to orthogonal polynomials, whose history started in the nineteenth century from the study of continued fractions. There are many excellent books on this topic, so we just recall the properties that will be useful in the other chapters. The important point for our purposes is that orthogonal polynomials satisfy three-term recurrences. We are also interested in some properties of the zeros of these polynomials. We give some examples of classical orthogonal polynomials like the Legendre, Chebyshev and Laguerre polynomials. Some of them will be used later in several algorithms and in numerical experiments. We also introduce a less classical topic, matrix orthogonal polynomials, that is, polynomials whose coefficients are square matrices. These polynomials satisfy block three-term recurrences and lead to consideration of block tridiagonal matrices. They will be useful for computing estimates of off-diagonal elements of functions of matrices.

Since tridiagonal matrices will play a prominent role in the algorithms described in this book, chapter 3 recalls properties of these matrices. We consider Cholesky-like factorizations of symmetric tridiagonal matrices and properties of the eigenvalues and eigenvectors. We will see that some elements of the inverse of tridiagonal matrices (particularly the $(1, 1)$ element) come into play for estimating bilinear forms involving the inverse of A. Hence, we give expressions of elements of the inverse obtained from Cholesky factorizations and algorithms to cheaply compute elements of the inverse. Finally, we describe the QD algorithm which was introduced by H. Rutishauser to compute eigenvalues of tridiagonal matrices and some of its variants. This algorithm will be used to solve inverse problems, namely reconstruction of symmetric tridiagonal matrices from their spectral properties.

Chapter 4 briefly describes the well-known Lanczos and conjugate gradient (CG) algorithms. The Lanczos algorithm will be used to generate the recurrence coefficients of orthogonal polynomials related to our problem. The conjugate gradient algorithm is closely linked to Gauss quadrature and we will see that quadrature rules can be used to obtain bounds or estimates of norms of the error during CG iterations when solving symmetric positive definite linear systems. We also describe the nonsymmetric Lanczos and the block Lanczos algorithms which will be useful to compute estimates of bilinear forms $u^T f(A)v$ when $u \neq v$. Another topic of interest in this chapter is the Golub–Kahan bidiagonalization algorithms that are useful when solving least squares problems.

Chapter 5 deals with the computation of the tridiagonal matrices containing the coefficients of the three-term recurrences satisfied by orthogonal polynomials. These matrices are called Jacobi matrices. There are many circumstances in which we have to compute the Jacobi matrices either from knowledge of the measure of a Riemann–Stieltjes integral or from the moments related to the measure. It is also important to be able to solve the inverse problem of reconstructing the Jacobi matrices from the nodes and weights of a quadrature formula which defines a discrete measure. We first describe the Stieltjes procedure, which dates back to the nineteenth century. It computes the coefficients from the measure which, in most cases, has to be approximated by a discrete measure. This algorithm can be considered as a predecessor of the Lanczos algorithm although it was not constructed to compute eigenvalues. Unfortunately there are cases for which the Stieltjes algorithm gives poor results due to a sensitivity to roundoff errors. Then we show how the nonzero entries of the Jacobi matrices are related to determinants of Hankel matrices constructed from the moments. These formulas are of little computational interest even though they have been used in some algorithms. More interesting is the modified

Chebyshev algorithm, which uses so-called modified moments to compute the Jacobi matrix. These modified moments are obtained from some known auxiliary orthogonal polynomials. The next section consider several algorithms for solving the problem of constructing the Jacobi matrix from the nodes and weights of a discrete measure. They are the eigenvalues and squares of the first elements of the eigenvectors. Hence, this is in fact an inverse eigenvalue problem of reconstructing a tridiagonal matrix from spectral information. Finally, we describe modification algorithms which compute the Jacobi matrices for measures that are given by a measure for which we know the coefficients of the three-term recurrence multiplied or divided by a polynomial.

The subject of chapter 6 is Gauss quadrature rules to obtain approximations or bounds for Riemann–Stieltjes integrals. The nodes and weights of these rules are related to the orthogonal polynomials associated with the measure and they can be computed using the eigenvalues and eigenvectors of the Jacobi matrix describing the three-term recurrence. With N nodes, the Gauss rule is exact for polynomials of order $2N - 1$. The Jacobi matrix has to be modified if one wants to fix a node at one end or at both ends of the integration interval. This gives respectively the Gauss–Radau and Gauss–Lobatto quadrature rules. We also consider the anti-Gauss quadrature rule devised by D. P. Laurie to obtain a rule whose error is the opposite of the error of the Gauss rule. This is useful to estimate the error of the Gauss quadrature rule. The Gauss–Kronrod quadrature rule uses $2N + 1$ nodes of which N are the Gauss rule nodes to obtain a rule that is exact for polynomials of degree $3N + 1$. It can also be used to estimate errors in Gauss rules. Then we turn to topics that may be less familiar to the reader. The first one is the nonsymmetric Gauss quadrature rule which uses two sets of orthogonal polynomials. The second one is block Gauss quadrature rules to handle the case where the measure is a symmetric matrix. This involves the matrix orthogonal polynomials that were studied in chapter 2.

Chapter 7 is, in a sense, a summary of the previous chapters. It shows how the theoretical results and the techniques presented before allow one to obtain bounds and estimates of bilinear forms $u^T f(A)v$ when A is a symmetric matrix and f a smooth function. First, we consider the case of a quadratic form with $u = v$. To solve this problem we use the Lanczos algorithm which provides a Jacobi matrix. Using the eigenvalues and eigenvectors of this matrix (eventually suitably modified) we can compute the nodes and weights of Gauss quadrature rules. This gives estimates or bounds (if the signs of the derivatives of f are constant over the interval of integration) of the quadratic form. When $u \neq v$ we use either the nonsymmetric Lanczos algorithm or the block Lanczos algorithm. With the former we can in some cases obtain bounds for the bilinear form whereas with the latter we obtain only estimates. However, the block Lanczos algorithm has the advantage of delivering estimates of several elements of $f(A)$ instead of just one for the nonsymmetric Lanczos algorithm.

Chapter 8 briefly describes extensions of the techniques summarized in chapter 7 to the case of a nonsymmetric matrix A. The biconjugate gradient and the Arnoldi algorithms have been used to compute estimates of $u^T f(A)v$ or $u^H f(A)v$ in the complex case. Some justifications of this can be obtained through the use of Gauss quadrature in the complex plane [293] or, more interestingly, the Vorobyev moment problem [316].

The first part of the book is ended by chapter 9 which is devoted to solving secular equations. We give some examples of problems for which it is useful to solve such equations. One example is computing the eigenvalues of a matrix A perturbed by a rank-one matrix cc^T where c is a given vector. To compute the eigenvalues μ we have to solve the equation $1 + c^T(A - \mu I)^{-1}c = 0$. Note that this equation involves a quadratic form. Using the spectral decomposition of A, this problem can be reduced to solving a secular equation. We review different numerical techniques to solve such equations. Most of them are based on use of rational interpolants.

The second part of the book describes applications and gives numerical examples of the algorithms and techniques developed in the first nine chapters.

Even though this is not the main topic of the book, chapter 10 gives examples of computation of Gauss quadrature rules. It amounts to computing eigenvalues and the first components of the eigenvectors. We compare the Golub and Welsch algorithm with other implementations of the QR or the QL algorithms. We also show some examples of computation of integrals and describe experiments with modification algorithms where one computes the Jacobi matrix associated with a known measure multiplied or divided by a polynomial.

Chapter 11 is concerned with the computation of bounds for elements of $f(A)$. The functions f we are interested in as examples are A^{-1}, $\exp(A)$ and $\sqrt{A}$. We start by giving analytical lower and upper bounds for elements of the inverse. This is obtained by doing "by hand" one or two iterations of the Lanczos algorithm. These results are then extended to any function f. We also show how to compute estimates of the trace of the inverse and of the determinant of A, a problem which does not exactly fit in the same framework. These algorithms are important for some applications in physics. Several numerical examples are provided to show the efficiency of our techniques for computing bounds and to analyze their accuracy.

Chapter 12 studies the close relationships of the conjugate gradient algorithm with Gauss quadrature. In fact, the square of the A-norm of the error at iteration k is the remainder of a k-point Gauss quadrature rule for computing $(r^0)^T A^{-1} r^0$ where r^0 is the initial residual. Bounds of the A-norm of the error can be computed during CG iterations by exploiting this relationship. If one is interested in the l_2 norm of the error, it can also be estimated during the CG iterations. This leads to the definition of reliable stopping criteria for the CG algorithm. These estimates have been used when solving finite element problems. One can define a stopping criterion such that the norm of the error with the solution of the continuous problem is at the level one can expect for a given mesh size. Numerous examples of computation of bounds of error norms are provided.

In chapter 13 we consider the least squares fit of some given data by polynomials. The solution to this problem can be expressed using the orthogonal polynomials related to the discrete inner product defined by the data. We are particularly interested in the updating and downdating operations where one adds or deletes data from the sample. This amounts to computing new Jacobi matrices from known ones. We review algorithms using orthogonal transformations to solve these problems, which are also linked to inverse eigenvalue problems. We also consider the problem of computing the backward error of a least squares solution. Use of the exact expression of the backward error is difficult because it amounts to computing the smallest eigenvalue of a rank-one modification of a singular matrix. However one can compute an approximation of the backward error with Gauss quadrature.

Given a matrix A and a right-hand side c, the method of Total Least Squares (TLS) looks for the solution of $(A + E)x = c + r$ where E and r are the smallest perturbations in the Frobenius norm such that $c + r$ is in the range of $A + E$. To compute the solution we need the smallest singular value of the matrix $(A \quad c)$. It is given as the solution of a secular equation. In chapter 14, approximations of this solution are obtained by using the Golub–Kahan bidiagonalization algorithm and Gauss quadrature.

Finally, chapter 15 considers the determination of the Tikhonov regularization parameter for discrete ill-posed problems. There are many criteria which have been devised to define good parameters. We mainly study generalized cross-validation (GCV) and the L-curve criteria. The computations of the "optimal" parameters for these methods involve the computation of quadratic forms which can be approximated using Gauss quadrature rules. We describe improvements of algorithms which have been proposed in the literature and we provide numerical experiments to compare the different criteria and the algorithms implementing them.

This book should be useful to researchers in numerical linear algebra and more generally to people interested in matrix computations. It can be of interest too to scientists and engineers solving problems in which computation of bilinear forms arises naturally.

Chapter Two

Orthogonal Polynomials

In this chapter, we briefly recall the properties of orthogonal polynomials which will be needed in the next chapters. We are mainly interested in polynomials of a real variable defined in an interval of the real line. For more details, see the book by Szegö [323] or the book by Chihara [64], and also the paper [65] for theoretical results on classical orthogonal polynomials and the nice book by Gautschi [131] for the computational aspects.

2.1 Definition of Orthogonal Polynomials

We will define orthogonal polynomials in either a finite or an infinite interval $[a, b]$ of the real line. We first have to define orthogonality. For our purposes this is done through the definition of an inner product for functions of a real variable by using Riemann–Stieltjes integrals.

DEFINITION 2.1 *A Riemann–Stieltjes integral of a real valued continuous function f of a real variable with respect to a real function α is denoted by*

$$\int_a^b f(\lambda) \, d\alpha(\lambda), \tag{2.1}$$

and is defined to be the limit (if it exists), as the mesh size of the partition π of the interval $[a, b]$ goes to zero, of the sums

$$\sum_{\{\lambda_i\} \in \pi} f(c_i)(\alpha(\lambda_{i+1}) - \alpha(\lambda_i)),$$

where $c_i \in [\lambda_i, \lambda_{i+1}]$.

Note that we obtain a Riemann integral if $d\alpha(\lambda) = d\lambda$. If α is continuously differentiable, the integral (2.1) is equal to

$$\int_a^b f(\lambda)\alpha'(\lambda) \, d\lambda.$$

See for instance Riesz and Nagy [283]. But this is not always the case since α may have jumps or may have a zero derivative almost everywhere. Then the Riemann–Stieltjes integral (2.1) cannot be reduced to a Riemann integral. However, in many cases Riemann–Stieltjes integrals are directly written as

$$\int_a^b f(\lambda) \, w(\lambda) d\lambda,$$

where w is called the weight function.

The simplest existence theorem for the type of integral (2.1) says that if f is continuous and α is of bounded variation on $[a, b]$ then the integral exists. The function α is of bounded variation if it is the difference of two nondecreasing functions. In particular, the integral exists if f is continuous and α is nondecreasing.

Let α be a nondecreasing function on the interval (a, b) having finite limits at $\pm\infty$ if $a = -\infty$ and/or $b = +\infty$ and infinitely many points of increase.

DEFINITION 2.2 *The numbers*

$$\mu_i = \int_a^b \lambda^i \, d\alpha(\lambda), \ i = 0, 1, \ldots \tag{2.2}$$

are called the **moments** *related to the measure* α.

This name was chosen by Stieltjes because of analogy with some definitions in mechanical problems. To be able to use the Riemann–Stieltjes integral for polynomials we assume that all the moments are finite. Let us define an inner product given by a Riemann–Stieltjes integral (2.1).

DEFINITION 2.3 *Let $\mathcal{P}$ be the space of real polynomials. We define an inner product (related to the measure α) of two polynomials p and $q \in \mathcal{P}$ as*

$$\langle p, q \rangle = \int_a^b p(\lambda)q(\lambda) \, d\alpha(\lambda). \tag{2.3}$$

The norm of p is defined as

$$\|p\|_\alpha = \left(\int_a^b p(\lambda)^2 \, d\alpha(\lambda) \right)^{\frac{1}{2}}. \tag{2.4}$$

Note that with our hypothesis for the moments the integral (2.3) exists. When it is necessary to refer to the measure α, we will also denote the inner product as $\langle \cdot, \cdot \rangle_\alpha$. We will consider also discrete inner products as

$$\langle p, q \rangle = \sum_{j=1}^m p(t_j)q(t_j)w_j^2. \tag{2.5}$$

The values t_j are referred to as points or nodes and the values w_j^2 are the weights. Several times in this book we will use the fact that the sum in equation (2.5) can be seen as an approximation of the integral (2.3). Conversely, it can be written as a Riemann–Stieltjes integral for a measure α which is piecewise constant and has jumps at the nodes t_j (that we assume to be distinct for simplicity):

$$\alpha(\lambda) = \begin{cases} 0, & \text{if } \lambda < t_1, \\ \sum_{j=1}^i [w_j]^2, & \text{if } t_i \leq \lambda < t_{i+1}, \ i = 1, \ldots, m-1, \\ \sum_{j=1}^m [w_j]^2, & \text{if } t_m \leq \lambda; \end{cases}$$

see Atkinson [13], Dahlquist, Eisenstat and Golub [75] and Dahlquist, Golub and Nash [76]. There are different ways to normalize polynomials. A polynomial p of exact degree k is said to be monic if the coefficient of the monomial of highest degree is 1, that is, it is defined as $p(\lambda) = \lambda^k + c_{k-1}\lambda^{k-1} + \ldots$.

DEFINITION 2.4 *The polynomials p and q are said to be orthogonal (with respect to inner products (2.3) or (2.5)) if* $\langle p, q \rangle = 0$. *The polynomials p in a set of polynomials are orthonormal if they are mutually orthogonal and if* $\langle p, p \rangle = 1$. *Polynomials in a set are said to be monic orthogonal polynomials if they are orthogonal, monic and their norms are strictly positive.*

Sometimes, polynomials are also normalized by fixing their value at 0, for example $p(0) = 1$. It is not so obvious to know when there exist orthogonal polynomials for a given measure and the corresponding inner product. The inner product $\langle \cdot, \cdot \rangle_\alpha$ is said to be positive definite if $\|p\|_\alpha > 0$ for all nonzero p in $\mathcal{P}$. A necessary and sufficient condition for having a positive definite inner product is that the determinants of the Hankel moment matrices are positive,

$$\det \begin{pmatrix} \mu_0 & \mu_1 & \cdots & \mu_{k-1} \\ \mu_1 & \mu_2 & \cdots & \mu_k \\ \vdots & \vdots & & \vdots \\ \mu_{k-1} & \mu_k & \cdots & \mu_{2k-2} \end{pmatrix} > 0, \; k = 1, 2, \ldots,$$

where μ_i are the moments of definition (2.2). This leads to a sufficient condition for the existence of the orthogonal polynomials.

THEOREM 2.5 *If the inner product* $\langle \cdot, \cdot \rangle_\alpha$ *is positive definite on* $\mathcal{P}$, *there exists a unique infinite sequence of monic orthogonal polynomials related to the measure* α.

Proof. See Gautschi [131]. □

Orthogonal polynomials have interesting minimization properties for the l_2 norm.

THEOREM 2.6 *If* q_k *is a monic polynomial of degree* k, *then*

$$\min_{q_k} \int_a^b q_k^2(\lambda) \, d\alpha(\lambda),$$

is attained if and only if q_k *is a constant times the orthogonal polynomial* p_k *related to* α.

Proof. See Szegö [323]. □

We have defined orthogonality relative to an inner product given by a Riemann–Stieltjes integral but, more generally, orthogonal polynomials can be defined relative to a linear functional L such that $L(\lambda^k) = \mu_k$. Two polynomials p and q are said to be orthogonal if $L(pq) = 0$. One obtains the same kind of existence result as in theorem 2.5; see the book by Brezinski [36].

2.2 Three-Term Recurrences

For our purposes in this book, the most important property of orthogonal polynomials is that they satisfy a three-term recurrence relation. The main ingredient to obtain this result is to have the following property for the inner product:

$$\langle \lambda p, q \rangle = \langle p, \lambda q \rangle.$$

This is obviously satisfied for the inner product defined in equation (2.3) using the Riemann–Stieltjes integral. Let us first consider monic orthogonal polynomials p_k of degree k.

THEOREM 2.7 *For monic orthogonal polynomials, there exist sequences of coefficients α_k, $k = 1, 2, \ldots$ and γ_k, $k = 1, 2, \ldots$ such that*

$$p_{k+1}(\lambda) = (\lambda - \alpha_{k+1})p_k(\lambda) - \gamma_k p_{k-1}(\lambda), \ k = 0, 1, \ldots \quad (2.6)$$

$$p_{-1}(\lambda) \equiv 0, \ p_0(\lambda) \equiv 1.$$

where

$$\alpha_{k+1} = \frac{\langle \lambda p_k, p_k \rangle}{\langle p_k, p_k \rangle}, \ k = 0, 1, \ldots,$$

$$\gamma_k = \frac{\langle p_k, p_k \rangle}{\langle p_{k-1}, p_{k-1} \rangle}, \ k = 1, 2, \ldots.$$

Proof. Notice that γ_0 does not need to be defined since $p_{-1} \equiv 0$. We follow the proof in Gautschi [131]. It is easy to prove that a set of monic orthogonal polynomials p_j is linearly independent and any polynomial p of degree k can be written as

$$p = \sum_{j=0}^{k} \omega_j p_j,$$

for some real numbers ω_j. Since we consider monic polynomials, the polynomial $p_{k+1} - \lambda p_k$ is of degree $\leq k$ and can be written as a linear combination of the orthogonal polynomials p_j, $j = 0, \ldots, k$. Let us write this as

$$p_{k+1} - \lambda p_k = -\alpha_{k+1} p_k - \gamma_k p_{k-1} + \sum_{j=0}^{k-2} \delta_j p_j, \quad (2.7)$$

where the coefficients δ_j are real numbers. We would like to prove that $\delta_j = 0$, $j = 0, \ldots, k - 2$. Taking the inner product of equation (2.7) with p_k and using orthogonality, we obtain

$$\langle \lambda p_k, p_k \rangle = \alpha_{k+1} \langle p_k, p_k \rangle.$$

Since $\langle p_k, p_k \rangle > 0$, this gives the value of α_{k+1}. Multiplying equation (2.7) by p_{k-1}, we have

$$\langle \lambda p_k, p_{k-1} \rangle = \gamma_k \langle p_{k-1}, p_{k-1} \rangle.$$

But, using equation (2.7) for the degree $k - 1$,

$$\langle \lambda p_k, p_{k-1} \rangle = \langle p_k, \lambda p_{k-1} \rangle = \langle p_k, p_k \rangle.$$

This gives the expression of γ_k. For the other terms, we multiply equation (2.7) with p_j, $j < k - 1$ obtaining

$$\langle \lambda p_k, p_j \rangle = \delta_j \langle p_j, p_j \rangle.$$

The left-hand side of the last equation vanishes. For this, the property $\langle \lambda p_k, p_j \rangle = \langle p_k, \lambda p_j \rangle$ is crucial. Since λp_j is of degree $< k$, the left-hand side is 0 and it implies $\delta_j = 0$, $j = 0, \ldots, k - 2$. $\square$

We remark that the coefficients γ_k are strictly positive. There is a converse to theorem 2.7. It is is attributed to J. Favard [104] whose paper was published in 1935, although this result had also been obtained by J. Shohat [298] at about the same time and it was known earlier to Stieltjes [313]; see [230]. Without all the technical details, the result of Favard is the following.

THEOREM 2.8 *If a sequence of monic orthogonal polynomials p_k, $k = 0, 1, \ldots$, satisfies a three-term recurrence relation such as equation (2.6) with real coefficients and $\gamma_k > 0$, then there exists a positive measure α such that the sequence p_k is orthogonal with respect to an inner product defined by a Riemann–Stieltjes integral for the measure α.*

Proof. For a proof in a more general setting, see Marcellán and Alvarez–Nodarse [230]. $\square$

Moreover, there are additional conditions on the coefficients of recurrence (2.6) which implies that the support of the measure is in $[0, \infty)$ or in a bounded interval.

Considering the recurrences for orthonormal polynomials, we have the following result.

THEOREM 2.9 *For orthonormal polynomials, there exist sequences of coefficients α_k, $k = 1, 2, \ldots$ and β_k, $k = 1, 2, \ldots$ such that*

$$\sqrt{\beta_{k+1}} p_{k+1}(\lambda) = (\lambda - \alpha_{k+1}) p_k(\lambda) - \sqrt{\beta_k} p_{k-1}(\lambda), \ k = 0, 1, \ldots, \quad (2.8)$$

$$p_{-1}(\lambda) \equiv 0, \ p_0(\lambda) \equiv 1/\sqrt{\beta_0}, \ \beta_0 = \int_a^b d\alpha,$$

where

$$\alpha_{k+1} = \langle \lambda p_k, p_k \rangle, \ k = 0, 1, \ldots$$

and β_k is computed such that $\|p_k\|_\alpha = 1$.

Proof. The proof is basically the same as for theorem 2.7. $\square$

Generally to avoid the square roots we will directly use $\eta_k = \sqrt{\beta_k}$ in the three-term recurrence. The square roots naturally arise from the relation between monic and orthonormal polynomials. If we assume that we have a system of monic polynomials p_k satisfying a three-term recurrence (2.6), then we can obtain orthonormal polynomials $\hat{p}_k$ by normalization

$$\hat{p}_k(\lambda) = \frac{p_k(\lambda)}{\langle p_k, p_k \rangle^{1/2}}.$$

Using equation (2.6) we have

$$\|p_{k+1}\|_\alpha \hat{p}_{k+1} = \left(\lambda \|p_k\|_\alpha - \frac{\langle \lambda p_k, p_k \rangle}{\|p_k\|_\alpha} \right) \hat{p}_k - \frac{\|p_k\|_\alpha^2}{\|p_{k-1}\|_\alpha} \hat{p}_{k-1}.$$

After some manipulations we obtain

$$\frac{\|p_{k+1}\|_\alpha}{\|p_k\|_\alpha}\hat{p}_{k+1} = (\lambda - \langle\lambda\hat{p}_k, \hat{p}_k\rangle)\hat{p}_k - \frac{\|p_k\|_\alpha}{\|p_{k-1}\|_\alpha}\hat{p}_{k-1}.$$

We note that

$$\langle\lambda\hat{p}_k, \hat{p}_k\rangle = \frac{\langle\lambda p_k, p_k\rangle}{\|p_k\|_\alpha^2},$$

and

$$\sqrt{\beta_{k+1}} = \frac{\|p_{k+1}\|_\alpha}{\|p_k\|_\alpha}.$$

Therefore the coefficients α_k are the same and $\beta_k = \gamma_k$. If we have the coefficients of monic orthogonal polynomials we just have to take the square root of γ_k to obtain the coefficients of the corresponding orthonormal polynomials.

If the orthonormal polynomials in theorem 2.9 exist for all k, there is an infinite symmetric tridiagonal matrix J_∞ associated with them,

$$J_\infty = \begin{pmatrix} \alpha_1 & \sqrt{\beta_1} & & \\ \sqrt{\beta_1} & \alpha_2 & \sqrt{\beta_2} & \\ & \sqrt{\beta_2} & \alpha_3 & \sqrt{\beta_3} \\ & & \ddots & \ddots & \ddots \end{pmatrix}.$$

Since it has positive subdiagonal elements, the matrix J_∞ is called an infinite Jacobi matrix. Its leading principal submatrix of order k is denoted as J_k.

THEOREM 2.10 *The Jacobi matrix J_k of dimension k related to α is uniquely determined by the first $2k$ moments of α.*

Proof. See Elhay, Golub and Kautsky [101]. Their proof relies on the fact (as we will see when studying quadrature rules) that the first $2k$ moments uniquely define a k-point Gauss quadrature. Since the polynomials of degree less than or equal to $2k - 1$ orthogonal with respect to the measure α are also orthogonal with respect to the discrete measure given by this quadrature rule, we obtain the result.   □

Note that a given Jacobi matrix of finite size k corresponds to an infinite set of normalized weight functions (with $\mu_0 = 1$) which all have the same first $2k$ moments. On this topic see also Kautsky [199].

Using the three-term recurrence relation satisfied by orthogonal polynomials one can also prove the following interesting and useful result. This is known as a Christoffel–Darboux formula.

THEOREM 2.11 *Let p_k, $k = 0, 1, \dots$ be orthonormal polynomials, then*

$$\sum_{i=0}^k p_i(\lambda)p_i(\mu) = \sqrt{\beta_{k+1}}\frac{p_{k+1}(\lambda)p_k(\mu) - p_k(\lambda)p_{k+1}(\mu)}{\lambda - \mu}, \text{ if } \lambda \neq \mu, \quad (2.9)$$

and

$$\sum_{i=0}^k p_i^2(\lambda) = \sqrt{\beta_{k+1}}[p_{k+1}'(\lambda)p_k(\lambda) - p_k'(\lambda)p_{k+1}(\lambda)].$$

Proof. Multiply the relation (2.8) by $p_k(\mu)$ and subtract this relation from a similar one with λ and μ interchanged. The other relation is obtained by letting $\lambda \to \mu$. $\square$

COROLLARY 2.12 *For monic orthogonal polynomials we have*

$$\sum_{i=0}^{k} \gamma_k \gamma_{k-1} \cdots \gamma_{i+1} p_i(\lambda) p_i(\mu) = \frac{p_{k+1}(\lambda) p_k(\mu) - p_k(\lambda) p_{k+1}(\mu)}{\lambda - \mu}, \text{ if } \lambda \neq \mu.$$

On the Christoffel–Darboux formula, see also Brezinski [35].

2.3 Properties of Zeros

Let us put the values at λ of the first k orthonormal polynomials in a vector and denote

$$P_k(\lambda) = \begin{pmatrix} p_0(\lambda) & p_1(\lambda) & \cdots & p_{k-1}(\lambda) \end{pmatrix}^T.$$

Then, in matrix form, the three-term recurrence is written as

$$\lambda P_k = J_k P_k + \eta_k p_k(\lambda) e^k, \tag{2.10}$$

where J_k is the Jacobi matrix of order k with subdiagonal elements η_i and e^k is the last column of the identity matrix of order k. This leads to the fundamental following result.

THEOREM 2.13 *The zeros $\theta_j^{(k)}$ of the orthonormal polynomial p_k are the eigenvalues of the Jacobi matrix J_k.*

Proof. If θ is a zero of p_k, from equation (2.10) we have

$$\theta P_k(\theta) = J_k P_k(\theta).$$

This shows that θ is an eigenvalue of J_k and $P_k(\theta)$ is a corresponding (unnormalized) eigenvector. $\square$

The matrix J_k being a symmetric tridiagonal matrix, its eigenvalues (the zeros of the orthogonal polynomial p_k) are real and distinct as we will see in chapter 3. Moreover, we have the following result about location of the zeros; see Szegö [323].

THEOREM 2.14 *The zeros of the orthogonal polynomials p_k associated with the measure α on $[a, b]$ are real, distinct and located in the interior of $[a, b]$.*

Proof. The first two assertions are proved as a consequence of theorem 2.13. The statement concerning the location follows from the minimization property in theorem 2.6. If there is a zero outside the interval $[a, b]$, then the value of the Riemann–Stieltjes integral can be decreased by moving this zero. But this is not possible since p_k, properly normalized, gives the minimum of the integral. $\square$

2.4 Historical Remarks

In 1894-1895, Stieltjes published a seminal paper: "Recherches sur les fractions continues" [313]. He proposed and solved the following problem.

Find a bounded nondecreasing function α in the interval $[0, \infty)$ such that its moments have a prescribed set of values μ_n,

$$\int_0^\infty \lambda^n \, d\alpha(\lambda) = \mu_n, \; n = 0, 1, 2, \dots .$$

The name "problem of moments" was chosen by Stieltjes in analogy with mechanical problems. In fact $d\alpha(\lambda)$ can be considered as a mass distributed over $[\lambda, \lambda+d\lambda]$. Then the integral

$$\int_0^x d\alpha(\lambda)$$

represents the mass over the segment $[0, x]$. This is why α is often called a distribution function. The integrals

$$\int_0^\infty \lambda \, d\alpha(\lambda), \quad \int_0^\infty \lambda^2 \, d\alpha(\lambda)$$

represent the first statical moment and the moment of inertia (with respect to 0) of the total mass distributed over $[0, \infty)$.

Stieltjes showed that a necessary and sufficient condition to have a solution is the positiveness of certain Hankel determinants given by the moments μ_n. The solution may be unique or there can be infinitely many solutions.

Prior to Stieltjes, Chebyshev in a series of papers started in 1855 studied related problems. He was interested in how far a sequence of given moments such that

$$\int_{-\infty}^\infty \lambda^n f(\lambda) \, d\lambda = \mu_n, \; n = 0, 1, 2, \dots ,$$

determine the function f. A. Markov, a student of Chebyshev, continued his work. Heine (1861, 1878, 1881) also worked on related problems before Stieltjes. H. Hamburger considered the moment problem on the whole real axis (1920, 1921). He gave a sufficient and necessary condition for the existence of a solution. This is again given by Hankel determinants. Hausdorff (1923) gave a criterion for the moment problem to have a solution in a finite interval.

Many of these early papers used the theory of continued fractions to solve moment problems. The study of the polynomials which are the denominators of convergents (that is truncated sums) of continued fractions was at the beginning of the modern theory of orthogonal polynomials. More recent papers studied the moment problem using the tools of functional analysis. For details on the moment problem, see the books by Shohat and Tamarkin [299] and Akhiezer [4].

2.5 Examples of Orthogonal Polynomials

There are many cases for which the coefficients of the three-term recurrence are explicitly known. Let us consider some classical examples where the measure is defined through a weight function w by $d\alpha(\lambda) = w(\lambda)\, d\lambda$. Let

$$a = -1,\ b = 1,\quad w(\lambda) = (1 - \lambda)^\delta (1 + \lambda)^\beta,\ \delta, \beta > -1.$$

The corresponding orthogonal polynomials are known as Jacobi polynomials. Some special cases of interest for different choices of the exponents δ and β are described in the next subsections.

2.5.1 Chebyshev Polynomials of the First Kind

These polynomials, denoted as C_k (or sometimes T_k depending on the spelling used, Chebyshev or Tchebicheff; see Davis [77]), are obtained by choosing $\delta = \beta = -1/2$. They are defined for $|\lambda| \le 1$ as

$$C_k(\lambda) = \cos(k \arccos \lambda).$$

It is not immediately obvious that this defines polynomials, but using trigonometric identities one can see that the functions C_k satisfy a three-term recurrence,

$$C_0(\lambda) \equiv 1,\ C_1(\lambda) \equiv \lambda,\quad C_{k+1}(\lambda) = 2\lambda C_k(\lambda) - C_{k-1}(\lambda).$$

Hence, they are indeed polynomials. The corresponding Jacobi matrix J_k is the tridiagonal matrix $(\,1/2\quad 0\quad 1/2\,)$ of order k with constant diagonals except for the first row which has 0 and 1 in positions $(1, 1)$ and $(1, 2)$. Its eigenvalues (and consequently the zeros of C_k) are

$$\lambda_{j+1} = \cos\left(\frac{2j + 1}{k}\frac{\pi}{2}\right),\ j = 0, 1, \ldots k - 1.$$

The polynomial C_k has $k + 1$ extremas in $[-1, 1]$,

$$\lambda_j' = \cos\left(\frac{j\pi}{k}\right),\ j = 0, 1, \ldots, k$$

and $C_k(\lambda_j') = (-1)^j$. The polynomials C_k, $k = 1, \ldots, 7$ are displayed in figure 2.1. We see that their absolute values grow very fast outside $[-1, 1]$. On $[-1, 1]$ the polynomials oscillate between -1 and 1.

For $k \ge 1$, C_k has a leading coefficient 2^{k-1}. The inner product related to the measure $\alpha(\lambda) = (1 - \lambda^2)^{-1/2}$ is such that

$$\langle C_i, C_j \rangle_\alpha = \begin{cases} 0, & i \ne j, \\ \frac{\pi}{2}, & i = j \ne 0, \\ \pi, & i = j = 0. \end{cases}$$

So these polynomials are neither monic nor orthonormal. But, of course, they can be normalized, and one of the most interesting properties of the Chebyshev polynomials is the following (see for instance Dahlquist and Björck [73] or the recent book [74]).

THEOREM 2.15 *For all monic polynomials of degree k, $C_k/2^{k-1}$ has the smallest maximum norm, equal to $1/2^{k-1}$.*

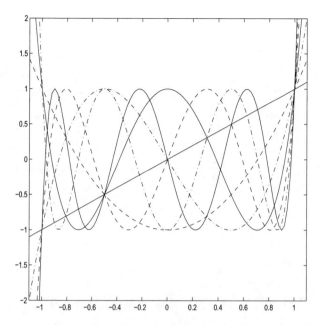

Figure 2.1 Chebyshev polynomials (first kind) C_k, $k = 1, \ldots, 7$ on $[-1.1, 1.1]$

Let $\pi_n^1 = \{$ polynomials of degree n in λ whose value is 1 for $\lambda = 0$ $\}$. Chebyshev polynomials provide the solution of the minimization problem

$$\min_{q_n \in \pi_n^1} \max_{\lambda \in [a,b]} |q_n(\lambda)|.$$

The solution is written as

$$\min_{q_n \in \pi_n^1} \max_{\lambda \in [a,b]} |q_n(\lambda)| = \max_{\lambda \in [a,b]} \left| \frac{C_n\left(\frac{2\lambda - (a+b)}{b-a}\right)}{C_n\left(\frac{a+b}{b-a}\right)} \right| = \left| \frac{1}{C_n\left(\frac{a+b}{b-a}\right)} \right|.$$

2.5.2 Chebyshev Polynomials of the Second Kind

In this case we have $\delta = \beta = 1/2$ and the polynomials U_k are defined as

$$U_k(\lambda) = \frac{\sin(k+1)\theta}{\sin \theta}, \quad \lambda = \cos \theta.$$

They satisfy the same three-term recurrence as the Chebyshev polynomials of the first kind but with initial conditions $U_0 \equiv 1$, $U_1 \equiv 2\lambda$. Of all monic polynomials q_k, $2^{-k} U_k$ gives the smallest L_1 norm,

$$\|q_k\|_1 = \int_{-1}^{1} |q_k(\lambda)| \, d\lambda.$$

The first polynomials U_k are displayed in figure 2.2.

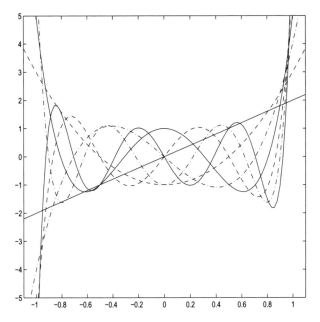

Figure 2.2 Chebyshev polynomials (second kind) U_k, $k = 1, \ldots, 7$ on $[-1.1, 1.1]$

2.5.3 Legendre Polynomials

The choice $\delta = \beta = 0$ (that is, a weight function equal to 1) gives the Legendre polynomials P_k. The three-term recurrence is

$$(k + 1)P_{k+1}(\lambda) = (2k + 1)\lambda P_k(\lambda) - kP_{k-1}(\lambda), \quad P_0(\lambda) \equiv 1, \ P_1(\lambda) \equiv \lambda.$$

The Legendre polynomial P_k is bounded by 1 on $[-1, 1]$. The first polynomials P_k are displayed in figure 2.3.

2.5.4 Laguerre and Hermite Polynomials

Other classical examples on different intervals are the Laguerre and Hermite polynomials. For Laguerre polynomials L_k, the interval is $[0, \infty)$ and the weight function is $e^{-\lambda}$. The recurrence relation is

$$(k + 1)L_{k+1} = (2k + 1 - \lambda)L_k - kL_{k-1}, \quad L_0 \equiv 1, \ L_1 \equiv 1 - \lambda.$$

The first polynomials L_k are displayed in figure 2.4. We will also consider as examples generalized Laguerre polynomials for which the interval is the same and the weight function is $e^{-\lambda}\lambda^\omega$.

For Hermite polynomials H_k, the interval is $(-\infty, \infty)$ and the weight function is $e^{-\lambda^2}$. The recurrence relation is

$$H_{k+1} = 2\lambda H_k - 2kH_{k-1}, \quad H_0 \equiv 1, \ H_1 \equiv 2\lambda.$$

The first polynomials H_k are displayed in figure 2.5.

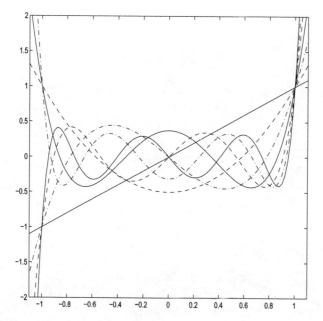

Figure 2.3 Legendre polynomials P_k, $k = 1, \ldots, 7$ on $[-1.1, 1.1]$

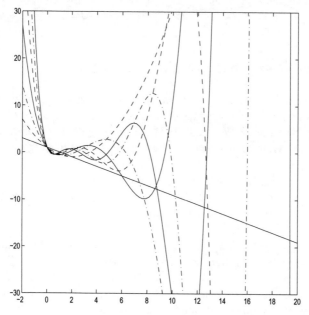

Figure 2.4 Laguerre polynomials L_k, $k = 1, \ldots, 7$ on $[-2, 20]$

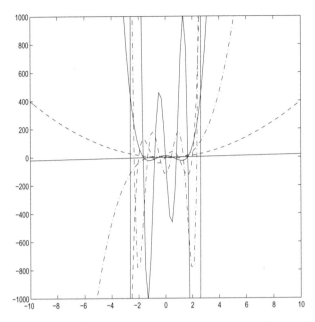

Figure 2.5 Hermite polynomials H_k, $k = 1, \ldots, 7$ on $[-10, 10]$

Other interesting polynomials are the Gegenbauer and Meixner–Pollaczek polynomials; see Szegö [323] or Gautschi [131]. Less classical examples, in particular polynomials of a discrete variable, are described in Gautschi [131].

2.6 Variable-Signed Weight Functions

So far we have assumed that the measure is positive. What happens if the measure is defined by a weight function which changes sign in the interval $[a, b]$? This problem was considered by G. W. Struble [321]. Regarding existence of the orthogonal polynomials related to such a measure, the result is the following.

THEOREM 2.16 *Assume that all the moments exist and are finite. For any $k > 0$, there exists a polynomial p_k of degree at most k such that p_k is orthogonal to all polynomials of degree $\leq k - 1$ with respect to w.*

The important words in this result are "of degree at most k". In some cases the polynomial p_k can be of degree less than k. If $C(k)$ denotes the set of polynomials of degree $\leq k$ orthogonal to all polynomials of degree $\leq k - 1$, $C(k)$ is called degenerate if it contains polynomials of degree less than k. If $C(k)$ is nondegenerate it contains one unique polynomial (up to a multiplicative constant). If $C(k)$ is nondegenerate, we may consider the next nondegenerate set $C(k+n)$, $n > 0$. The following result from [321] characterizes the sets in between.

THEOREM 2.17 *Let $C(k)$ be nondegenerate with a polynomial p_k. Assume $C(k+n)$, $n > 0$ is the next nondegenerate set. Then p_k is the unique (up to a multiplicative constant) polynomial of lowest degree in $C(k + m)$, $m = 1, \ldots, n - 1$.*

Each polynomial in $C(k+m)$, $m = 1, \ldots, n - 1$ is divisible by p_k. If $C(k+1)$ is degenerate, p_k is orthogonal to itself. Notice that this is not possible when the weight function is positive.

If one considers the set of polynomials p_k of degree d_k belonging to the sets $C(k)$ which are nondegenerate, they satisfy a three-term recurrence but with different coefficients as in the positive case. We have for $k = 2, 3, \ldots$,

$$p_k(\lambda) = \left(\alpha_k \lambda^{d_k - d_{k-1}} + \sum_{i=0}^{d_k - d_{k-1} - 1} \beta_{k,i} \lambda^i \right) p_{k-1}(\lambda) - \gamma_{k-1} p_{k-2}(\lambda) \quad (2.11)$$

and

$$p_0(\lambda) \equiv 1, \quad p_1(\lambda) = \left(\alpha_1 \lambda^{d_1} + \sum_{i=0}^{d_1 - 1} \beta_{1,i} \lambda^i \right) p_0(\lambda).$$

The coefficient of p_{k-1} contains powers of λ depending on the difference of the degrees of the polynomials in the nondegenerate cases. The coefficients α_k and γ_{k-1} have to be nonzero. Reciprocally, given the sequence of degrees and the coefficients such that α_k and γ_{k-1} are different from zero, Struble [321] constructed a weight function w for which the orthogonal polynomials satisfy equation (2.11). On degenerate orthogonal polynomials and the longer recurrence relation they satisfy, see also Draux [90].

2.7 Matrix Orthogonal Polynomials

In the previous sections, the coefficients of the polynomials were real numbers. We would like to generalize this to have matrices as coefficients. There are several ways to define orthogonal polynomials whose coefficients are square matrices. Orthogonal matrix polynomials on the real line were considered by M. G. Krein [208] a long time ago. Here, we follow the development of [149] published in 1994; see also [7], [135], [305], [306], [304], [231], [80], [81] and [82]. The coefficients to be considered are 2×2 matrices since this is sufficient for the applications we have in mind. But these results can be easily generalized to square matrices of any order.

DEFINITION 2.18 *For λ real, a matrix polynomial $p_i(\lambda)$ which is, in our case, a 2×2 matrix is defined as*

$$p_i(\lambda) = \sum_{j=0}^{i} \lambda^j C_j^{(i)},$$

where the coefficients $C_j^{(i)}$ are given 2×2 real matrices. If the leading coefficient is the identity matrix, the matrix polynomial is said to be monic.

The measure $\alpha(\lambda)$ is now a matrix of order 2 that we suppose to be symmetric and positive semidefinite. Moreover, we assume that if $\lambda_1 < \lambda_2$, then $\alpha(\lambda_2) - \alpha(\lambda_1)$ is positive semidefinite. The integral $\int_a^b f(\lambda)d\alpha(\lambda)$ is a 2×2 symmetric matrix. We also assume that the (matrix) moments

$$M_k = \int_a^b \lambda^k \, d\alpha(\lambda) \tag{2.12}$$

exist for all k.

The "inner product" of two matrix polynomials p and q is defined as

$$\langle p, q \rangle = \int_a^b p(\lambda) \, d\alpha(\lambda) q(\lambda)^T. \tag{2.13}$$

Note that this defines a matrix and we have to be careful about the order for the matrix multiplications under the integral sign. Two matrix polynomials in a sequence p_k, $k = 0, 1, \ldots$ are said to be orthonormal if

$$\langle p_i, p_j \rangle = \delta_{i,j} I_2, \tag{2.14}$$

where $\delta_{i,j}$ is the Kronecker symbol and I_2 the identity matrix of order 2.

THEOREM 2.19 *Sequences of matrix orthogonal polynomials satisfy a block three-term recurrence,*

$$p_j(\lambda)\Gamma_j = \lambda p_{j-1}(\lambda) - p_{j-1}(\lambda)\Omega_j - p_{j-2}(\lambda)\Gamma_{j-1}^T, \tag{2.15}$$

$$p_0(\lambda) \equiv I_2, \quad p_{-1}(\lambda) \equiv 0,$$

where Γ_j, Ω_j are 2×2 matrices and the matrices Ω_j are symmetric.

Proof. The proof is essentially the same as in the scalar case.   □

The block three-term recurrence can be written in matrix form as

$$\lambda[p_0(\lambda), \ldots, p_{k-1}(\lambda)] = [p_0(\lambda), \ldots, p_{k-1}(\lambda)]J_k + [0, \ldots, 0, p_k(\lambda)\Gamma_k], \tag{2.16}$$

where

$$J_k = \begin{pmatrix} \Omega_1 & \Gamma_1^T & & & \\ \Gamma_1 & \Omega_2 & \Gamma_2^T & & \\ & \ddots & \ddots & \ddots & \\ & & \Gamma_{k-2} & \Omega_{k-1} & \Gamma_{k-1}^T \\ & & & \Gamma_{k-1} & \Omega_k \end{pmatrix}$$

is a block tridiagonal matrix of order $2k$ with 2×2 blocks. Let us put the k first matrix polynomials at λ in $P(\lambda) = [p_0(\lambda), \ldots, p_{k-1}(\lambda)]^T$. Because of the symmetry of J_k, by transposing equation (2.16) we have

$$J_k P(\lambda) = \lambda P(\lambda) - [0, \ldots, 0, p_k(\lambda)\Gamma_k]^T.$$

We note that if θ_r is an eigenvalue of J_k and if we choose $u = u_r$ to be a vector of length 2 whose components are the first two components of the corresponding eigenvector, then $P(\theta_r)u$ is this eigenvector (because of the relations that are satisfied) and if Γ_k is nonsingular, $p_k^T(\theta_r)u = 0$.

There is also a matrix analog of the Favard theorem. If a sequence of matrix polynomials satisfies a three-term block recurrence then there exists a matrix measure for which they are orthonormal; see Aptekarev and Nikishin [7]. For more details, see Dette and Studden [80]. Some of the following results on the properties of the matrix polynomials were derived in [149].

THEOREM 2.20 *The eigenvalues of J_k are the zeros of* $\det[p_k(\lambda)]$.

Proof. Let θ be a zero of $\det[p_k(\lambda)]$. This implies that the rows of $p_k(\theta)$ are linearly dependent and there exists a vector v with two components such that

$$v^T p_k(\theta) = 0. \tag{2.17}$$

Using the matrix three-term recurrence we have

$$\theta[v^T p_0(\theta), \ldots, v^T p_{k-1}(\theta)] = [v^T p_0(\theta), \ldots, v^T p_{k-1}(\theta)] J_k.$$

Therefore θ is an eigenvalue of J_k. The determinant of $p_k(\lambda)$ is a polynomial of degree $2k$ in λ. Hence, there exists $2k$ zeros of the determinant and therefore all eigenvalues are zeros of $\det[p_k(\lambda)]$. □

THEOREM 2.21 *For λ and μ real, we have the matrix analog of the Christoffel–Darboux identity,*

$$(\lambda - \mu) \sum_{j=0}^{k-1} p_j(\mu) p_j^T(\lambda) = p_{k-1}(\mu) \Gamma_k^T p_k^T(\lambda) - p_k(\mu) \Gamma_k p_{k-1}^T(\lambda). \tag{2.18}$$

Proof. Using the three-term recurrence (2.15), we have

$$\Gamma_{j+1}^T p_{j+1}^T(\lambda) = \lambda p_j^T(\lambda) - \Omega_{j+1} p_j^T(\lambda) - \Gamma_j p_{j-1}^T(\lambda) \tag{2.19}$$

and

$$p_{j+1}(\mu) \Gamma_{j+1} = \mu p_j(\mu) - p_j(\mu) \Omega_{j+1} - p_{j-1}(\mu) \Gamma_j^T. \tag{2.20}$$

Multiplying equation (2.19) on the left by $p_j(\mu)$ and equation (2.20) on the right by $p_j^T(\lambda)$ gives

$$p_j(\mu) \Gamma_{j+1}^T p_{j+1}^T(\lambda) - p_{j+1}(\mu) \Gamma_{j+1} p_j^T(\lambda) =$$

$$(\lambda - \mu) p_j(\mu) p_j^T(\lambda) - p_j(\mu) \Gamma_j p_{j-1}^T(\lambda) + p_{j-1}(\mu) \Gamma_j^T p_j^T(\lambda).$$

Summing these equalities over j, some terms cancel and we obtain the desired result. □

Chapter Three

Properties of Tridiagonal Matrices

We have seen that the tridiagonal Jacobi matrices are closely linked to orthogonal polynomials since they describe the three-term recurrence satisfied by these polynomials. We will see in chapter 4 that they are also key ingredients in the Lanczos and conjugate gradient algorithms. In this chapter we summarize some properties of tridiagonal matrices that will be useful in the next chapters.

3.1 Similarity

Let us consider a nonsymmetric tridiagonal matrix of order k, with real coefficients

$$T_k = \begin{pmatrix} \alpha_1 & \omega_1 & & & \\ \beta_1 & \alpha_2 & \omega_2 & & \\ & \ddots & \ddots & \ddots & \\ & & \beta_{k-2} & \alpha_{k-1} & \omega_{k-1} \\ & & & \beta_{k-1} & \alpha_k \end{pmatrix},$$

and $\beta_i \neq \omega_i$, $i = 1, \ldots, k - 1$. Then we have the following result.

PROPOSITION 3.1 *Assume that the coefficients ω_j, $j = 1, \ldots, k - 1$ are different from zero and the products $\beta_j \omega_j$ are positive. Then, the matrix T_k is similar to a symmetric tridiagonal matrix. Therefore, its eigenvalues are real.*

Proof. Let γ_j, $j = 1, \ldots, k$ be the diagonal elements of a diagonal matrix D_k. Then, we consider $D_k^{-1} T_k D_k$ which is similar to T_k. The diagonal coefficients of this matrix are α_j. If we want to have the $(2, 1)$ element equal to the $(1, 2)$ element, we need to have

$$\frac{\gamma_1}{\gamma_2} \beta_1 = \frac{\gamma_2}{\gamma_1} \omega_1.$$

If we take, for instance, $\gamma_1 = 1$, we have

$$\gamma_2^2 = \frac{\beta_1}{\omega_1}.$$

To have the symmetry of all the nondiagonal coefficients, we find by induction

$$\gamma_j^2 = \frac{\beta_{j-1} \cdots \beta_1}{\omega_{j-1} \cdots \omega_1}, \quad j = 2, \ldots k.$$

With the hypothesis, we see that we can compute real values γ_j. $\qquad \square$

Using this result, we consider only symmetric tridiagonal matrices in this chapter.

3.2 Cholesky Factorizations of a Tridiagonal Matrix

Let us consider a symmetric tridiagonal matrix of order k,

$$
J_k = \begin{pmatrix}
\alpha_1 & \beta_1 & & & & \\
\beta_1 & \alpha_2 & \beta_2 & & & \\
& \ddots & \ddots & \ddots & & \\
& & \beta_{k-2} & \alpha_{k-1} & \beta_{k-1} \\
& & & \beta_{k-1} & \alpha_k
\end{pmatrix},
$$

where the values $\beta_j, j = 1, \ldots, k - 1$ are assumed to be nonzero. We remark that the determinant of J_k verifies a three-term recurrence.

LEMMA 3.2

$$
\det(J_{k+1}) = \alpha_{k+1} \det(J_k) - \beta_k^2 \det(J_{k-1})
$$

with initial conditions

$$
\det(J_1) = \alpha_1, \quad \det(J_2) = \alpha_1 \alpha_2 - \beta_1^2.
$$

Proof. This is obtained by expanding the determinant of J_{k+1} along the last row or column of J_{k+1}. □

The eigenvalues of J_k are the zeros of $\det(J_k - \lambda I)$. From lemma 3.2, we see that the zeros do not depend on the signs of the coefficients $\beta_j, j = 1, \ldots, k - 1$. Hence, we can choose the sign of these coefficients to our convenience. Let us assume they are positive and thus that J_k is a Jacobi matrix.

We would like to find a factorization of this matrix. We consider a Cholesky-like factorization of J_k. Let Δ_k be a diagonal matrix with diagonal elements $\delta_j, j = 1, \ldots, k$ and let L_k be a lower triangular (bidiagonal) matrix to be determined,

$$
L_k = \begin{pmatrix}
1 & & & & \\
l_1 & 1 & & & \\
& \ddots & \ddots & & \\
& & l_{k-2} & 1 & \\
& & & l_{k-1} & 1
\end{pmatrix}.
$$

Let the factorization of J_k be $J_k = L_k \Delta_k L_k^T$. By identification it is easy to see that we have

$$
\delta_1 = \alpha_1, \quad l_1 = \beta_1/\delta_1,
$$

$$
\delta_j = \alpha_j - \frac{\beta_{j-1}^2}{\delta_{j-1}}, \, j = 2, \ldots, k, \quad l_j = \beta_j/\delta_j, \, j = 2, \ldots, k - 1
$$

The factorization can be completed if no δ_j is zero for $j = 1, \ldots, k - 1$. This does not happen if the matrix J_k is positive definite. In such a case, all the elements δ_j are positive and the genuine Cholesky factorization (see for instance Golub and Van Loan [154]) can be obtained from Δ_k and L_k by writing $\Delta_k = \Delta_k^{1/2} \Delta_k^{1/2}$.

Using this transformation we have $J_k = L_k^C(L_k^C)^T$ with $L_k^C = L_k \Delta_k^{1/2}$, which is

$$
L_k^C =
\begin{pmatrix}
\sqrt{\delta_1} & & & & \\
\frac{\beta_1}{\sqrt{\delta_1}} & \sqrt{\delta_2} & & & \\
& \ddots & \ddots & & \\
& & \frac{\beta_{k-2}}{\sqrt{\delta_{k-2}}} & \sqrt{\delta_{k-1}} & \\
& & & \frac{\beta_{k-1}}{\sqrt{\delta_{k-1}}} & \sqrt{\delta_k}
\end{pmatrix}.
$$

The factorization can also be written as $J_k = L_k^D \Delta_k^{-1} (L_k^D)^T$ with

$$
L_k^D =
\begin{pmatrix}
\delta_1 & & & & \\
\beta_1 & \delta_2 & & & \\
& \ddots & \ddots & & \\
& & \beta_{k-2} & \delta_{k-1} & \\
& & & \beta_{k-1} & \delta_k
\end{pmatrix}.
$$

Clearly, we see that the only elements we have to compute and store are the diagonal elements δ_j, $j = 1, \ldots, k$. The last factorization is more interesting computationally since we do not need to compute square roots or the elements l_i. If we want to solve a linear system $J_k x = c$, we successively solve

$$
L_k^D y = c, \quad (L_k^D)^T x = \Delta_k y.
$$

Looking at the components, we have

$$
y_1 = \frac{c_1}{\delta_1}, \quad y_j = \frac{c_j - \beta_{j-1} y_{j-1}}{\delta_j}, \, j = 2, \ldots, k,
$$

$$
x_k = y_k, \quad x_j = y_j - \frac{\beta_j}{\delta_j} x_{j+1}, \, j = k - 1, \ldots, 1.
$$

Note that it is better to store Δ_k^{-1} instead of Δ_k if we have several linear systems to solve with the same matrix and different right-hand sides.

The matrices J_j, $j < k$ are leading matrices of J_k. We introduce also the trailing matrices of J_k,

$$
J_{j,k} =
\begin{pmatrix}
\alpha_j & \beta_j & & & \\
\beta_j & \alpha_{j+1} & \beta_{j+1} & & \\
& \ddots & \ddots & \ddots & \\
& & \beta_{k-2} & \alpha_{k-1} & \beta_{k-1} \\
& & & \beta_{k-1} & \alpha_k
\end{pmatrix}.
$$

The determinants of these matrices satisfy a three-term recurrence,

$$
\det(J_{j,k}) = \alpha_j \det(J_{j+1,k}) - \beta_j^2 \det(J_{j+2,k}).
$$

The previous Cholesky-like factorizations proceed from top to bottom giving a lower triangular matrix L_k. One can also proceed from bottom to top, obtaining an upper triangular matrix. The UL factorization from the bottom to the top is written as $J_k = \bar{L}_k^T D_k^{-1} \bar{L}_k$, with $\bar{L}_k$ a lower bidiagonal matrix

$$
\bar{L}_k = \begin{pmatrix}
d_1^{(k)} & & & & & \\
\beta_1 & d_2^{(k)} & & & & \\
& \ddots & \ddots & & & \\
& & \beta_{k-2} & d_{k-1}^{(k)} & \\
& & & \beta_{k-1} & d_k^{(k)}
\end{pmatrix},
$$

and D_k a diagonal matrix whose diagonal elements are $d_j^{(k)}$. Assuming the decomposition exists, it is easy to see that

$$
d_k^{(k)} = \alpha_k, \quad d_j^{(k)} = \alpha_j - \frac{\beta_j^2}{d_{j+1}^{(k)}}, \; j = k-1,\dots,1.
$$

The diagonal elements of the UL factorization are denoted with an upper index (k) because when we augment the matrix from J_k to J_{k+1} all the diagonal elements change, contrary to the LU factorization for which it is enough to compute δ_{k+1} from the previous elements δ_j obtained from the factorization of J_k.

From the LU and UL factorizations we can obtain all the so-called "twisted" factorizations of J_k. A twisted factorization starts both at the top and at the bottom of the matrix. The forward and backward steps meet at some given index l, $1 \leq l \leq k$. Then, $J_k = M_k \Omega_k M_k^T$, M_k being lower bidiagonal at the top for rows whose index is smaller than l and upper bidiagonal at the bottom for rows whose index is larger than l. The elements ω_j of the diagonal matrix Ω_k are given by

$$
\omega_1 = \alpha_1, \quad \omega_j = \alpha_j - \frac{\beta_{j-1}^2}{\omega_{j-1}}, \; j = 2,\dots,l-1,
$$

$$
\omega_k = \alpha_k, \quad \omega_j = \alpha_j - \frac{\beta_j^2}{\omega_{j+1}}, \; j = k-1,\dots,l+1,
$$

$$
\omega_l = \alpha_l - \frac{\beta_{l-1}^2}{\omega_{l-1}} - \frac{\beta_l^2}{\omega_{l+1}}.
$$

The twisted factorizations are useful to establish some theoretical results on the inverse of a tridiagonal matrix but also computationally to solve tridiagonal linear systems. The importance of Cholesky-like factorizations for tridiagonal matrices have been emphasized in Parlett [267].

3.3 Eigenvalues and Eigenvectors

The eigenvalues of J_k are the zeros of $\det(J_k - \lambda I)$. We can consider the LU and UL factorizations of $J_k - \lambda I$ when they exist, that is, when λ is different from the eigenvalues of J_k. From the previous section we obtain functions $\delta_j(\lambda)$ and $d_j^{(k)}(\lambda)$ and

$$
\det(J_k - \lambda I) = \delta_1(\lambda) \cdots \delta_k(\lambda) = d_1^{(k)}(\lambda) \cdots d_k^{(k)}(\lambda).
$$

This shows that

$$\delta_k(\lambda) = \frac{\det(J_k - \lambda I)}{\det(J_{k-1} - \lambda I)}, \quad d_1^{(k)}(\lambda) = \frac{\det(J_k - \lambda I)}{\det(J_{2,k} - \lambda I)}.$$

Hence, both $\delta_k(\lambda)$ and $d_1^{(k)}(\lambda)$ are rational functions of λ. The poles of $\delta_k(\lambda)$ are the eigenvalues of J_{k-1} and the poles of $d_1^{(k)}(\lambda)$ are the eigenvalues of $J_{2,k}$. Let us denote by $\theta_j^{(k)}$ the eigenvalues of J_k that are real numbers. The previous results lead to a proof of the famous Cauchy interlacing property.

THEOREM 3.3 *Let us denote by $\theta_j^{(k)}$ the eigenvalues of J_k that are real numbers. The eigenvalues of J_{k+1} strictly interlace the eigenvalues of J_k,*

$$\theta_1^{(k+1)} < \theta_1^{(k)} < \theta_2^{(k+1)} < \theta_2^{(k)} < \cdots < \theta_k^{(k)} < \theta_{k+1}^{(k+1)}.$$

The proof of this result can also be obtained by writing the eigenvector x corresponding to an eigenvalue θ of J_{k+1} as $x = (y \ \zeta)^T$ where y is a vector of length k, ζ is a real number and writing the equations satisfied by the components of x,

$$J_k y + \beta_k \zeta e^k = \theta y,$$

$$\beta_k y_k + \alpha_{k+1} \zeta = \theta \zeta.$$

Eliminating y from these relations, we obtain

$$(\alpha_{k+1} - \beta_k^2 ((e^k)^T (J_k - \theta I)^{-1} e^k)) \zeta = \theta \zeta.$$

The real number ζ is different from zero. Otherwise, θ would be an eigenvalue of J_k. Therefore, we have the following equation for θ,

$$\alpha_{k+1} - \beta_k^2 \sum_{j=1}^k \frac{\xi_j^2}{\theta_j^{(k)} - \theta} = \theta,$$

where ξ_j is the last component of the jth eigenvector of J_k. An equation like this one is called a "secular" equation. The function is monotone in each interval defined by the poles $\theta_j^{(k)}$. There is only one root in each interval and this proves the result. We will study secular equations in more details in chapter 9. Note that the quadratic form $(e^k)^T (J_k - \theta I)^{-1} e^k$ is an essential part of this equation. For bounds on the eigenvalues of tridiagonal matrices, see Golub [138].

Later in this book we will need some components of the eigenvectors of J_k, particularly the first and the last ones. We recall the following results whose proof can be found, for instance, in [239].

PROPOSITION 3.4 *Let $\chi_{j,k}(\lambda)$ be the determinant of $J_{j,k} - \lambda I$. The first components of the eigenvectors z^i of J_k are*

$$(z_1^i)^2 = \left| \frac{\chi_{2,k}(\theta_i^{(k)})}{\chi_{1,k}'(\theta_i^{(k)})} \right|,$$

that is

$$(z_1^i)^2 = \frac{\theta_i^{(k)} - \theta_1^{(2,k)}}{\theta_i^{(k)} - \theta_1^{(k)}} \cdots \frac{\theta_i^{(k)} - \theta_{i-1}^{(2,k)}}{\theta_i^{(k)} - \theta_{i-1}^{(k)}} \frac{\theta_i^{(2,k)} - \theta_i^{(k)}}{\theta_{i+1}^{(k)} - \theta_i^{(k)}} \cdots \frac{\theta_{k-1}^{(2,k)} - \theta_i^{(k)}}{\theta_k^{(k)} - \theta_i^{(k)}}.$$

The last components of the eigenvectors z^i of J_k satisfy

$$(z_k^i)^2 = \left| \frac{\chi_{1,k-1}(\theta_i^{(k)})}{\chi_{1,k}'(\theta_i^{(k)})} \right|,$$

that is,

$$(z_k^i)^2 = \frac{\theta_i^{(k)} - \theta_1^{(k-1)}}{\theta_i^{(k)} - \theta_1^{(k)}} \cdots \frac{\theta_i^{(k)} - \theta_{i-1}^{(k-1)}}{\theta_i^{(k)} - \theta_{i-1}^{(k)}} \frac{\theta_{i-1}^{(k-1)} - \theta_i^{(k)}}{\theta_{i+1}^{(k)} - \theta_i^{(k)}} \cdots \frac{\theta_{k-1}^{(k-1)} - \theta_i^{(k)}}{\theta_k^{(k)} - \theta_i^{(k)}}.$$

The components of the eigenvectors are also related to (the derivatives of) the functions $\delta_j(\lambda)$ and $d_j^{(k)}(\lambda)$.

PROPOSITION 3.5 *The first components of the eigenvectors of J_k are given by*

$$(z_1^i)^2 = \left| \frac{1}{[d_1^{(k)}]'(\theta_i^{(k)})} \right|.$$

For the last components we have

$$(z_k^i)^2 = \left| \frac{1}{\delta_k'(\theta_i^{(k)})} \right|.$$

3.4 Elements of the Inverse

We will see in the next chapters that we are also interested in some elements of the inverse of J_k, particularly the $(1, 1)$ element. Therefore, we now recall some results about the inverse of a tridiagonal matrix.

From Baranger and Duc-Jacquet [20], Meurant [234] and the references therein, it is known that there exist two sequences of numbers $\{u_i\}, \{v_i\}, i = 1, \ldots, k$ such that

$$J_k^{-1} = \begin{pmatrix} u_1 v_1 & u_1 v_2 & u_1 v_3 & \cdots & u_1 v_k \\ u_1 v_2 & u_2 v_2 & u_2 v_3 & \cdots & u_2 v_k \\ u_1 v_3 & u_2 v_3 & u_3 v_3 & \cdots & u_3 v_k \\ \vdots & \vdots & \vdots & \ddots & \vdots \\ u_1 v_k & u_2 v_k & u_3 v_k & \cdots & u_k v_k \end{pmatrix}.$$

Moreover, u_1 can be chosen arbitrarily, for instance $u_1 = 1$. From this, we see that to have all the elements of the inverse it is enough to compute the first column of the inverse (that is, $J_k^{-1} e^1$) to obtain the sequence $\{v_j\}$ and then the last column of the inverse (that is, $J_k^{-1} e^k$) to obtain the sequence $\{u_j\}$.

To solve $J_k v = e^1$, it is natural to use the UL factorization of J_k. In the first phase we have to solve $\bar{L}_k^T y = e^1$. All the components of the vector y are zero, except the first one, $y_1 = 1/d_1^{(k)}$. Going forward (and down) we obtain

$$v_1 = \frac{1}{d_1^{(k)}}, \quad v_j = (-1)^{j-1} \frac{\beta_1 \cdots \beta_{j-1}}{d_1^{(k)} \cdots d_j^{(k)}}, \ j = 2, \ldots, k.$$

To solve $v_k J_k u = e^k$, we use the LU factorization of J_k. In the first (forward) phase we solve $L_k^D y = e^k$. All the components of y are zero except the last one, $y_k = 1/(\delta_k v_k)$. Going backward (and up) we obtain

$$u_k = \frac{1}{\delta_k v_k}, \quad u_{k-j} = (-1)^j \frac{\beta_{k-j} \cdots \beta_{k-1}}{\delta_{k-j} \cdots \delta_k v_k}, \quad j = 1, \ldots, k-1.$$

This leads to the following result.

THEOREM 3.6 *The inverse of the symmetric tridiagonal matrix J_k is characterized as*

$$(J_k^{-1})_{i,j} = (-1)^{j-i} \beta_i \cdots \beta_{j-1} \frac{d_{j+1}^{(k)} \cdots d_k^{(k)}}{\delta_i \cdots \delta_k}, \quad \forall i, \forall j > i,$$

$$(J_k^{-1})_{i,i} = \frac{d_{i+1}^{(k)} \cdots d_k^{(k)}}{\delta_i \cdots \delta_k}, \quad \forall i,$$

where δ_j and $d_j^{(k)}$, $j = 1, \ldots, k$ are the diagonal elements of the LU and UL factorizations of J_k.

Proof. From the previous results, we have

$$u_i = (-1)^{-(i+1)} \frac{1}{\beta_1 \cdots \beta_{i-1}} \frac{d_1^{(k)} \cdots d_k^{(k)}}{\delta_i \cdots \delta_k}.$$

Since, for $j \geq i$, we have $(J_k^{-1})_{i,j} = u_i v_j$, we obtain the result.   □

The diagonal elements of the inverse of J_k can also be obtained using twisted factorizations.

THEOREM 3.7 *Let l be a fixed index and ω_j the diagonal elements of the corresponding twisted factorization of J_k. Then,*

$$(J_k^{-1})_{l,l} = \frac{1}{\omega_l}.$$

Proof. This is obtained by solving $J_k y = e^l$ and looking at the lth element of the solution. Since all the components of e^l are zero except the lth one, starting from the top and the bottom, all the components of the solution of the first phase are zero except for the lth one which is $1/\omega_l$. The second phase fills all the components but this is not a concern.   □

As we said before, we are particularly interested in the $(1, 1)$ element of the inverse. During the course of the previous proofs we have seen that

$$(J_k^{-1})_{1,1} = \frac{1}{d_1^{(k)}}.$$

So, if we want to know what $(J_{k+1}^{-1})_{1,1}$ is in relation with $(J_k^{-1})_{1,1}$, we have to find the relation between $d_1^{(k)}$ and $d_1^{(k+1)}$. This has been done in Meurant [239] in the proof of theorem 2.14. The result is the following.

PROPOSITION 3.8

$$d_1^{(k)} - d_1^{(k+1)} = \frac{(\beta_1 \cdots \beta_k)^2 d_1^{(k)} d_1^{(k+1)}}{\det(J_k) \det(J_{k+1})}.$$

This gives

$$\frac{1}{d_1^{(k+1)}} - \frac{1}{d_1^{(k)}} = \frac{(\beta_1 \cdots \beta_k)^2}{\det(J_k) \det(J_{k+1})}.$$

The relation between $(J_{k+1}^{-1})_{1,1}$ and $(J_k^{-1})_{1,1}$ can also be obtained by writing the matrix J_{k+1} in block form as

$$J_{k+1} = \begin{pmatrix} J_k & \beta_k e^k \\ \beta_k (e^k)^T & \alpha_{k+1} \end{pmatrix}.$$

The upper left block of J_{k+1}^{-1} (which contains the $(1,1)$ element) is the inverse of the Schur complement, that is,

$$\left(J_k - \frac{\beta_k^2}{\alpha_{k+1}} e^k (e^k)^T \right)^{-1}.$$

The matrix within parenthesis is a rank-one modification of J_k. The only term of J_k which is modified is the element (k, k). We use the Sherman–Morrison formula (see for instance Golub and Van Loan [154]) to obtain

$$\left(J_k - \frac{\beta_k^2}{\alpha_{k+1}} e^k (e^k)^T \right)^{-1} = J_k^{-1} + \frac{(J_k^{-1} e^k)((e^k)^T J_k^{-1})}{\frac{\alpha_{k+1}}{\beta_k^2} - (e^k)^T J_k^{-1} e^k}.$$

Let $l^k = J_k^{-1} e^k$ be the last column of the inverse of J_k. From the last relation, we have

$$(J_{k+1}^{-1})_{1,1} = (J_k^{-1})_{1,1} + \frac{\beta_k^2 (l_1^k)^2}{\alpha_{k+1} - \beta_k^2 l_k^k}. \tag{3.1}$$

It remains to compute l_1^k and l_k^k. This is done by using the LU factorization of J_k. We obtain

$$l_1^k = (-1)^{k-1} \frac{\beta_1 \cdots \beta_{k-1}}{\delta_1 \cdots \delta_k}, \quad l_k^k = \frac{1}{\delta_k}.$$

To simplify the formulas, we note that

$$\alpha_{k+1} - \beta_k^2 l_k^k = \alpha_{k+1} - \frac{\beta_k^2}{\delta_k} = \delta_{k+1}.$$

Therefore, using either proposition 3.8 or the previous derivation, we have the following result.

THEOREM 3.9

$$(J_{k+1}^{-1})_{1,1} = (J_k^{-1})_{1,1} + \frac{(\beta_1 \cdots \beta_k)^2}{(\delta_1 \cdots \delta_k)^2 \delta_{k+1}}.$$

Proof. This is given by relation (3.1) and also by proposition 3.8 since it gives

$$\frac{1}{d_1^{(k+1)}} - \frac{1}{d_1^{(k)}} = \frac{(\beta_1 \cdots \beta_k)^2}{\det(J_k)\det(J_{k+1})},$$

and $\det(J_k) = \delta_1 \cdots \delta_k$. □

Hence, by computing the diagonal elements δ_k of the Cholesky-like factorization, we can compute incrementally the $(1,1)$ element of the inverse of the Jacobi matrix. We start with $(J_1^{-1})_{1,1} = 1/\alpha_1$ and $c_1 = 1$. From the previous steps, we compute δ_{k+1} and c_{k+1} by

$$\delta_{k+1} = \alpha_{k+1} - \frac{\beta_k^2}{\delta_k}, \quad c_{k+1} = c_k \frac{\beta_k^2}{\delta_k} \frac{1}{\delta_k}.$$

Then,

$$(J_{k+1}^{-1})_{1,1} = (J_k^{-1})_{1,1} + \frac{c_{k+1}}{\delta_{k+1}}.$$

In these formulas we see that, in fact, we can compute and store $\pi_k = 1/\delta_k$. Then,

$$t = \beta_k^2 \pi_k, \quad \delta_{k+1} = \alpha_{k+1} - t, \quad \pi_{k+1} = \frac{1}{\delta_{k+1}}, \quad c_{k+1} = c_k t \pi_k.$$

This gives

$$(J_{k+1}^{-1})_{1,1} = (J_k^{-1})_{1,1} + c_{k+1}\pi_{k+1},$$

and we can store $c_{k+1}\pi_{k+1}$ for the next step. Therefore, updating the $(1,1)$ element of the inverse of a symmetric tridiagonal matrix is particularly cheap. It costs only seven floating point operations (four multiplications, two additions and one division) and moreover we obtain the Cholesky-like factorization of the Jacobi matrices.

3.5 The QD Algorithm

In this section we review the QD algorithm since it can be used to solve some inverse problems in which we will be interested later. The QD algorithm is a method introduced by Heinz Rutishauser [286] to compute the eigenvalues of a tridiagonal matrix. For details on the QD algorithms, see Stiefel [311], Henrici [186] [185], Fernando and Parlett [106], Parlett [265] and Laurie [220]. The QD algorithm is also related to Padé-type approximation; see Brezinski [34]. It involves the Cholesky factorizations that we have reviewed in the previous sections.

An orthogonal QD algorithm has been developed by U. von Matt [337]. For a matrix QD algorithm and applications, see Dette and Studden [81].

3.5.1 The Basic QD Algorithm

If starting from the Cholesky factorization $J_k = L_k L_k^T$ of the tridiagonal positive definite matrix J_k we compute $\hat{J}_k = L_k^T L_k$ we have $\hat{J}_k = L_k^{-1} J_k L_k$. This shows that the matrix $\hat{J}_k$ is similar to the matrix J_k and therefore has the same eigenvalues. Now, we can compute the Cholesky factorization of $\hat{J}_k$ and iterate the process, obtaining a series of matrices $J_k^{(i)}$ with $J_k^{(0)} = J_k$, $J_k^{(1)} = \hat{J}_k, \ldots$. This is the basis of the LR algorithm of H. Rutishauser [287]. The off-diagonal elements tend to zero and one obtains in the limit the eigenvalues of J_k on the diagonal.

To use this algorithm we are faced with the problem of computing the Cholesky factorization $\hat{L}_k \hat{L}_k^T$ of $\hat{J}_k = L_k^T L_k$. Let us see if we can achieve this without explicitly computing $\hat{J}_k$. We have

$$
\hat{J}_k = \begin{pmatrix}
\delta_1 + \frac{\beta_1^2}{\delta_1} & \beta_1 \sqrt{\frac{\delta_2}{\delta_1}} & & & \\
\beta_1 \sqrt{\frac{\delta_2}{\delta_1}} & \delta_2 + \frac{\beta_2^2}{\delta_2} & \beta_2 \sqrt{\frac{\delta_3}{\delta_2}} & & \\
& \ddots & \ddots & \ddots & \\
& & \beta_{k-2} \sqrt{\frac{\delta_{k-1}}{\delta_{k-2}}} & \delta_{k-1} + \frac{\beta_{k-1}^2}{\delta_{k-1}} & \beta_{k-1} \sqrt{\frac{\delta_k}{\delta_{k-1}}} \\
& & & \beta_{k-1} \sqrt{\frac{\delta_k}{\delta_{k-1}}} & \delta_k
\end{pmatrix}.
$$

Let us denote the subdiagonal entries of $L_k = L_k^C$ by $\sqrt{\epsilon_j}$. Therefore, $\epsilon_j = \beta_j^2 / \delta_j$. The diagonal elements of the Cholesky-like factorization of $\hat{J}_k$ are given by

$$
\hat{\delta}_1 = \delta_1 + \frac{\beta_1^2}{\delta_1} = \delta_1 + \epsilon_1,
$$

$$
\hat{\delta}_j = \delta_j + \frac{\beta_j^2}{\delta_j} - \frac{\beta_{j-1}^2 \delta_j / \delta_{j-1}}{\hat{\delta}_{j-1}} = \delta_j + \epsilon_j - \frac{\epsilon_{j-1} \delta_j}{\hat{\delta}_{j-1}}, \quad j = 2, \ldots, k-1,
$$

$$
\hat{\delta}_k = \delta_k - \frac{\beta_{k-1}^2 \delta_k / \delta_{k-1}}{\hat{\delta}_{k-1}} = \delta_k - \frac{\epsilon_{k-1} \delta_k}{\hat{\delta}_{k-1}}.
$$

Let $\hat{\epsilon}_j = \beta_j^2 \delta_{j+1} / (\delta_j \hat{\delta}_j)$. The diagonal entries of $\hat{L}_k$ are $\sqrt{\hat{\delta}_j}$, the subdiagonal entries are $\sqrt{\hat{\epsilon}_j}$ and we have

$$
\hat{\epsilon}_j = \epsilon_j \frac{\delta_{j+1}}{\hat{\delta}_j}.
$$

The expression for $\hat{\delta}_j$ can be written as

$$
\hat{\delta}_j = \delta_j + \epsilon_j - \hat{\epsilon}_{j-1}.
$$

Therefore, in pseudocode the QD algorithm is the following; given δ_j and ϵ_j:

```
ê₀ = 0
for j=1:k-1
    δ̂ⱼ = (δⱼ − ε̂ⱼ₋₁) + εⱼ
    ε̂ⱼ = εⱼδⱼ₊₁/δ̂ⱼ
end
δ̂ₖ = δₖ − ε̂ₖ₋₁
```

Then, in the LR algorithm, we do $\delta_j = \hat{\delta}_j$, $\epsilon_j = \hat{\epsilon}_j$ and we iterate until the off-diagonal elements are small enough. The QD algorithm is particularly simple and elegant.

3.5.2 The Differential QD Algorithm

The QD algorithm can be modified (or improved) to remove subtractions since this is supposed to improve the stability. We remark that

$$\hat{\delta}_j = \frac{\delta_j}{\hat{\delta}_{j-1}}(\hat{\delta}_{j-1} - \epsilon_{j-1}) + \epsilon_j.$$

Let us introduce a new variable $\hat{t}_j = \hat{\delta}_j - \epsilon_j$. The previous equation shows that

$$\hat{t}_j = \hat{t}_{j-1}\frac{\delta_j}{\hat{\delta}_{j-1}},$$

and obviously, we have $\hat{\delta}_j = \hat{t}_j + \epsilon_j$. The differential QD algorithm (dqd) is

```
t = δ₁
for j=1:k-1
    δ̂ⱼ = t + εⱼ
    f = δⱼ₊₁/δ̂ⱼ
    ε̂ⱼ = fεⱼ
    t = ft
end
δ̂ₖ = t
```

3.5.3 The QD Algorithms with Shift

When using the QD algorithms to compute eigenvalues it is important to introduce shifts (and also deflation) to speed up convergence. This is relatively easy to do. The QD algorithm with a shift μ (qds(μ)) is

```
ε̂₀ = 0
for j=1:k-1
    δ̂ⱼ = (δⱼ - ε̂ⱼ₋₁) + εⱼ - μ
    ε̂ⱼ = εⱼδⱼ₊₁/δ̂ⱼ
end
δ̂ₖ = δₖ - ε̂ₖ₋₁ - μ
```

The differential QD algorithm with shift (dqds(μ)) is:

```
t = δ₁ - μ
for j=1:k-1
    δ̂ⱼ = t + εⱼ
    f = δⱼ₊₁/δ̂ⱼ
    ε̂ⱼ = fεⱼ
    t = ft - μ
end
δ̂ₖ = t
```

Of course, shifting introduces subtractions even in the differential QD algorithm. For an implementation, see Parlett and Marques [269].

3.5.4 Cholesky Factorization of a Shifted Tridiagonal Matrix

To start the QD algorithm from a tridiagonal matrix J_k, it is necessary to compute the Cholesky factorization of the shifted matrix. The same problem arises when using one of the QD algorithms to compute eigenvalues. Therefore, let us consider the $\hat{L}\hat{D}\hat{L}^T$ factorization of the shifted matrix $J_k - \mu I$ when we know the factorization of $J_k = LDL^T$. Using the elements of J_k (which we may eventually not want to compute explicitly) we have

$$\hat{\delta}_1 = \alpha_1 - \mu, \ \hat{\delta}_j = \alpha_j - \mu - \frac{\beta_{j-1}^2}{\hat{\delta}_{j-1}}, \ j = 2, \ldots, k, \quad \hat{l}_j = \frac{\beta_j}{\hat{\delta}_j}, \ j = 1, \ldots, k-1.$$

But we have that $\alpha_1 = \delta_1$, $\alpha_j = \delta_j + \delta_{j-1}l_{j-1}$, $j = 2, \ldots, k$, $\beta_j = \delta_j l_j$, $j = 1, \ldots, k-1$. We can eliminate α_j and β_j from the formulas giving $\hat{\delta}_j$ and $\hat{l}_j$,

$$\hat{\delta}_1 = \delta_1 - \mu, \ \hat{\delta}_j = \delta_j - \mu + \delta_{j-1}l_{j-1}^2 - \frac{\delta_{j-1}^2 l_{j-1}^2}{\hat{\delta}_{j-1}}, \ j = 2, \ldots, k$$

and

$$\hat{l}_j = \frac{\delta_j l_j}{\hat{\delta}_j}, \ j = 1, \ldots, k-1.$$

Finally,

$$\hat{\delta}_j = \delta_j - \mu + \delta_{j-1}l_{j-1}^2 - \delta_{j-1}l_{j-1}\hat{l}_{j-1}, \ j = 2, \ldots, k.$$

This algorithm, which computes the new factorization of the shifted matrix from the factorization of J_k, has been named stqds by Dhillon and Parlett [84], even though we have seen that it is nothing other than the Cholesky factorization. The last formula can be rearranged to remove one subtraction. Let us introduce a new variable $s_j = \hat{\delta}_j - \delta_j$; then

$$s_j = l_{j-1}\hat{l}_{j-1}\left(\frac{l_{j-1}\delta_{j-1}}{\hat{l}_{j-1}} - \delta_{j-1}\right) - \mu$$

$$= l_{j-1}\hat{l}_{j-1}(\hat{\delta}_{j-1} - \delta_{j-1}) - \mu$$

$$= l_{j-1}\hat{l}_{j-1}s_{j-1} - \mu.$$

This algorithm is said to be the differential form of stqds and denoted by dstqds. It starts from the factorization of J_k and gives the factorization of $J_k - \mu I$:

```
s = -μ
for j=1:k-1
    δ̂_j = s + δ_j
    l̂_j = (δ_j l_j)/δ̂_j
    s = l̂_j l_j s - μ
end
δ̂_k = s + δ_k
```

One can also develop algorithms using the UL instead of the LU factorization; see Dhillon [83].

3.5.5 Relation to Finding the Poles of a Function from its Taylor Series

The following application comes from the exposition of Parlett [264]; see also Henrici [186]. We describe it because of its own interest and also because it is linked to moments, has an interesting relationship to the QD algorithm and has some connection with our main topic, which is estimation of bilinear forms.

Rutishauser's QD algorithm is linked to a classical problem of finding the singularities (poles) of a meromorphic function from the coefficients of its series at a regular point. In complex analysis, a meromorphic function on an open subset Ω of the complex plane is a function that is holomorphic on all Ω except for a set of isolated points, which are poles for the function. Every meromorphic function on Ω can be expressed as the ratio between two holomorphic functions, the poles being the zeros of the denominator. Holomorphic functions are functions defined on an open subset of the complex plane with complex values that are complex-differentiable at every point.

For two given vectors x and y, one considers the following rational function of the complex variable z:

$$f(z) = y^*(I - zA)^{-1}x,$$

where the star denotes the conjugate transpose. We are mainly interested in real symmetric matrices for which we can consider having $x = y$ and real vectors. The poles of f are the inverses of the eigenvalues of A (which are real if A is symmetric). If we are in the real symmetric case, we have the spectral decomposition $A = Q\Lambda Q^T$ where Q is orthogonal and Λ diagonal. Assuming we choose $x = y = e^1$, e^1 being the first column of the identity matrix, we have

$$f(z) = (e^1)^T Q(I - z\Lambda)^{-1}Q^T e^1 = \sum_{j=1}^{n} \frac{(q_1^j)^2}{1 - z\lambda_j},$$

where q_1^j are the first components of the eigenvectors of A. The function f is determined by its Taylor expansion at the origin which converges for $|z| \leq |\lambda_{\max}|^{-1}$. Let

$$f(z) = \sum_{j=0}^{\infty} \mu_j z^j.$$

The coefficients μ_j are $\mu_j = y^* A^j x$ which are the moments if $y = x$. A problem that has been considered at least since the nineteenth century is to find the poles of f from the coefficients of the Taylor series (moments). This problem was solved theoretically by J. Hadamard in 1892 [168]. The solution is given using Hankel determinants. Let $H_0^j = 1$ and

$$H_k^j = \det \begin{pmatrix} \mu_j & \mu_{j+1} & \cdots & \mu_{j+k-1} \\ \mu_{j+1} & \mu_{j+2} & \cdots & \mu_{j+k} \\ \vdots & \vdots & & \vdots \\ \mu_{j+k-1} & \mu_{j+k} & \cdots & \mu_{j+2k-2} \end{pmatrix}, \ j = 0, 1, \ldots, \ k = 1, 2, \ldots.$$

The results are that, for j large enough, $H_k^j \neq 0$ and as $j \to \infty$ the ratio $H_k^{j+1}/H_k^j \to \lambda_1 \cdots \lambda_k$ where the λ_i are the poles in decreasing modulus order. Therefore, the solution of the problem is

$$\lambda_k = \lim_{j \to \infty} \left(\frac{H_k^{j+1} H_{k-1}^j}{H_k^j H_{k-1}^{j+1}} \right).$$

Of course, this cannot be used in practice for the computation of the poles. A step towards the solution was obtained by A. C. Aitken in the 1920s when he was working on methods for finding all the zeros of a polynomial. The Hankel determinants can be arranged in a two-dimensional table:

$$
\begin{array}{cccccc}
1 \\
1 & H_1^0 \\
1 & H_1^1 & H_2^0 \\
1 & H_1^2 & H_2^1 & H_3^0 \\
1 & H_1^3 & H_2^2 & H_3^1 & H_4^0 \\
1 & H_1^4 & H_2^3 & H_3^2 & H_4^1 & H_5^0 \\
\end{array}
$$

If the function f has only a finite number N of poles, the H-table has only $N+1$ columns since all the other coefficients can be shown to be zero. Aitken proves the following relation between Hankel determinants:

$$(H_k^j)^2 - H_k^{j-1} H_k^{j+1} + H_{k+1}^{j-1} H_{k-1}^{j+1} = 0.$$

Note that this is nothing other than Sylvester's identity for determinants; see Gantmacher [122]. As noted by Parlett, for a point P in the table, this is $P^2 = NS - WE$ where N, S, W and E refer to the north, south, west, and east neighbors of P. To compute the determinants using this formula it is better to proceed by diagonals rather than by columns.

However, the Hankel determinants are not the most interesting variables in this problem. It was noted by H. Rutishauser that the proper variables are

$$q_k^j = \frac{H_k^{j+1} H_{k-1}^j}{H_k^j H_{k-1}^{j+1}}.$$

As we have seen, we have $q_k^j \to \lambda_k$ as $j \to \infty$. If one introduces the auxiliary quantity

$$e_k^j = \frac{H_{k-1}^{j+1} H_{k+1}^j}{H_k^j H_k^{j+1}},$$

then the relation between Hankel determinants shows that

$$q_k^j + e_k^j = q_k^{j+1} + e_{k-1}^{j+1}.$$

Moreover, we have

$$q_k^{j+1} e_k^{j+1} = q_{k+1}^j e_k^j.$$

We recognize that, even though the notations are different, these are the relations we have in the QD algorithm. They had been called the Rhombus rules by E. Stiefel. Putting the values of q_k^j and e_k^j in a so-called QD table,

$$0 = e_0^0$$
$$q_1^0$$
$$0 = e_0^1 \qquad e_1^0$$
$$q_1^1 \qquad q_2^0$$
$$0 = e_0^2 \qquad e_1^1 \qquad e_2^0$$
$$q_1^2 \qquad q_2^1 \qquad q_3^0$$
$$0 = e_0^3 \qquad e_1^2 \qquad e_2^1 \qquad e_3^0$$
$$q_1^3 \qquad q_2^2 \qquad q_3^1 \qquad q_4^0$$

$$\cdot \quad \cdot \quad \cdot \quad \cdot \quad \cdot \quad \cdot \quad \cdot \quad \cdot$$

we can compute the elements either columnwise moving to the right or diagonal-wise moving down. We can use the algorithms qd or dqd provided that we know the first diagonal $q_1^0, e_1^0, q_2^0, e_2^0, \dots$. It turns out that, in exact arithmetic, the first diagonal can be obtained by the nonsymmetric Lanczos algorithm (which we will study in chapter 4) with initial vectors x and y. If $x = y$ this reduces to the (symmetric) Lanczos algorithm. In fact, the elements of the first diagonal are given by the Cholesky factorization of the Jacobi matrix given by the Lanczos algorithm. This is almost obvious when we remember the way we have derived the QD algorithm from the Cholesky factorization of a Jacobi matrix.

Chapter Four

The Lanczos and Conjugate Gradient Algorithms

In this chapter we introduce the Lanczos algorithm for symmetric matrices as well as its block version and also the nonsymmetric Lanczos algorithm. These algorithms, which were devised to compute eigenvalues or to solve linear systems, are closely related to the moment problem and they will be used to compute quadrature formulas and to estimate bilinear forms. The Lanczos algorithm provides examples of orthonormal polynomials related to a discrete (usually unknown) measure. Moreover, we describe the Golub–Kahan bidiagonalization algorithms which are special versions of the Lanczos algorithm for matrices AA^T or $A^T A$. This is useful when solving least squares problems. We also consider the conjugate gradient (CG) algorithm since it can be derived from the Lanczos algorithm and it is the most used algorithm for solving positive definite symmetric linear systems. As we will see, CG is closely linked to the remainder of Gauss quadrature. Conversely, Gauss quadrature rules can be used to estimate norms of the error between the CG approximate solutions and the exact solution during the iterations. This also gives a reliable and cheap way to compute stopping criteria for the CG iterations.

4.1 The Lanczos Algorithm

Let A be a real symmetric matrix of order n. We introduce the Lanczos algorithm as a means of computing an orthogonal basis of a Krylov subspace. Let v be a given vector and

$$K_k = (\, v, \quad Av, \quad \cdots, \quad A^{k-1}v \,) \tag{4.1}$$

be the Krylov matrix of dimension $n \times k$. The subspace that is spanned by the columns of the matrix K_k is called a Krylov subspace and denoted by $\mathcal{K}_k(A, v)$ or $\mathcal{K}(A, v)$ when no confusion is possible. There is a maximal dimension $k = m \leq n$ for which the rank of K_k is k. For any v this maximal dimension is always less than the degree of the minimal polynomial of A.

The algorithm that is now called the Lanczos algorithm was introduced by C. Lanczos in 1950 [216] to construct a basis of the Krylov subspace (see also [217] for the solution of linear systems). It can be considered as a particular form of the Stieltjes algorithm that we will study in chapter 5. The natural basis of the Krylov subspace $\mathcal{K}(A, v)$ given by the columns of the Krylov matrix K_k is badly conditioned when k is large. In fact, when k increases, the vectors $A^k v$ tend to align with the eigenvector corresponding to the eigenvalue of A of largest modulus. Numerically, the Krylov vectors may lose their independence before the maximal dimension is reached. Note that the elements of the matrix $K_k^T A K_k$ are of the form $v^T A^{i+j} v$. They are moments corresponding to an unknown measure which depends on the eigenvalues of A.

The Lanczos algorithm constructs an orthonormal basis of the Krylov subspace $\mathcal{K}(A, v)$. We start the derivation of the algorithm by applying a variant of the Gram–Schmidt orthogonalization process (see for instance Golub and Van Loan [154]) to the Krylov basis (the columns of K_k) without the assumption that the matrix A is symmetric. Consider the set of vectors $v^{(j+1)} = A v^{(j)}$ with $v^{(1)} = v$; then $\mathcal{K}(A, v)$ is spanned by the vectors $v^{(j)}, j = 1, \dots, k$. For constructing orthogonal basis vectors v^j, instead of orthogonalizing $A^j v$ against the previous vectors, we can orthogonalize $A v^j$. Starting from $v^1 = v$ (normalized if necessary), the algorithm for computing the $(j + 1)$st vector of the basis using the previous vectors is

$$h_{i,j} = (A v^j, v^i), \quad i = 1, \dots, j,$$

$$\bar{v}^j = A v^j - \sum_{i=1}^{j} h_{i,j} v^i,$$

$$h_{j+1,j} = \|\bar{v}^j\|, \quad \text{if } h_{j+1,j} = 0 \text{ then stop},$$

$$v^{j+1} = \frac{\bar{v}^j}{h_{j+1,j}}.$$

The second step is the subtraction of the components of $A v^j$ on the previous basis vectors from $A v^j$. Then the resulting vector is normalized if this is possible. It is easy to verify that the vectors v^j span the Krylov subspace and that they are orthonormal. This orthogonalization process is known as the Arnoldi algorithm [11]. If we collect the vectors $v^j, j = 1, \dots, k$ in a matrix V_k, the relations defining the vector v^{k+1} can be written in matrix form as

$$A V_k = V_k H_k + h_{k+1,k} v^{k+1} (e^k)^T, \tag{4.2}$$

where H_k is an upper Hessenberg matrix with elements $h_{i,j}$; note that $h_{i,j} = 0, j = 1, \dots, i - 2, i > 2$. As before, the vector e^k is the kth column of the identity matrix of order k. If we suppose that the matrix A is symmetric, then the matrix H_k is also symmetric since, by multiplying equation (4.2) by V_k^T and using orthogonality, we have $H_k = V_k^T A V_k$. Clearly, a symmetric Hessenberg matrix is tridiagonal. Therefore, we denote H_k by J_k (since this is a Jacobi matrix, the elements in the sub- and superdiagonals being strictly positive) and we have $h_{i,j} = 0, j = i + 2, \dots, k$. This implies that $\bar{v}^k$ and hence the new vector v^{k+1} can be computed by using only the two previous vectors v^k and v^{k-1}. This describes the Lanczos algorithm. In fact, V_k is the orthonormal matrix (that is, such that $V_k^T V_k = I$) involved in a QR factorization of the Krylov matrix K_k, and the matrix $K_k^T A K_k$ is similar to $J_k = V_k^T A V_k$.

The relation for the matrix V_k whose columns are the orthogonal basis vectors can be written as

$$AV_k = V_k J_k + \eta_k v^{k+1}(e^k)^T, \tag{4.3}$$

where η_i denotes the nonzero off-diagonal entries of J_k. We also have $AV_n = V_n J_n$, if no v^j is zero before step n, since $v^{n+1} = 0$ because v^{n+1} is a vector orthogonal to a set of n orthogonal vectors in a space of dimension n. Otherwise there exists an $m < n$ for which $AV_m = V_m J_m$ and the algorithm has found an invariant subspace of A, the eigenvalues of J_m being eigenvalues of A. Equation (4.3) describes in matrix form the elegant Lanczos algorithm, which is written, starting from a nonzero vector $v^1 = v/\|v\|$, $\alpha_1 = (Av^1, v^1)$, $\tilde{v}^2 = Av^1 - \alpha_1 v^1$ and then, for $k = 2, 3, \ldots$,

$$\eta_{k-1} = \|\tilde{v}^k\|,$$

$$v^k = \frac{\tilde{v}^k}{\eta_{k-1}},$$

$$\alpha_k = (v^k, Av^k) = (v^k)^T Av^k,$$

$$\tilde{v}^{k+1} = Av^k - \alpha_k v^k - \eta_{k-1} v^{k-1}.$$

The real numbers α_j and η_j are the nonzero coefficients of the tridiagonal matrix J_k.

A variant of the Lanczos algorithm has been proposed by Paige [255] to improve the local orthogonality (with the previous vector) in finite precision computations. It replaces the third and fourth steps by

$$\alpha_k = (v^k)^T (Av^k - \eta_{k-1} v^{k-1}),$$

$$\tilde{v}^{k+1} = (Av^k - \eta_{k-1} v^{k-1}) - \alpha_k v^k.$$

Note that this variant can be implemented by using only two vectors of storage instead of three for the basic formulation. It corresponds to using the modified Gram–Schmidt orthogonalization process; see Golub and Van Loan [154]. In exact arithmetic both versions are mathematically equivalent but the modified variant better preserves the local orthogonality in finite precision arithmetic.

Since we can suppose that $\eta_i \neq 0$, the tridiagonal Jacobi matrix J_k has real and simple eigenvalues which we denote by $\theta_j^{(k)}$. They are known as the Ritz values and are the approximations of the eigenvalues of A given by the Lanczos algorithm. For our purposes, the most important property of the Lanczos algorithm is that the Lanczos vectors v^k are given as a polynomial in A applied to the initial vector v^1 as stated in the following theorem.

THEOREM 4.1 *Let $\chi_k(\lambda)$ be the determinant of $J_k - \lambda I$ (which is a monic polynomial); then*

$$v^k = p_k(A)v^1, \quad p_k(\lambda) = (-1)^{k-1} \frac{\chi_{k-1}(\lambda)}{\eta_1 \cdots \eta_{k-1}}, \quad k > 1, \quad p_1 \equiv 1.$$

The polynomials p_k of degree $k-1$ are called the normalized Lanczos polynomials.

Proof. As we have seen in chapter 3, the determinant of the tridiagonal matrix J_k satisfies a three-term recurrence relation,

$$\det(J_{k+1}) = \alpha_{k+1}\det(J_k) - \eta_k^2\det(J_{k-1}),$$

with initial conditions

$$\det(J_1) = \alpha_1, \quad \det(J_2) = \alpha_1\alpha_2 - \eta_1^2.$$

Comparison of the three-term recurrence for v^k and for the determinant $\chi_k(\lambda)$ of $J_k - \lambda I$ gives the result.   □

Obviously, the polynomials p_k satisfy a scalar three-term recurrence,

$$\eta_k p_{k+1}(\lambda) = (\lambda - \alpha_k)p_k(\lambda) - \eta_{k-1}p_{k-1}(\lambda), \ k = 1, 2, \ldots$$

with initial conditions $p_0 \equiv 0$, $p_1 \equiv 1$. Therefore, by the Favard theorem we have that there exists a measure for which these polynomials are orthogonal. It turns out that we are able to explicitly write down the measure, unfortunately in terms of unknown quantities.

THEOREM 4.2 *Consider the Lanczos vectors v^k. There exists a measure α (defined in the proof) such that*

$$(v^k, v^l) = \langle p_k, p_l \rangle = \int_a^b p_k(\lambda)p_l(\lambda)d\alpha(\lambda),$$

where $a \leq \lambda_1 = \lambda_{\min}$ and $b \geq \lambda_n = \lambda_{\max}$, $\lambda_{\min}$ and $\lambda_{\max}$ being the smallest and largest eigenvalues of A, and p_i are the Lanczos polynomials associated with A and v^1.

Proof. The matrix A being symmetric, let $A = Q\Lambda Q^T$ be the spectral decomposition of A, with Q orthonormal and Λ being the diagonal matrix of the eigenvalues λ_i such that

$$\lambda_1 \leq \lambda_2 \leq \cdots \leq \lambda_n.$$

Since the vectors v^j are orthonormal and $p_k(A) = Qp_k(\Lambda)Q^T$, we have

$$\begin{aligned}
(v^k, v^l) &= (v^1)^T p_k(A)^T p_l(A)v^1 \\
&= (v^1)^T Q p_k(\Lambda)Q^T Q p_l(\Lambda)Q^T v^1 \\
&= (v^1)^T Q p_k(\Lambda)p_l(\Lambda)Q^T v^1 \\
&= \sum_{j=1}^n p_k(\lambda_j)p_l(\lambda_j)[\hat{v}_j]^2,
\end{aligned}$$

where $\hat{v} = Q^T v^1$. This describes a discrete inner product for the polynomials p_k and p_l. The last sum can be written as an integral for a measure α which is piecewise constant (here we suppose for the sake of simplicity that the eigenvalues of A are distinct):

$$\alpha(\lambda) = \begin{cases} 0, & \text{if } \lambda < \lambda_1, \\ \sum_{j=1}^i [\hat{v}_j]^2, & \text{if } \lambda_i \leq \lambda < \lambda_{i+1}, \\ \sum_{j=1}^n [\hat{v}_j]^2, & \text{if } \lambda_n \leq \lambda. \end{cases}$$

The measure α has a finite number of points of increase at the (unknown) eigenvalues of A. □

The Lanczos algorithm provides an example of orthonormal polynomials for an unknown measure. The polynomials are described by their recurrence coefficients (which depend also on the starting vector v^1) in the Jacobi matrix J_k. Note that the inner product of two polynomials p_k and p_l can be computed by the inner product of the two Lanczos vectors v^k and v^l.

We note that the Lanczos algorithm can be used also to solve linear systems $Ax = c$ when A is symmetric and c is a given vector; see Lanczos [217]. Let x^0 be a given starting vector and $r^0 = c - Ax^0$ be the corresponding residual. Then one defines the first Lanczos vector as $v = v^1 = r^0/\|r^0\|$. The approximate solution x^k at iteration k is given by

$$x^k = x^0 + V_k y^k,$$

where the vector y^k of dimension k is obtained by requesting the residual $r^k = c - Ax^k$ to be orthogonal to the Krylov subspace of dimension k. This gives the condition $V_k^T r^k = 0$. By using the definition of the residual vector, we have

$$V_k^T r^k = V_k^T c - V_k^T Ax^0 - V_k^T AV_k y^k = V_k^T r^0 - J_k y^k.$$

But $r^0 = \|r^0\|v^1$ and, because of the orthogonality of the Lanczos vectors v^j, we have $V_k^T r^0 = \|r^0\|e^1$. Therefore, the vector y^k is obtained by solving a tridiagonal linear system

$$J_k y^k = \|r^0\|e^1,$$

at every iteration. In theory, the Lanczos algorithm constructs an orthogonal basis of the Krylov subspace. However, when using this algorithm on a computer we often lose the orthogonality of the computed Lanczos vectors. This problem arises because of rounding errors. Moreover, even though the eigenvalues of the Jacobi matrices must be simple, eigenvalues which are very close to already computed ones appear again and again during the computation. The study of these problems was started by Chris Paige in his Ph.D. thesis [255] in 1971. For a summary of his results and also more recent insights into the convergence of the Ritz values to the eigenvalues of A, see Meurant [239] and Meurant and Strakoš [242]. In fact, the rounding errors begin to increase when a Ritz value starts converging to an eigenvalue. The appearance of multiple copies delays the convergence toward the other eigenvalues which are not yet approximated. A remedy for these problems is to reorthogonalize the Lanczos vectors. A complete reorthogonalization can be done at every iteration but this is very costly. Better and cheaper strategies have been developed by Parlett [270] and Simon [303].

4.2 The Nonsymmetric Lanczos Algorithm

When the matrix A is not symmetric we cannot generally construct a vector v^{k+1} orthogonal to all the previous basis vectors by using only the two previous vectors

v^k and v^{k-1}. In other words, we lose the nice property of having a short recurrence. To work around this problem, an algorithm for nonsymmetric matrices was introduced by C. Lanczos in 1950 [216]. Its goal is to construct two biorthogonal sequences of vectors for a nonsymmetric matrix A. Fortunately, this can be done using short recurrences. The drawback is that the algorithm may break down. The standard development of the Lanczos algorithm for nonsymmetric matrices depends upon using the matrix A and its transpose A^T.

We choose two starting vectors v^1 and $\tilde{v}^1$ with $(v^1, \tilde{v}^1) \neq 0$ normalized such that $(v^1, \tilde{v}^1) = 1$. We set $v^0 = \tilde{v}^0 = 0$. Then for $k = 1, 2, \ldots$

$$z^k = Av^k - \omega_k v^k - \eta_{k-1} v^{k-1},$$
$$w^k = A^T \tilde{v}^k - \omega_k \tilde{v}^k - \tilde{\eta}_{k-1} \tilde{v}^{k-1},$$

the coefficient ω_k being computed as

$$\omega_k = (\tilde{v}^k, Av^k).$$

The other coefficients η_k and $\tilde{\eta}_k$ are chosen (provided $(z^k, w^k) \neq 0$) such that

$$\eta_k \tilde{\eta}_k = (z^k, w^k),$$

and the new vectors at step $k + 1$ are given by

$$v^{k+1} = \frac{z^k}{\tilde{\eta}_k}, \quad \tilde{v}^{k+1} = \frac{w^k}{\eta_k}.$$

These relations can be written in matrix form. Let

$$J_k = \begin{pmatrix} \omega_1 & \eta_1 & & & \\ \tilde{\eta}_1 & \omega_2 & \eta_2 & & \\ & \ddots & \ddots & \ddots & \\ & & \tilde{\eta}_{k-2} & \omega_{k-1} & \eta_{k-1} \\ & & & \tilde{\eta}_{k-1} & \omega_k \end{pmatrix}$$

and

$$V_k = [v^1 \cdots v^k], \quad \tilde{V}_k = [\tilde{v}^1 \cdots \tilde{v}^k].$$

Then

$$AV_k = V_k J_k + \tilde{\eta}_k v^{k+1} (e^k)^T,$$
$$A^T \tilde{V}_k = \tilde{V}_k J_k^T + \eta_k \tilde{v}^{k+1} (e^k)^T.$$

THEOREM 4.3 *If the nonsymmetric Lanczos algorithm does not break down with $\eta_k \tilde{\eta}_k$ being zero, the algorithm yields biorthogonal vectors such that*

$$(\tilde{v}^i, v^j) = 0, \, i \neq j, \quad i, j = 1, 2, \ldots.$$

The vectors $v^1, \ldots, v^k$ span $\mathcal{K}_k(A, v^1)$ and $\tilde{v}^1, \ldots, \tilde{v}^k$ span $\mathcal{K}_k(A^T, \tilde{v}^1)$. The two sequences of vectors can be written as

$$v^k = p_k(A)v^1, \quad \tilde{v}^k = \tilde{p}_k(A^T)\tilde{v}^1,$$

where p_k and $\tilde{p}_k$ are polynomials of degree $k - 1$.

Proof. These properties follow straightforwardly from the definition of the vectors v^k and $\tilde{v}^k$. □

The polynomials p_k and $\tilde{p}_k$ satisfy three-term recurrences

$$\tilde{\eta}_k p_{k+1} = (\lambda - \omega_k) p_k - \eta_{k-1} p_{k-1},$$

$$\eta_k \tilde{p}_{k+1} = (\lambda - \omega_k) \tilde{p}_k - \tilde{\eta}_{k-1} \tilde{p}_{k-1}.$$

The algorithm breaks down if at some step we have $(z^k, w^k) = 0$. There are two different cases.

a) $z^k = 0$ and/or $w^k = 0$. In both cases we have found an invariant subspace. If $z^k = 0$ we can compute the eigenvalues or the solution of the linear system $Ax = c$. If $z^k \neq 0$ and $w^k = 0$, the only way to deal with this situation is to restart the algorithm with another vector $\tilde{v}^1$. Usually use of a random initial vector is enough to avoid this kind of breakdown.

b) The more dramatic situation (which is called a "serious breakdown") is when $(z^k, w^k) = 0$ with z^k and $w^k \neq 0$. Then, a way to solve this problem is to use a look-ahead strategy. The solution is to construct the vectors v^{k+1} and $\tilde{v}^{k+1}$ at step k maintaining biorthogonality only in a blockwise sense. If this is not possible, one tries to construct also vectors v^{k+2} and $\tilde{v}^{k+2}$ and so on. The worst case is when we reach the order of the matrix A without having been able to return to the normal situation. This is known as an incurable breakdown.

In finite precision arithmetic, it is unlikely that we get $(z^k, w^k) = 0$ with z^k and $w^k \neq 0$. However, it may happen that (z^k, w^k) is small. This is known as a near breakdown and it is really this problem that look-ahead strategies must deal with; see Freund, Gutknecht and Nachtigal [115] and Brezinski, Redivo-Zaglia and Sadok [38].

In the next chapters, for computational purposes, we will use the nonsymmetric Lanczos algorithm even for a symmetric matrix $A = A^T$. In this particular application it is possible to choose η_k and $\tilde{\eta}_k$ such that

$$\eta_k = \pm\tilde{\eta}_k = \pm\sqrt{|(z^k, w^k)|},$$

with, for instance, $\eta_k \geq 0$ and $\tilde{\eta}_k = \text{sgn}[(z^k, w^k)]\,\eta_k$. Then

$$\tilde{p}_k = \pm p_k.$$

Note that if the off-diagonal elements are different from zero, the matrix J_k is diagonally similar to a symmetric matrix.

4.3 The Golub–Kahan Bidiagonalization Algorithms

When we wish to solve a linear system of the form $A^T A x = A^T c$, where A is an $m \times n$ matrix, we can apply the Lanczos algorithm with the symmetric matrix $K = A^T A$. However, we can use to our advantage the fact that the matrix K is such a product. This was done in a paper by Golub and Kahan [144] for the purpose of computing the singular values of A.

Two different algorithms were defined. The first one, which is denoted as "Lanczos bidiagonalization I" in Golub and von Matt [159], reduces A to an upper bidiagonal form and is the following.

Let $q^0 = c/\|c\|$, $r^0 = Aq^0$, $\delta_1 = \|r^0\|$, $p^0 = r^0/\delta_1$; then for $k = 1, 2, \ldots$

$$u^k = A^T p^{k-1} - \delta_k q^{k-1},$$

$$\gamma_k = \|u^k\|,$$

$$q^k = u^k/\gamma_k,$$

$$r^k = Aq^k - \gamma_k p^{k-1},$$

$$\delta_{k+1} = \|r^k\|,$$

$$p^k = r^k/\delta_{k+1}.$$

In this algorithm there is one multiplication with A and one with A^T. If we denote

$$P_k = (p^0 \quad \cdots \quad p^{k-1}), \quad Q_k = (q^0 \quad \cdots \quad q^{k-1}),$$

and

$$B_k = \begin{pmatrix} \delta_1 & \gamma_1 & & \\ & \ddots & \ddots & \\ & & \delta_{k-1} & \gamma_{k-1} \\ & & & \delta_k \end{pmatrix},$$

then P_k and Q_k, which is an orthogonal matrix, satisfy the equations

$$AQ_k = P_k B_k,$$
$$A^T P_k = Q_k B_k^T + \gamma_k q^k (e^k)^T.$$

Of course, by eliminating P_k in these equations we obtain

$$A^T AQ_k = Q_k B_k^T B_k + \gamma_k \delta_k q^k (e^k)^T.$$

This shows that $B_k^T B_k = J_k$ is the Lanczos Jacobi matrix corresponding to $A^T A$. Hence, we have directly computed the Cholesky factorization of J_k.

The previous algorithm was devised to compute the singular values of A (which are the square roots of the eigenvalues of $A^T A$). For this purpose the starting vector is not really important. For solving linear systems or least squares problems, this algorithm (using a different starting vector $q^0 = A^T c/\|A^T c\|$ as it is necessary to solve $A^T Ax = A^T c$) was considered also in a paper by Paige and Saunders [256], [257] under the name "Bidiag 2".

The second algorithm (named "Lanczos bidiagonalization II" in [159] and "Bidiag 1" in [256]) reduces A to lower bidiagonal form. The steps are the following, the coefficients δ_k and γ_k being different from the ones in the previous algorithm.

Let $p^0 = c/\|c\|$, $u^0 = A^T p^0$, $\gamma_1 = \|u^0\|$, $q^0 = u^0/\gamma_1$, $r^1 = Aq^0 - \gamma_1 p^0$, $\delta_1 = \|r^1\|$, $p^1 = r^1/\delta_1$; then for $k = 2, 3, \ldots$

$$u^{k-1} = A^T p^{k-1} - \delta_{k-1} q^{k-2},$$

$$\gamma_k = \|u^{k-1}\|,$$

$$q^{k-1} = u^{k-1}/\gamma_k,$$

$$r^k = Aq^{k-1} - \gamma_k p^{k-1},$$

$$\delta_k = \|r^k\|,$$

$$p^k = r^k/\delta_k.$$

If we denote

$$P_{k+1} = (p^0 \quad \cdots \quad p^k), \quad Q_k = (q^0 \quad \cdots \quad q^{k-1}),$$

and

$$C_k = \begin{pmatrix} \gamma_1 & & & \\ \delta_1 & \ddots & & \\ & \ddots & \ddots & \\ & & \ddots & \gamma_k \\ & & & \delta_k \end{pmatrix},$$

a $(k+1) \times k$ matrix, then P_k and Q_k, which is an orthogonal matrix, satisfy the equations

$$AQ_k = P_{k+1}C_k,$$
$$A^T P_{k+1} = Q_k C_k^T + \gamma_{k+1} q^k (e^{k+1})^T.$$

Of course, by eliminating P_{k+1} in these equations we obtain

$$A^T AQ_k = Q_k C_k^T C_k + \gamma_{k+1}\delta_k q^k (e^k)^T.$$

Therefore, we have

$$C_k^T C_k = B_k^T B_k = J_k,$$

and B_k is also the Cholesky factor of the product $C_k^T C_k$. This last algorithm is the basis for the algorithm LSQR of Paige and Saunders [256], who used an incremental reduction of the matrix C_k to upper triangular form by using Givens rotations.

4.4 The Block Lanczos Algorithm

In this section we consider the block Lanczos algorithm that was proposed by Golub and Underwood; see [152]. A block conjugate gradient has also been developed by O'Leary [249]. We restrict ourselves to the case of 2×2 blocks. Let X_0 be an $n \times 2$ given matrix, such that $X_0^T X_0 = I_2$ where I_2 is the 2×2 identity matrix. Let $X_{-1} = 0$ be an $n \times 2$ matrix. Then, for $k = 1, 2, \ldots$

$$\Omega_k = X_{k-1}^T AX_{k-1},$$

$$R_k = AX_{k-1} - X_{k-1}\Omega_k - X_{k-2}\Gamma_{k-1}^T, \tag{4.4}$$

$$X_k \Gamma_k = R_k.$$

The last step of the algorithm is the QR factorization of R_k (see Golub and Van Loan [154]) such that X_k is $n \times 2$ with $X_k^T X_k = I_2$. The matrix Γ_k is 2×2 upper triangular. The other symmetric coefficient matrix Ω_k is 2×2. The matrix R_k can eventually be rank deficient and in that case Γ_k is singular. The solution of this problem is given in [152]. One of the columns of X_k can be chosen arbitrarily. To complete the algorithm, we choose this column to be orthogonal with the previous block vectors X_j. We can, for instance, choose (randomly) another vector and orthogonalize it against the previous ones. The block tridiagonal matrix that is produced by the algorithm has the same structure as in equation (2.16). Note that it can be considered also as a band matrix.

The block Lanczos algorithm generates a sequence of matrices such that

$$X_j^T X_i = \delta_{ij} I_2,$$

where δ_{ij} is the Kronecker symbol. We can relate the iterates X_i to a matrix polynomial p_k, see chapter 2.

PROPOSITION 4.4

$$X_i = \sum_{k=0}^{i} A^k X_0 C_k^{(i)},$$

where $C_k^{(i)}$ are 2×2 matrices.

Proof. The proof is easily obtained by induction. □

As in the scalar case, the matrix Lanczos polynomials satisfy a three-term block recurrence.

THEOREM 4.5 *The matrix valued polynomials p_k satisfy*

$$p_k(\lambda)\Gamma_k = \lambda p_{k-1}(\lambda) - p_{k-1}(\lambda)\Omega_k - p_{k-2}(\lambda)\Gamma_{k-1}^T,$$

$$p_{-1}(\lambda) \equiv 0, \quad p_0(\lambda) \equiv I_2,$$

where λ is a scalar and $p_k(\lambda) = \sum_{j=0}^{k} \lambda^j X_0 C_j^{(k)}$.

Proof. From the previous definition of the algorithm, we show by induction that p_k can be generated by the given (matrix) recursion. □

This block three-term recurrence can be written as

$$\lambda[p_0(\lambda), \ldots, p_{N-1}(\lambda)] = [p_0(\lambda), \ldots, p_{N-1}(\lambda)]J_N + [0, \ldots, 0, p_N(\lambda)\Gamma_N],$$

and as $P(\lambda) = [p_0(\lambda), \ldots, p_{N-1}(\lambda)]^T$,

$$J_N P(\lambda) = \lambda P(\lambda) - [0, \ldots, 0, p_N(\lambda)\Gamma_N]^T,$$

with J_N defined by equation (2.16).

As in the scalar case, the inner product of the matrices X_i can be related to an integral.

THEOREM 4.6 *Considering the matrices X_k, there exists a matrix measure α (defined in the proof) such that*

$$X_i^T X_j = \int_a^b p_i(\lambda)^T d\alpha(\lambda) p_j(\lambda) = \delta_{ij} I_2,$$

where $a \leq \lambda_1 = \lambda_{\min}$ and $b \geq \lambda_n = \lambda_{\max}$.

Proof. Using the orthogonality of the X_i's and the spectral decomposition of A, we can write

$$
\begin{aligned}
\delta_{ij} I_2 = X_i^T X_j &= \left(\sum_{k=0}^i (C_k^{(i)})^T X_0^T A^k \right) \left(\sum_{l=0}^j A^l X_0 C_l^{(j)} \right) \\
&= \sum_{k,l} (C_k^{(i)})^T X_0^T Q \Lambda^{k+l} Q^T X_0 C_l^{(j)} \\
&= \sum_{k,l} (C_k^{(i)})^T \hat{X} \Lambda^{k+l} \hat{X}^T C_l^{(j)} \\
&= \sum_{k,l} (C_k^{(i)})^T \left(\sum_{m=1}^n \lambda_m^{k+l} \hat{X}_m \hat{X}_m^T \right) C_l^{(j)} \\
&= \sum_{m=1}^n \left(\sum_k \lambda_m^k (C_k^{(i)})^T \right) \hat{X}_m \hat{X}_m^T \left(\sum_l \lambda_m^l C_l^{(j)} \right),
\end{aligned}
$$

where $\hat{X}_m$ are the columns of $\hat{X} = X_0^T Q$, which is a $2 \times n$ matrix. Therefore,

$$X_i^T X_j = \sum_{m=1}^n p_i(\lambda_m)^T \hat{X}_m \hat{X}_m^T p_j(\lambda_m).$$

The sum in the right-hand side can be written as an integral for a 2×2 matrix measure,

$$
\alpha(\lambda) = \begin{cases}
0, & \text{if } \lambda < \lambda_1, \\
\sum_{j=1}^i \hat{X}_j \hat{X}_j^T, & \text{if } \lambda_i \leq \lambda < \lambda_{i+1}, \\
\sum_{j=1}^n \hat{X}_j \hat{X}_j^T, & \text{if } \lambda_n \leq \lambda.
\end{cases}
$$

Then,

$$X_i^T X_j = \int_a^b p_i(\lambda)^T d\alpha(\lambda) p_j(\lambda).$$

$\square$

The p_j^T's are matrix orthogonal polynomials for the matrix measure α.

4.5 The Conjugate Gradient Algorithm

The conjugate gradient (CG) algorithm is an iterative method to solve linear systems $Ax = c$ where the matrix A is symmetric positive definite. It was introduced at the beginning of the 1950s by Magnus Hestenes and Eduard Stiefel [187]. It can be derived from the Lanczos algorithm, which can also be used for indefinite matrices; see, for instance, Householder [193] and Meurant [239]. However, we have seen that to solve a linear system with the Lanczos algorithm we have to store (or recompute) all the Lanczos vectors. In contrast, the storage for the CG algorithm will be only a few vectors. A way to derive the CG algorithm is to consider an LU factorization of the Jacobi matrix J_k obtained in the Lanczos algorithm. Even though this is not the most frequently used form of the algorithm, we will define the three-term recurrence variant of CG since one can see more clearly the relations with the Lanczos algorithm and also with the Chebyshev semi-iterative method; see Golub and Varga [155], [156]. This particular form of the CG algorithm was popularized by Concus, Golub and O'Leary [68]. Therefore, we consider iterates x^k defined by

$$x^{k+1} = \nu_{k+1}(\mu_k r^k + x^k - x^{k-1}) + x^{k-1}, \tag{4.5}$$

where ν_{k+1} and μ_k are parameters to be determined by orthogonality constraints. Equation (4.5) gives us a relation for the residual vectors $r^k = c - Ax^k$,

$$r^{k+1} = r^{k-1} - \nu_{k+1}(\mu_k Ar^k - r^k + r^{k-1}). \tag{4.6}$$

The parameters ν_{k+1} and μ_k are computed by requiring that the residual vector r^{k+1} is orthogonal to r^k and r^{k-1}.

PROPOSITION 4.7 *If the parameter μ_k is chosen as*

$$\mu_k = \frac{(r^k, r^k)}{(r^k, Ar^k)}, \tag{4.7}$$

then $(r^k, r^{k+1}) = 0$. If the parameter ν_{k+1} is chosen as

$$\nu_{k+1} = \frac{1}{1 + \mu_k \dfrac{(r^{k-1}, Ar^k)}{(r^{k-1}, r^{k-1})}}, \tag{4.8}$$

then $(r^{k-1}, r^{k+1}) = 0$.

Proof. See Concus, Golub and O'Leary [68] or Meurant [237].   □

As we will see, this choice of parameters guarantees that $(r^l, r^k) = 0, l \leq k$, that is, the residual vectors are mutually orthogonal. Fortunately (see for instance [237]) there is an alternate expression for ν_{k+1},

$$\nu_{k+1} = \frac{1}{1 - \dfrac{\mu_k(r^k, r^k)}{\nu_k \mu_{k-1}(r^{k-1}, r^{k-1})}}.$$

This last formula is computationally more efficient than the formula (4.8) in proposition 4.7 since for computing both coefficients μ_k and ν_{k+1} we only have to compute two inner products instead of three with the previous formula. The iterations are started by taking $\nu_1 = 1$. Then

$$x^1 = \mu_0 r^0 + x^0,$$

and we need only to define x^0 and then $r^0 = c - Ax^0$. Now, the key point is to show that we have global orthogonality as we said before, that is, the new vector r^{k+1} is orthogonal not only to the last two vectors r^k and r^{k-1}, but to all the previous vectors: $(r^{k+1}, r^j) = 0$, $0 \leq j < k - 1$. This is similar to what happens in the Lanczos algorithm and is proved by induction supposing the property is true up to step k. Multiplying equation (4.6) by r^j, $0 \leq j < k - 1$, we have

$$(r^j, r^{k+1}) = (r^j, r^{k-1}) - \nu_{k+1}[\mu_k(r^j, Ar^k) - (r^j, r^k) + (r^j, r^{k-1})].$$

But, since $j < k - 1$, some terms are zero and

$$(r^j, r^{k+1}) = \nu_{k+1}\mu_k(r^j, Ar^k).$$

Writing the definition of r^{j+1} we obtain

$$r^{j+1} = r^{j-1} - \nu_{j+1}(\mu_j Ar^j - r^j + r^{j-1}).$$

Multiplying this equation by r^k and taking into account that $j + 1 < k$, we have

$$\nu_{j+1}\mu_j(r^k, Ar^j) = 0.$$

Because of the symmetry of A we obtain $(Ar^k, r^j) = 0$. This shows that we have $(r^j, r^{k+1}) = 0$ for all j such that $j < k - 1$. Therefore, as in the Lanczos algorithm and because A is symmetric, the local orthogonality with r^k and r^{k-1} implies the global orthogonality with all r^j, $j = k - 2, \ldots, 0$. Since A is assumed to be positive definite, the algorithm cannot break down. If $\|r^k\| = 0$ or $(r^k, Ar^k) = 0$, the algorithm has found the solution.

The standard two-term form of CG is obtained by using an LU factorization of the Jacobi matrix of the Lanczos algorithm. It uses two-term recurrences and is the following, starting from a given x^0 and $r^0 = c - Ax^0$:

for $k = 0, 1, \ldots$ until convergence,

$$\beta_k = \frac{(r^k, r^k)}{(r^{k-1}, r^{k-1})}, \quad \beta_0 = 0,$$

$$p^k = r^k + \beta_k p^{k-1},$$

$$\gamma_k = \frac{(r^k, r^k)}{(Ap^k, p^k)},$$

$$x^{k+1} = x^k + \gamma_k p^k,$$

$$r^{k+1} = r^k - \gamma_k Ap^k.$$

Of course, there are some relations between the coefficients of the two-term and three-term CG recurrences. They are obtained by eliminating p^k in the two-term recurrence equations,

$$\nu_{k+1} = 1 + \frac{\gamma_k \beta_k}{\gamma_{k-1}},$$

$$\mu_k = \frac{\gamma_k}{1 + \frac{\gamma_k \beta_k}{\gamma_{k-1}}}.$$

The relations between the Lanczos algorithm coefficients α_k and η_k and those of the two-term form of CG are

$$\alpha_k = \frac{1}{\gamma_{k-1}} + \frac{\beta_{k-1}}{\gamma_{k-2}}, \ \beta_0 = 0, \ \gamma_{-1} = 1, \tag{4.9}$$

$$\eta_k = \frac{\sqrt{\beta_k}}{\gamma_{k-1}}. \tag{4.10}$$

There are also relations between the three-term recurrence CG coefficients and those of the Lanczos algorithm. We write the three-term recurrence for the residuals as

$$r^{k+1} = -\nu_{k+1}\mu_k A r^k + \nu_{k+1} r^k + (1 - \nu_{k+1}) r^{k-1}.$$

There is a relation between the residuals and the Lanczos basis vectors $v^{k+1} = (-1)^k r^k / \|r^k\|$; see Meurant [239]. This leads to

$$\mu_k = \frac{1}{\alpha_{k+1}}, \quad \nu_{k+1} = 1 + \frac{\eta_k}{\eta_{k+1}} \frac{\|r^{k+1}\|}{\|r^{k-1}\|}.$$

One of the most interesting features of the CG algorithm is that it has several optimality properties. The first step in proving this is to show that CG is a polynomial method as the Lanczos algorithm. Using the relation of the residuals and basis vectors, we have

$$r^k = (-1)^k \frac{\|r^k\|}{\|r^0\|} p_{k+1}(A) r^0.$$

From the three-term CG recurrence we can show the following.

PROPOSITION 4.8 *The residual vector r^{k+1} is a polynomial in A,*

$$r^{k+1} = [I - A s_k(A)] r^0,$$

where s_k is a kth degree polynomial satisfying a three-term recurrence

$$s_k(\lambda) = \mu_k \nu_{k+1} + \nu_{k+1}(1 - \mu_k \lambda) s_{k-1}(\lambda) - (\nu_{k+1} - 1) s_{k-2}(\lambda),$$

$$s_0(\lambda) = \mu_0, \quad s_1(\lambda) = \nu_2(\mu_0 + \mu_1 - \mu_0 \mu_1 \lambda).$$

Proof. The proof is obtained by induction on k.   □

This gives a relation between the Lanczos and the CG polynomials

$$1 - \lambda s_{k-1}(\lambda) = (-1)^k \frac{1}{|p_{k+1}(0)|} p_{k+1}(\lambda).$$

Remember that the Lanczos polynomial p_{k+1} is of exact degree k.

PROPOSITION 4.9 *Let s_k be the polynomial defined in proposition 4.8. The CG iterates are given by*

$$x^{k+1} = x^0 + s_k(A) r^0.$$

Proof. We have

$$x^{k+1} = x^{k-1} + \nu_{k+1}(\mu_k z^k + x^k - x^{k-1}).$$

By induction and with the help of proposition 4.8 this is written as

$$x^{k+1} = x^0 + s_{k-2}(A)r^0 + \nu_{k+1}(\mu_k[I - As_{k-1}(A)]r^0 + s_{k-1}(A)r^0 - s_{k-2}(A)r^0).$$

Hence

$$x^{k+1} = x^0 + s_k(A)r^0,$$

because of the recurrence relation satisfied by s_k. □

For CG, the most interesting measure of the error $\epsilon^k = x - x^k$ is the A-norm.

DEFINITION 4.10 *Let A be a symmetric positive definite matrix. The A-norm of a vector ϵ^k is defined as*

$$\|\epsilon^k\|_A = (A\epsilon^k, \epsilon^k)^{1/2}.$$

The CG optimality property involving the A-norm is the following.

THEOREM 4.11 *Consider all the iterative methods that can be written as*

$$\overline{x}^{k+1} = \overline{x}^0 + q_k(A)\overline{r}^0, \quad \overline{x}^0 = x^0, \quad \overline{r}^0 = c - A\overline{x}^0,$$

where q_k is a polynomial of degree k. Of all these methods, CG is the one which minimizes $\|\epsilon^k\|_A$ at each iteration.

Proof. See Meurant [237] for a proof. □

This optimality result allows us to obtain bounds on the A-norm of the error.

THEOREM 4.12

$$\|\epsilon^k\|_A^2 \leq \max_{1 \leq i \leq n}(t_k(\lambda_i))^2 \|\epsilon^0\|_A^2,$$

for all polynomials t_k of degree k such that $t_k(0) = 1$.

Proof. From theorem 4.11, we know that the CG polynomial minimizes $\|\epsilon^k\|_A$. Thus, if we replace the CG polynomial s_{k-1} by any other polynomial of degree $k-1$, we obtain an upper bound for the A-norm of the error. This can be written as

$$\|\epsilon^k\|_A^2 \leq \sum_{i=1}^n (t_k(\lambda_i))^2 (\overline{\epsilon}_i^0)^2,$$

where $\overline{\epsilon}^j = \Lambda^{1/2} Q^T \epsilon^j$, Q being the orthonormal matrix whose columns are the eigenvectors of A and Λ being the diagonal matrix of the eigenvalues. This result holds for all polynomials t_k of degree k, such that $t_k(0) = 1$, equality holding only if $t_k(\lambda) = 1 - \lambda s_{k-1}(\lambda)$. Therefore,

$$\|\epsilon^k\|_A^2 \leq \max_{1 \leq i \leq n}(t_k(\lambda_i))^2 \sum_{i=1}^n (\overline{\epsilon}_i^0)^2.$$

But, $(\overline{\epsilon}^0, \overline{\epsilon}^0) = \|\epsilon^0\|_A^2$, which proves the result. □

In theorem 4.12 we are free to choose the polynomial in the right-hand side. The only constraint is that it must have a value of 1 at 0. This leads to the most well known a priori bound for the CG A-norm of the error.

THEOREM 4.13

$$\|\epsilon^k\|_A \le 2 \left(\frac{\sqrt{\kappa} - 1}{\sqrt{\kappa} + 1} \right)^k \|\epsilon^0\|_A,$$

where $\kappa = \frac{\lambda_n}{\lambda_1}$ is the condition number of A.

Proof. The right-hand side in theorem 4.12 $\max_{1 \le i \le n}(t_k(\lambda_i))^2$ is bounded by $\max_{\lambda_1 \le \lambda \le \lambda_n}(t_k(\lambda))^2$. For t_k we choose the polynomial of degree k such that $t_k(0) = 1$, which minimizes the maximum. The solution to this problem is given by the shifted Chebyshev polynomials (see chapter 2),

$$t_k(\lambda) = \frac{C_k \left(\frac{\lambda_1 + \lambda_n - 2\lambda}{\lambda_n - \lambda_1} \right)}{C_k \left(\frac{\lambda_1 + \lambda_n}{\lambda_n - \lambda_1} \right)}.$$

By the properties of the Chebyshev polynomials,

$$\max_{\lambda_1 \le \lambda \le \lambda_n} |t_k(\lambda)| \le 2 \left(\frac{\sqrt{\kappa} - 1}{\sqrt{\kappa} + 1} \right)^k.$$

This proves the theorem.   □

There are many cases for which this bound is overly pessimistic. We will see in chapter 12 that CG convergence depends not only on the condition number but on the distribution of all the eigenvalues of A, and that good estimates of the A-norm of the error can be obtained during the iterations using the relations of CG with Gauss quadrature.

Chapter Five

Computation of the Jacobi Matrices

We have seen in chapter 2 that we know the coefficients of the three-term recurrence for the classical orthogonal polynomials. In other cases, we have to compute these coefficients from some other information sources. There are many circumstances in which one wants to determine the coefficients of the three-term recurrence (that is, the Jacobi matrix J_k) of a family of orthogonal polynomials given either the measure α, the moments μ_k defined in equation (2.2) or the nodes and weights of a quadrature formula. We will see some examples of applications later in this book.

5.1 The Stieltjes Procedure

A way to compute the coefficients of the three-term recurrence given the measure α is to approximate it by a discrete measure and to compute the coefficients of the recurrence corresponding to the discrete measure. If the discretizations are done properly, the process will converge; see Gautschi [131]. The problem (which arises also directly if the given measure is discrete) is now to compute the coefficients of the recurrence. Probably the simplest way to do this is to use the Stieltjes procedure [312] (or algorithm) which dates back to the nineteenth century. With a discrete inner product, sums like

$$\langle p, q \rangle = \sum_{j=1}^{m} p(t_j) q(t_j) w_j^2$$

are trivial to compute given the nodes t_j and the weights w_j^2. The coefficients of the three-term recurrence are given by

$$\alpha_{k+1} = \frac{\langle \lambda p_k, p_k \rangle}{\langle p_k, p_k \rangle}, \quad \gamma_k = \frac{\langle p_k, p_k \rangle}{\langle p_{k-1}, p_{k-1} \rangle}$$

for a monic polynomial p_k having a recurrence relation (2.6) or

$$\alpha_{k+1} = \langle \lambda p_k, p_k \rangle, \quad \beta_{k+1} = \langle \lambda p_k, \lambda p_k \rangle - \alpha_{k+1}^2 - \beta_k$$

for an orthonormal polynomial with a recurrence (2.8). This is obtained by taking the inner product of the recurrence relation with either p_k or p_{k-1} or by expressing p_{k+1} in terms of p_k and p_{k-1} using the three-term recurrence. For instance to compute γ_k we need to compute $\langle p_k, p_k \rangle$ assuming $\langle p_{k-1}, p_{k-1} \rangle$ is already known. The values $p_i(t_j)$ of the polynomials at the nodes needed in the inner products can be computed by recurrences (2.6) or (2.8) and from this, the coefficients are computed provided that the nodes are not the roots of the orthogonal polynomials.

So we can intertwine the computation of the coefficients and of the values of the polynomials at the nodes. The algorithm starts by computing α_1 from p_0 which is identically 1. Then one can compute the values of p_1 at the nodes t_j and then γ_1 and α_2. This, in turn, allows to compute the values of p_2 at the nodes, and so on.

This seems a very simple and elegant algorithm. However, in finite precision arithmetic, the Stieltjes procedure can be sensitive to roundoff errors. In fact, the Stieltjes procedure can be seen as a predecessor of the Lanczos algorithm with a different inner product (see chapter 4). Unfortunately, in many circumstances, the Lanczos algorithm may have a large growth of rounding errors (see [239], [242]) and it is the same for the Stieltjes algorithm. Therefore, the Jacobi matrix computed by the Stieltjes algorithm in finite precision arithmetic may sometimes be far from the exact one. We will review some other algorithms for computing the Jacobi matrix from the nodes and weights of a discrete inner product in the next sections.

5.2 Computing the Coefficients from the Moments

Assume that we do not know the measure α but that we know the moments related to it. There are expressions directly relating the moments to the polynomial coefficients. Let us quote two results which use Hankel matrices; see Szegö [323] or Gautschi [131]. Let

$$\Delta_0 = 1, \quad \Delta_k = \det(H_k), \quad H_k = \begin{pmatrix} \mu_0 & \mu_1 & \cdots & \mu_{k-1} \\ \mu_1 & \mu_2 & \cdots & \mu_k \\ \vdots & \vdots & & \vdots \\ \mu_{k-1} & \mu_k & \cdots & \mu_{2k-2} \end{pmatrix}, \ k = 1, 2, \ldots$$

and

$$\Delta_0' = 0, \ \Delta_1' = \mu_1, \ \Delta_k' = \det \begin{pmatrix} \mu_0 & \mu_1 & \cdots & \mu_{k-2} & \mu_k \\ \mu_1 & \mu_2 & \cdots & \mu_{k-1} & \mu_{k+1} \\ \vdots & \vdots & & \vdots & \vdots \\ \mu_{k-1} & \mu_k & \cdots & \mu_{2k-3} & \mu_{2k-1} \end{pmatrix}, \ k = 2, 3, \ldots .$$

THEOREM 5.1 *The monic orthogonal polynomial π_k of degree k associated with the moments μ_j, $j = 0, \ldots, 2k - 1$ is*

$$\pi_k(\lambda) = \frac{1}{\Delta_k} \det \begin{pmatrix} \mu_0 & \mu_1 & \cdots & \mu_k \\ \mu_1 & \mu_2 & \cdots & \mu_{k+1} \\ \vdots & \vdots & & \vdots \\ \mu_{k-1} & \mu_k & \cdots & \mu_{2k-1} \\ 1 & \lambda & \cdots & \lambda^k \end{pmatrix}, \ k = 1, 2, \ldots .$$

THEOREM 5.2 *The recursion coefficients of the three-term recurrence for the polynomial π_k of theorem 5.1,*

$$\pi_{k+1}(\lambda) = (\lambda - \alpha_{k+1})\pi_k(\lambda) - \gamma_k \pi_{k-1}(\lambda), \quad \pi_{-1}(\lambda) \equiv 0, \ \pi_0(\lambda) \equiv 1,$$

are given by

$$\alpha_{k+1} = \frac{\Delta'_{k+1}}{\Delta_{k+1}} - \frac{\Delta'_k}{\Delta_k}, \ k = 0, 1, \ldots, \tag{5.1}$$

$$\gamma_k = \frac{\Delta_{k+1}\Delta_{k-1}}{\Delta_k^2}, \ k = 1, 2, \ldots . \tag{5.2}$$

These results are mainly of theoretical interest since the map giving the coefficients as a function of the moments is badly conditioned. The condition number of this map has been studied by Gautschi (see [131]). In [328] Tyrtyshnikov proved that the condition number of any real positive Hankel matrix of order n is larger than $3 \cdot 2^{n-6}$. Fasino [103] proved that the condition number of any Hankel matrix generated by moments of positive functions is essentially the same as the condition number of the Hilbert matrix of the same size. Hilbert matrices are notoriously badly conditioned. On these topics, see also Beckermann [24].

The coefficients of the orthogonal polynomial can also be found from the moments by an algorithm proposed by Gautschi which was described in Golub and Welsch [160]. The main tool is the factorization of the Hankel matrix H_k. This is more convenient than working with determinants.

PROPOSITION 5.3 *Assume the measure α is defined by a positive weight function w; then the Hankel matrix H_k is positive definite.*

Proof. The inner product defined by the measure α is positive definite. This implies that all the principal minors of the Hankel matrix are positive.   □

Therefore, we can consider the Cholesky factorization $H_k = R_k^T R_k$, R_k being an upper triangular matrix. Let $s_{i,j}$, $j \geq i$ be the nonzero entries of R_k^{-1}. Let $q(\lambda)$ be a vector defined with components

$$q_j(\lambda) = s_{1,j} + s_{2,j}\lambda + \cdots + s_{j,j}\lambda^{j-1}, \ j = 1, \ldots, k.$$

Then,

$$q(\lambda) = R_k^{-T} \begin{pmatrix} 1 \\ \lambda \\ \vdots \\ \lambda^{k-1} \end{pmatrix} = R_k^{-T}\phi_k(\lambda).$$

PROPOSITION 5.4 *Let $\langle q(\lambda), q^T(\lambda) \rangle$ be the matrix whose entries are*

$$\langle q_i(\lambda), q_j(\lambda) \rangle = \int_a^b q_i(\lambda)q_j(\lambda)\, d\alpha.$$

We have

$$\langle q(\lambda), q^T(\lambda) \rangle = I.$$

Proof. From the definition of $q(\lambda)$, we have

$$\langle q(\lambda), q^T(\lambda) \rangle = R_k^{-T}\langle \phi_k(\lambda), \phi_k(\lambda)^T \rangle R_k^{-1} = R_k^{-T} H_k R_k^{-1} = I.$$

This proves the result.   □

The last proposition shows that $p_{j-1}(\lambda) = q_j(\lambda)$ is the orthonormal polynomial related to the measure α. This polynomial satisfies a three-term recurrence

$$\eta_{k+1}p_{k+1}(\lambda) = (\lambda - \alpha_{k+1})p_k(\lambda) - \eta_k p_{k-1}(\lambda). \qquad (5.3)$$

By comparing the coefficients of λ^j and λ^{j-1}, we have the relations

$$s_{j,j} = \eta_j s_{j+1j+1}, \quad s_{j-1,j} = \alpha_j s_{j,j} + \eta_j s_{j,j+1}.$$

From these relations, we obtain the values of the coefficients α_j, η_j from the entries of R_k^{-1}. Now, we can write the entries $s_{i,j}$ of R_k^{-1} we need as functions of the entries $r_{i,j}$ of R_k,

$$s_{j,j} = \frac{1}{r_{j,j}}, \quad s_{j,j+1} = -\frac{r_{j,j+1}}{r_{j,j}r_{j+1,j+1}}.$$

This leads to the following result.

THEOREM 5.5 *Let $H_k = R_k^T R_k$ be the Cholesky factorization of the moment matrix. The coefficients of the orthonormal polynomial satisfying the three-term recurrence (5.3) are given by*

$$\eta_j = \frac{r_{j+1,j+1}}{r_{j,j}}, j = 1,\ldots,k-1 \quad \alpha_1 = r_{1,2}, \ \alpha_j = \frac{r_{j,j+1}}{r_{j,j}} - \frac{r_{j-1,j}}{r_{j-1,j-1}}, \ j = 2,\ldots,k.$$

Proof. See Golub and Welsch [160].   □

It is interesting to remark that these formulas are similar to what is obtained when doing a QR factorization of the Krylov matrix defined in equation (4.1) in the Lanczos algorithm described in chapter 4, see Meurant [239], section 1.1. Let $K_k = V_k R_k$ be a QR factorization of the Krylov matrix where V_k is $n \times k$ and orthonormal ($V_k^T V_k = I$, the identity matrix) and R_k is $k \times k$ nonsingular and upper triangular (say with positive elements on the diagonal). It is easy to see that V_k is the matrix whose columns are the Lanczos vectors v^j. The matrix R_k is also the Cholesky factor of the moment matrix $K_k^T K_k$. That is,

$$R_k^T R_k = \begin{pmatrix} 1 & (v, Av) & \cdots & (v, A^{k-1}v) \\ (v, Av) & (v, A^2v) & \cdots & (v, A^kv) \\ \vdots & \vdots & & \vdots \\ (v, A^{k-1}v) & & \cdots & (v, A^{2k-2}v) \end{pmatrix},$$

where v is the first Lanczos vector. The elements $r_{i,j}$ of R_k are related to those of the tridiagonal Lanczos matrix J_k by formulas similar to those of theorem 5.5.

5.3 The Modified Chebyshev Algorithm

Since using the moments μ_k to compute the recurrence coefficients may not be numerically safe (see Gautschi [131]), it is often wiser to use so-called modified moments defined by using another family of orthogonal polynomials p_k for which we know the recurrence coefficients. Then, instead of integrating the monomials λ^i, we consider the integration of the polynomials p_k. This is a sort of change of basis functions.

DEFINITION 5.6 *The* **modified moments** *(using known orthogonal polynomials* p_k*) are*

$$m_k = \int_a^b p_k(\lambda) \, d\alpha. \tag{5.4}$$

The modified Chebyshev algorithm was developed by J. Wheeler in 1974 [348] (see Gautschi [127]) from an algorithm due to P. Chebyshev in 1859 [63]. It applies the Chebyshev algorithm to modified moments instead of ordinary moments; see also Sack and Donovan [289]. For the exposition, we follow Gautschi [126], [131]. An interesting paper on modified moments is Beckermann and Bourreau [25]. For transforming one polynomial expansion into another, see Salzer [290].

Let us consider monic orthogonal polynomials satisfying

$$p_{k+1}(\lambda) = (\lambda - a_{k+1})p_k(\lambda) - c_k p_{k-1}(\lambda), \quad p_{-1}(\lambda) \equiv 0, \, p_0(\lambda) \equiv 1, \tag{5.5}$$

whose coefficients a_{k+1} and c_k are supposed to be known. The Chebyshev algorithm is obtained by using $p_k(\lambda) = \lambda^k, \forall k$. We also assume that we know the modified moments m_k, $k = 0, \dots, 2m - 1$ defined by equation (5.4). We would like to determine the coefficients α_{k+1} and η_k of the three-term recurrence

$$\pi_{k+1}(\lambda) = (\lambda - \alpha_{k+1})\pi_k(\lambda) - \eta_k \pi_{k-1}(\lambda), \quad \pi_{-1}(\lambda) \equiv 0, \, \pi_0(\lambda) \equiv 1, \tag{5.6}$$

where π_k are the unknown monic orthogonal polynomials associated with the measure α. To do this, we introduce mixed moments.

DEFINITION 5.7 *The* **mixed moments** *related to* p_l *and* α *are*

$$\sigma_{k,l} = \int_a^b \pi_k(\lambda) p_l(\lambda) \, d\alpha(\lambda).$$

We will derive relations allowing us to compute these mixed moments. By orthogonality, we have $\sigma_{k,l} = 0$, $k > l$. Moreover, since $\lambda p_{k-1}(\lambda)$ (which is a polynomial of degree k) can be written as the sum of π_k and a polynomial of degree strictly less than k, we can write

$$\sigma_{k,k} = \int_a^b \pi_k(\lambda) \lambda p_{k-1}(\lambda) \, d\alpha(\lambda) = \int_a^b \pi_k^2(\lambda) \, d\alpha(\lambda).$$

Multiplying equation (5.6) by p_{k-1} and integrating gives the relation

$$\sigma_{k,k} - \eta_k \sigma_{k-1,k-1} = 0,$$

and therefore

$$\eta_k = \frac{\sigma_{k,k}}{\sigma_{k-1,k-1}}.$$

Multiplying equation (5.6) by p_k and integrating gives

$$\alpha_{k+1}\sigma_{k,k} + \eta_k \sigma_{k-1,k} = \int_a^b \pi_k(\lambda) \lambda p_k(\lambda) \, d\alpha(\lambda).$$

We now use equation (5.5) to express $\lambda p_k(\lambda)$ in the previous equation. This gives the relation

$$\sigma_{k,k+1} + (a_{k+1} - \alpha_{k+1})\sigma_{k,k} - \eta_k \sigma_{k-1,k} = 0.$$

Using again equations (5.5) and (5.6) with different indices, one obtains

$$\sigma_{k,l} = \sigma_{k-1,l+1} - (\alpha_k - a_{l+1})\sigma_{k-1,l} - \eta_{k-1}\sigma_{k-2,l} + c_l\sigma_{k-1,l-1}.$$

With these relations we compute the first $2m - 1$ unknown recursion coefficients α_k, η_k from the first $2m$ modified moments m_l, since it is obvious that $\sigma_{0,k} = m_k$. The modified Chebyshev algorithm for monic polynomials is the following:

$$\sigma_{-1,l} = 0, \, l = 1, \ldots, 2m - 2, \quad \sigma_{0,l} = m_l, \, l = 0, 1, \ldots, 2m - 1$$

$$\alpha_1 = a_1 + \frac{m_1}{m_0},$$

and for $k = 1, \ldots, m - 1$,

$$\sigma_{k,l} = \sigma_{k-1,l+1} + (a_{l+1} - \alpha_k)\sigma_{k-1,l} + c_l\sigma_{k-1,l-1} - \eta_{k-1}\sigma_{k-2,l}, \, l = k, \ldots, 2m-k-1,$$

$$\alpha_{k+1} = a_{k+1} + \frac{\sigma_{k,k+1}}{\sigma_{k,k}} - \frac{\sigma_{k-1,k}}{\sigma_{k-1,k-1}},$$

$$\eta_k = \frac{\sigma_{k,k}}{\sigma_{k-1,k-1}}.$$

Note the similarity of these formulas with the ones given in theorem 5.5. The initial condition of the algorithm for $k = 0$ and $l = 0, \ldots, 2m-1$ is given by the modified moments m_l. Then in the plane l, k we proceed by going up in k. For each k we compute a new element $\sigma_{k,l}$ by using four previous values for $k - 1$ and $k - 2$. In a five-point discretization stencil (k, l) is the north point and it is computed from the central point and the west, south and east points. Note that when we increase k by 1 the number of mixed moments to compute decreases by 2. Even though we have to compute all the $\sigma_{k,l}$ for a given k, only the first two ones $\sigma_{k,k}$ and $\sigma_{k,k+1}$ are used to compute the coefficients. Of course, the modified Chebyshev algorithm depends on the knowledge of the modified moments; see Gautschi [131]. If the modified moments cannot be computed accurately, then the modified Chebyshev algorithm is not really useful.

A similar procedure can be used for orthonormal polynomials π_k. More generally, if the recurrence relation for the unknown polynomials is

$$\gamma_{k+1}\pi_{k+1}(\lambda) = (\lambda - \alpha_{k+1})\pi_k(\lambda) - \eta_k\pi_{k-1}(\lambda), \quad \pi_{-1}(\lambda) \equiv 0, \, \pi_0(\lambda) \equiv \pi_0, \tag{5.7}$$

and the known polynomials satisfy

$$b_{k+1}p_{k+1}(\lambda) = (\lambda - a_{k+1})p_k(\lambda) - c_k p_{k-1}(\lambda), \quad p_{-1}(\lambda) \equiv 0, \, p_0(\lambda) \equiv p_0, \tag{5.8}$$

then the modified moments are

$$m_l = \frac{\sigma_{0,l}}{\pi_0} = \int_a^b p_l(\lambda) \, d\alpha.$$

The value π_0 can be chosen arbitrarily. Then the modified Chebyshev algorithm is the following:

$$\sigma_{-1,l} = 0, \, l = 1, \ldots, 2m - 2, \quad \sigma_{0,l} = m_l\pi_0, \, l = 0, 1, \ldots, 2m - 1,$$

$$\alpha_1 = a_1 + b_1 \frac{m_1}{m_0},$$

and for $k = 1, \ldots, m - 1$,

we choose the normalization parameter $\gamma_k > 0$ and
for $l = k, \ldots, 2m - k - 1$,

$$\sigma_{k,l} = \frac{1}{\gamma_k}[b_{l+1}\sigma_{k-1,l+1} + (a_{l+1} - \alpha_k)\sigma_{k-1,l} + c_l\sigma_{k-1,l-1} - \eta_{k-1}\sigma_{k-2,l}],$$

then

$$\alpha_{k+1} = a_{k+1} + b_{k+1}\frac{\sigma_{k,k+1}}{\sigma_{k,k}} - b_k\frac{\sigma_{k-1,k}}{\sigma_{k-1,k-1}},$$

$$\eta_k = b_k\frac{\sigma_{k,k}}{\sigma_{k-1,k-1}}.$$

For orthonormal polynomials the coefficients γ_k are chosen to have a norm equal to 1.

5.4 The Modified Chebyshev Algorithm for Indefinite Weight Functions

In [143] Golub and Gutknecht extended the modified Chebyshev algorithm to the case of indefinite weight functions. Then one has to use formal orthogonal polynomials. They gave also a matrix interpretation of the modified Chebyshev algorithm. Let J be the (infinite) tridiagonal matrix of the coefficients we are looking for with 1's on the lower subdiagonal and H be the upper Hessenberg matrix (with 1's on the first subdiagonal) of the coefficients of the auxiliary polynomials p_k which are not supposed to satisfy a three-term recurrence. Moreover, let S be the lower triangular matrix of the mixed moments $\sigma_{k,l}$ and D be a diagonal matrix with the same diagonal as S. Using these infinite matrices, the recurrence relations for the two sets of polynomials can be written as

$$\lambda\Pi(\lambda) = \Pi(\lambda)J, \quad \lambda P(\lambda) = P(\lambda)H,$$

where $\Pi = [\pi_0, \pi_1, \ldots]$ and $P = [p_0, p_1, \ldots]$. If we denote by $\varphi(\lambda^i)$ the integral of λ^i with the measure α, φ is a linear functional on the set of polynomials and we have

$$\varphi(\Pi^T\Pi) = D, \quad \varphi(P^T\Pi) = S.$$

Using the linearity of φ, we obtain the matrix relation

$$SJ = H^TS.$$

The matrices on the left and on the right-hand sides are lower Hessenberg. Therefore, if no $\sigma_{l,l}$ is zero,

$$JS^{-1} = S^{-1}H^T,$$

which again is an equality between Hessenberg matrices. Let $q \equiv q(\lambda)$ be the vector of the monomials; then the moment (Hankel) matrix is $M = \varphi(q^Tq)$ and the Gramian matrix of the polynomials p_k is $G = \varphi(P^T P)$. There exist unit upper triangular matrices Z and R such that

$$q(\lambda) = \Pi(\lambda)Z, \quad P(\lambda) = \Pi(\lambda)R.$$

This gives $M = Z^T D Z$ and $S = R^T D$. It leads to $G = R^T D R = SR$. The matrix S is lower triangular and R is upper triangular. Therefore, we have an LU factorization of the Gramian matrix. The matrices $Z^T D Z$ and $R^T D R$ are the LDU factorizations of M and G. Finally, we have $JR = RH$.

In the modified Chebyshev algorithm, the elements of J are computed from those of S and H using $J = S^{-1} H^T S$ and S can be generated from the modified moments. Golub and Gutknecht proposed also an inverse Chebyshev algorithm to compute S from J. From the coefficients of J, one can compute the diagonal and first diagonal of S from which the whole matrix S can be built. The knowledge of S allows one to compute the modified moments. As we have seen before, S can also be generated from the diagonal and subdiagonal of G, and from this, one can compute J using two diagonals of S^{-1} as done by Golub and Welsch [160].

Golub and Gutknecht generalized the previous relations to the general case where φ may be an arbitrary complex linear functional. Then the formal orthogonal polynomials may or may not exist for all degrees. The matrix J is now a block tridiagonal matrix and the matrix S is block lower triangular. Block equivalents of the previous algorithms can then be derived. As far as we know the stability properties of these algorithms have not been investigated yet.

5.5 Relations between the Lanczos and Chebyshev Semi-Iterative Algorithms

We consider this topic in this chapter because it is related to the computation of modified moments. Several iterative methods for solving linear systems $Ax = c$ with a symmetric positive definite matrix A can be written as

$$x^{k+1} = x^{k-1} + \omega_{k+1}(\delta_k z^k + x^k - x^{k-1}), \tag{5.9}$$

with parameters ω_{k+1} and δ_k depending on the given method and the vector z^k given by solving

$$M z^k = r^k,$$

where r^k is the residual vector. The symmetric positive definite matrix M is the preconditioner whose role is to speed up convergence. Depending on the choice of parameters, equation (5.9) describes the Chebyshev semi-iterative (CSI) method (see Golub and Varga [155], [156]), the Richardson second-order method and the conjugate gradient (CG) method (see chapter 4).

The first two methods depend on having estimates of the extreme eigenvalues of $M^{-1}A$ and this can be seen as a disadvantage. However, they have the advantage over the conjugate gradient algorithm of not requiring any inner products. This can be important on parallel computers for which computing inner products is often a bottleneck. Of course, it is also necessary that the rate of convergence is not much slower than for CG. This is why we will concentrate on the Chebyshev semi-iterative method since the Richardson method is not very attractive in this respect.

As we said, the computation of the parameters δ_k and ω_k needs estimates, a and b, of the smallest and largest eigenvalues of $M^{-1}A$. Let $\mu = (b-a)/(b+a)$; the parameters of the CSI method are given (see [155], [156], [237]) by

$$\delta_k = \delta = \frac{2}{b+a}, \quad \omega_{k+1} = \frac{1}{1 - \frac{\mu^2}{4}\omega_k} \text{ with } \omega_1 = 1, \ \omega_2 = \frac{1}{1 - \frac{\mu^2}{2}}.$$

The generalized residual vectors z^k are given by polynomials. This is summarized in the following result.

PROPOSITION 5.8 *The vectors z^k of the CSI method defined by equation (5.9) with $\delta_k \equiv \delta$ are given by*

$$z^k = p_k(B)z^0, \quad B = I - \delta M^{-1}A,$$

where p_k is a polynomial of degree k satisfying a three-term recurrence

$$p_{k+1}(\lambda) = \omega_{k+1}\lambda p_k(\lambda) + (1 - \omega_{k+1})p_{k-1}(\lambda), \quad p_{-1}(\lambda) \equiv 0, \ p_0(\lambda) \equiv 1.$$

Proof. Multiplying equation (5.9) by A, one obtains an equation for the residual vectors r^k,

$$r^{k+1} = r^{k-1} - \omega_{k+1}(\delta Az^k - r^k + r^{k-1}).$$

But $Mz^k = r^k$ and multiplying by M^{-1} we have a relation for the generalized residual vectors,

$$z^{k+1} = z^{k-1} - \omega_{k+1}(\delta M^{-1}Az^k - z^k + z^{k-1}),$$

which can be written as

$$z^{k+1} = (1 - \omega_{k+1})z^{k-1} + \omega_{k+1}(I - \delta M^{-1}A)z^k.$$

Iterating this relation and using the spectral decomposition of the matrix B (which is similar to a symmetric matrix), it is obvious that $z^k = p_k(B)z^0$, where the polynomial p_k satisfies the three-term recurrence of the proposition.   □

It is well known how to obtain estimates of the extreme eigenvalues during CG iterations. This uses the relation between the CG and Lanczos algorithms. The paper [146] by Golub and Kent shows how to obtain estimates of the needed eigenvalues during the CSI iterations using modified moments. This is yet another application of the use of modified moments in a linear algebra problem.

Consider the matrix $C = M^{-1}A$. Then $B = I - \delta C$. The matrix C is similar to a symmetric matrix since

$$C = M^{-1/2}(M^{-1/2}AM^{-1/2})M^{1/2}.$$

Let $A = Q\Lambda Q^T$ be the spectral factorization of the symmetric matrix A with Q orthogonal and Λ diagonal and let $S = Q^T M^{1/2}$. Then

$$B = I - \delta C = I - \delta S^{-1}\Lambda S = S^{-1}(I - \delta\Lambda)S = S^{-1}\tilde{\Lambda}S.$$

The columns of the matrix S^{-1} are the unnormalized eigenvectors of B. Since $S^{-T}MS^{-1} = I$, they are M-orthonormal. Then $p_k(B) = S^{-1}p_k(\tilde{\Lambda})S$. If we denote by s^i the columns of the eigenvector matrix S^{-1}, we can decompose the initial generalized residual z^0 on the eigenvectors of B,

$$z^0 = \sum_{i=1}^{n} \bar{\alpha}_i s^i.$$

Then

$$z^k = \sum_{i=1}^{n} \bar{\alpha}_i p_k(\tilde{\lambda}_i) s^i,$$

where the elements $\tilde{\lambda}_i = 1 - \delta\lambda_i$ are the eigenvalues of B. The inner product of two vectors is given by

$$\langle z^k, z^l \rangle = (z^k, Mz^l) = \sum_{i=1}^{n} \bar{\alpha}_i^2 p_k(\lambda_i) p_l(\lambda_i).$$

As we know this sum can be written as a Riemann–Stieltjes integral with a measure α,

$$\langle z^k, z^l \rangle = \int_a^b p_k(\lambda) p_l(\lambda) \, d\alpha(\lambda).$$

Associated with the measure α there is a set of orthogonal polynomials ψ_k, $k = 1, \ldots, n$. The modified moments m_k are defined as the integral of the polynomials p_k

$$m_k = \langle z^k, z^0 \rangle = \int_a^b p_k(\lambda) \, d\alpha(\lambda).$$

The coefficients of the three-term recurrence for ψ_k can be computed by a slight variation of the modified Chebyshev algorithm. Let us write the three-term recurrences for both polynomials p_k and ψ_k as

$$\lambda p_k(\lambda) = b_{k+1} p_{k+1}(\lambda) + a_{k+1} p_k(\lambda) + c_k p_{k-1}(\lambda),$$

with $b_{k+1} = 1/\omega_{k+1}$, $a_{k+1} = 0$, $c_k = (\omega_{k+1} - 1)/\omega_{k+1}$ and

$$\lambda \psi_k(\lambda) = \beta_{k+1} \psi_{k+1}(\lambda) + \alpha_{k+1} \psi_k(\lambda) + \gamma_k \psi_{k-1}(\lambda).$$

Given an integer m, as we have already seen, the coefficients are computed through

$$\sigma_{-1,l} = 0, \ \sigma_{0,l} = m_l, \ l = 0, 1, \ldots,$$

$$\alpha_1 = a_1 + b_1 \frac{\sigma_{0,1}}{\sigma_{0,0}}, \ \gamma_0 = 0,$$

and for $k = 1, \ldots, m - 1$

$$\sigma_{k,l} = \frac{1}{\beta_k}[b_{l+1}\sigma_{k-1,l+1} + (a_{l+1} - \alpha_k)\sigma_{k-1,l} + c_l\sigma_{k-1,l-1} - \gamma_{k-1}\sigma_{k-2,l}]$$

$$\text{for } l = k, \ldots, 2m - k - 1,$$

$$\alpha_{k+1} = a_{k+1} + b_{k+1}\frac{\sigma_{k,k+1}}{\sigma_{k,k}} - b_k\frac{\sigma_{k-1,k}}{\sigma_{k-1,k-1}},$$

$$\gamma_k = b_k \frac{\sigma_{k,k}}{\sigma_{k-1,k-1}}.$$

The coefficients β_k can be chosen to scale the polynomials ψ_k. The choice made in [146] is $\beta_k = b_k$. The smallest and largest eigenvalues of the tridiagonal matrix

$$J_m = \begin{pmatrix} \alpha_1 & \beta_1 & & & & \\ \gamma_1 & \alpha_2 & \beta_2 & & & \\ & \ddots & \ddots & \ddots & & \\ & & \gamma_{m-2} & \alpha_{m-1} & \beta_{m-1} \\ & & & \gamma_{m-1} & \alpha_m \end{pmatrix}$$

are the estimates we are looking for. Even though the computation of the modified moments also require inner products, only one inner product per iteration is needed. Moreover, it is not necessary to compute $\langle z^k, z^0 \rangle$ at iteration k. We only need these inner products when we want to compute estimates of the extreme eigenvalues. Therefore, on parallel computers some inner products can be computed in parallel.

In the previous algorithm J_m is known only after $2m$ CSI iterations because we need $2m$ moments. However (see [146]), one can take advantage of the properties of the Chebyshev polynomials to obtain the modified moments required for the computation of J_m only after m iterations. The polynomial of the CSI method is given by

$$p_k(\lambda) = \frac{C_k(\lambda/\mu)}{C_k(1/\mu)},$$

where C_k is the Chebyshev polynomial of the first kind of degree k. Golub and Kent used the relation

$$C_{k+l} = 2C_k C_l - C_{|k-l|},$$

which arises from the trigonometric identity

$$\cos(k+l)\theta = 2\cos k\theta \cos l\theta - \cos(k-l)\theta.$$

This gives the two relations

$$C_{2k} = 2C_k^2 - C_0, \quad C_{2k+1} = 2C_k C_{k+1} - C_1.$$

Using these relations one obtains

$$p_{2k}(\lambda) = p_k^2(\lambda) + \frac{1}{C_{2k}(1/\mu)}[p_k^2(\lambda) - 1],$$

and something similar for p_{2k+1}. Integrating this relation gives two modified moments

$$\nu_{2k} = \langle z^k, z^k \rangle + \frac{1}{C_{2k}(1/\mu)}(\langle z^k, z^k \rangle - \nu_0),$$

$$\nu_{2k+1} = \langle z^k, z^{k+1} \rangle + \frac{1}{\mu C_{2k}(1/\mu)}(\langle z^k, z^{k+1} \rangle - \nu_1).$$

One may wonder why one would use this variant of the algorithm instead of CG to solve a symmetric positive linear system, since to obtain the eigenvalue estimates one has also to compute two inner products per iteration. However, we do not need to compute the estimates at every iteration, and the inner products need not be computed as soon as the vectors are computed as in the CG algorithm. Therefore, the CSI algorithm with computation of the eigenvalues offers more flexibility than CG on parallel computers. On the other hand, CG takes into account the distribution of all eigenvalues.

5.6 Inverse Eigenvalue Problems

When we have a discrete measure and therefore a discrete inner product, we may want to compute the recurrence coefficients from the nodes t_j and weights w_j. As we will see the nodes and weights are related to the eigenpairs of the Jacobi matrix. Hence, we have an inverse eigenvalue problem in which one wants to reconstruct a symmetric tridiagonal matrix from its eigenvalues and the first components of its eigenvectors.

We have seen that this can be done with the Stieltjes procedure but this algorithm may suffer from rounding errors. This inverse problem has been considered in the paper by de Boor and Golub [79] using the Lanczos algorithm. But the Lanczos algorithm also suffers from rounding errors. Gragg and Harrod [164] gave a more stable algorithm based on a paper by Rutishauser [288]. Their algorithm uses orthogonal transformations, namely Givens rotations. This kind of algorithm has also been used by Reichel [278], [279]. Laurie [220] proposed to use variants of the QD algorithm; see chapter 3. On inverse eigenvalue problems see also Boutry [32]. Let us now review these algorithms.

5.6.1 Solution Using the Lanczos Algorithm

In exact arithmetic the problem defined above of reconstructing the Jacobi matrix from the nodes and weights can be solved by the Lanczos algorithm, see chapter 4. The Lanczos vectors v^j of the orthonormal basis of the Krylov subspace are constructed by three-term recurrences because of the symmetry of A. We have seen that the basis vectors satisfy

$$v^k = p_k(A)v^1, \quad v^1 = v,$$

where p_k is a polynomial of degree $k - 1$. In our case we choose $A = \Lambda = \Lambda_m$ a diagonal matrix of order m whose diagonal elements are $t_1, \ldots, t_m$ the given nodes (or eigenvalues). Therefore, we have

$$(v^i, v^j) = (p_j(\Lambda_m)v, p_i(\Lambda_m)v) = \sum_{l=1}^{m} p_j(t_l)p_i(t_l)v_l^2 = \delta_{i,j}.$$

Hence, if the initial vector v is chosen as the vector of the square roots of the weights w^2, the Lanczos polynomials are orthogonal for the given discrete inner product and the Jacobi matrix that is sought is the tridiagonal matrix generated by the Lanczos algorithm. Even though it might seem strange to start from a diagonal matrix to end up with a tridiagonal matrix, this should solve the inverse problem defined above. Since the matrix Λ_m has distinct eigenvalues we should do exactly m Lanczos iterations.

Moreover, things can also be seen in a different way. Let K_m be the Krylov matrix $K_m = \begin{pmatrix} v & \Lambda v & \cdots & \Lambda^{m-1}v \end{pmatrix}$ and $K_m = V_m R_m$ be a QR factorization where V_m is an orthonormal matrix of order m and R_m an upper triangular matrix of order m with positive elements on the diagonal. It turns out that V_m is the matrix whose columns are the Lanczos vectors v^j. Moreover, R_m is the Cholesky factor of the moment matrix $K_m^T K_m$. That is,

$$R_m^T R_m = \begin{pmatrix} 1 & (v, \Lambda v) & \cdots & (v, \Lambda^{m-1} v) \\ (v, \Lambda v) & (v, \Lambda^2 v) & \cdots & (v, \Lambda^m v) \\ \vdots & \vdots & & \vdots \\ (v, \Lambda^{m-1} v) & \cdots & & (v, \Lambda^{2m-2} v) \end{pmatrix}.$$

The inner products involved in this matrix are equal to

$$(v, \Lambda^j v) = \sum_{l=1}^m t_l^j w_l^2,$$

that is, they are the moments computed with the monomials t^j. In fact, the Lanczos algorithm progressively constructs matrices V_k which are $m \times k$, $k \leq m$ such that $K_k = V_k R_k$ where R_k is upper triangular of order k. Matrices R_k are extended at each step by one row and one column. The elements $r_{i,j}$ of the successive matrices R_k can also be related to those of the Jacobi matrix J_k. If we denote the matrix J_k by

$$J_k = \begin{pmatrix} \alpha_1 & \eta_1 & & & \\ \eta_1 & \alpha_2 & \eta_2 & & \\ & \ddots & \ddots & \ddots & \\ & & \eta_{k-2} & \alpha_{k-2} & \eta_{k-1} \\ & & & \eta_{k-1} & \alpha_k \end{pmatrix},$$

we have $r_{1,1} = 1$ and

$$\eta_{i-1} = \frac{r_{i,i}}{r_{i-1,i-1}}.$$

This shows that

$$\eta_1 \cdots \eta_{k-1} = r_{k,k}.$$

We also know that

$$\alpha_1 = r_{1,2}, \quad \alpha_i = \frac{r_{i,i+1}}{r_{i,i}} - \frac{r_{i-1,i}}{r_{i-1,i-1}}, \quad i = 2, \cdots, k-1, \quad \alpha_k = w_k^k - \frac{r_{k-1,k}}{r_{k-1,k-1}},$$

$w_k^k = (v^k, \Lambda_m^k v)/r_{k,k}$ and $r_{k,k+1} = (v^k, \Lambda_m^k v)$. So, in principle, we can compute the Jacobi matrix from the moment matrix. The Lanczos algorithm does this for us in a simple and convenient way without having to use the Cholesky factorization of the moment matrix.

However, as we have seen in chapter 4, the Lanczos algorithm can suffer badly from rounding errors. When the rounding errors start to grow, the Lanczos vectors v^j lose their orthogonality. Moreover, multiple copies of some eigenvalues t_j appear among the Ritz values and this delays the convergence to the other eigenvalues. Hence, without reorthogonalization, it is not always feasible to solve the inverse eigenvalue problem using the Lanczos algorithm.

5.6.2 Solution Using the Stieltjes Algorithm

As we have seen before the Stieltjes algorithm can be used to solve the inverse eigenvalue problem. Remember that if we consider monic orthogonal polynomials p_k the coefficients of the tridiagonal matrix are

$$\alpha_{k+1} = \frac{(tp_k, p_k)}{(p_k, p_k)}, \quad \eta_k = \frac{(p_k, p_k)}{(p_{k-1}, p_{k-1})},$$

α_k (resp. η_k) being the diagonal (resp. off-diagonal) elements. The inner products can be easily computed using the nodes and weights. If we want to obtain the Jacobi matrix corresponding to orthonormal polynomials, we just have to take the square roots of the off-diagonal elements.

It is clear that the above procedure is nothing other than the Lanczos algorithm with a different normalization. Therefore, in floating point arithmetic it must suffer also from rounding error problems. Moreover, there is a potential danger of overflow.

5.6.3 Solution Using Rotations

In [164] Gragg and Harrod considered the reconstruction of Jacobi matrices from the spectral data. This terse but nice paper summarizes the relation between Jacobi matrices, orthogonal polynomials, continued fractions, Padé approximation and Gauss quadrature. The main theme of this paper is to consider that in finite precision arithmetic the Lanczos algorithm is sensitive to rounding errors and cannot reliably solve the problem of the computation of the Jacobi matrix. However, the authors suggested the use of a rational variant of the Lanczos algorithm designed for a diagonal matrix.

Let d be the vector whose elements are β_0 times the given first components. Then, the algorithm is the following,

$p^{-1} = 0$, $p^0 = d$, $\rho_{-1}^2 = 1$,

for $k = 0, 1, \ldots, n - 1$

$\quad w^k = p^k . * p^k$, $\rho_k^2 = e^T w^k$

$\quad \beta_k^2 = \rho_k^2 / \rho_{k-1}^2$, $\alpha_{k+1} = l^T w^k / \rho_k^2$

$\quad p^{k+1} = (l - \alpha_{k+1} e) . * p^k - p^{k-1} \beta_k^2$

end

where $.*$ denotes the element-by-element multiplication, e is the vector of all 1's, $l = \Lambda_m e$ and β_k are the off-diagonal elements.

The recommended algorithm in [164] uses orthogonal transformations to compute the Jacobi matrix. Let us describe it and denote $\Lambda = \Lambda_m$ for simplicity. Assume that

$$\begin{pmatrix} 1 & \\ & Q^T \end{pmatrix} \begin{pmatrix} \alpha_0 & d^T \\ d & \Lambda \end{pmatrix} \begin{pmatrix} 1 & \\ & Q \end{pmatrix} = \begin{pmatrix} \alpha_0 & \beta_0(e^1)^T \\ \beta_0 e^1 & J_n \end{pmatrix},$$

with Q an orthogonal matrix. This construction is done incrementally. Let us add (δ, λ) to the data (d, Λ). We have

$$\begin{pmatrix} 1 & & \\ & Q^T & \\ & & 1 \end{pmatrix} \begin{pmatrix} \alpha_0 & d^T & \delta \\ d & \Lambda & 0 \\ \delta & 0 & \lambda \end{pmatrix} \begin{pmatrix} 1 & & \\ & Q & \\ & & 1 \end{pmatrix} = \begin{pmatrix} \alpha_0 & \beta_0(e^1)^T & \delta \\ \beta_0 e^1 & J_n & 0 \\ \delta & 0 & \lambda \end{pmatrix}.$$

To tridiagonalize the matrix in the right-hand side, we use rotations to chase the element δ in the last row and column toward the diagonal, without changing α_0, which is not needed in the algorithm. At some intermediate stage, we obtain

$$\begin{pmatrix} \bar{\alpha}_{k-1} & \beta'_{k-1} & 0 & \delta_{k-1} \\ \beta'_{k-1} & \alpha_k & \beta_k(e^1)^T & \bar{\delta}_{k-1} \\ 0 & \beta_k e^1 & J_{k+1,n} & 0 \\ \delta_{k-1} & \bar{\delta}_{k-1} & 0 & \lambda_{k-1} \end{pmatrix},$$

with $\lambda_{k-1} = \lambda + \tau_{k-1}$. We denote

$$\begin{pmatrix} \beta'_{k-1} & \bar{\delta}_{k-1} \\ \delta_{k-1} & \tau_{k-1} \end{pmatrix} = \begin{pmatrix} \beta_{k-1} \\ \pi_{k-1} \end{pmatrix} \begin{pmatrix} \gamma_{k-1} & \sigma_{k-1} \end{pmatrix},$$

where π_k is not to be confused with the polynomial π_n and $\gamma_{k-1}^2 + \sigma_{k-1}^2 = 1$. We choose γ_k and σ_k to annihilate δ_{k-1} with a rotation between the second and last rows. This gives $\gamma_k^2 + \sigma_k^2 = 1$ and

$$\begin{pmatrix} \gamma_k & -\sigma_k \\ \sigma_k & \gamma_k \end{pmatrix} \begin{pmatrix} \beta'_{k-1} \\ \delta_{k-1} \end{pmatrix} = \begin{pmatrix} \bar{\beta}_{k-1} \\ 0 \end{pmatrix},$$

which is

$$\begin{pmatrix} \gamma_k & -\sigma_k \\ \sigma_k & \gamma_k \end{pmatrix} \begin{pmatrix} \beta_{k-1} \\ \pi_{k-1} \end{pmatrix} \gamma_{k-1} = \begin{pmatrix} \bar{\beta}_{k-1} \\ 0 \end{pmatrix}.$$

Let $\rho_k = (\beta_{k-1}^2 + \pi_{k-1}^2)^{1/2}$; then

$$\gamma_k = \frac{\beta_{k-1}}{\rho_k}, \quad \sigma_k = -\frac{\pi_{k-1}}{\rho_k}, \text{ if } \rho_k > 0,$$

and $\gamma_k = 1$, $\sigma_k = 0$ if $\rho_k = 0$. Hence,

$$\bar{\beta}_{k-1} = \gamma_{k-1}\rho_k, \quad \sigma_k \beta_{k-1} + \gamma_k \pi_{k-1} = 0.$$

Let

$$\begin{pmatrix} \beta'_k \\ \bar{\delta}_k \end{pmatrix} = \begin{pmatrix} \gamma_k & -\sigma_k \\ \sigma_k & \gamma_k \end{pmatrix} \begin{pmatrix} \beta_k \\ 0 \end{pmatrix}.$$

The result is

$$\begin{pmatrix} \bar{\alpha}_{k-1} & \beta'_{k-1} & 0 & 0 \\ \bar{\beta}_{k-1} & \bar{\alpha}_k & \beta'_k(e^1)^T & \bar{\delta}_{k-1} \\ 0 & \beta'_k e^1 & J_{k+1,n} & \bar{\delta}_k e^1 \\ 0 & \delta_k & \bar{\delta}_k(e^1)^T & \lambda_k \end{pmatrix},$$

with $\lambda_k = \lambda + \tau_k$ and

$$(\delta_k \quad \tau_k) = \pi_k(\gamma_k \quad \sigma_k), \quad \pi_k = \sigma_k(\alpha_k - \lambda) + \gamma_k \sigma_{k-1}\beta_{k-1},$$

$$\tau_k = \sigma_k^2(\alpha_k - \lambda) - \gamma_k^2 \tau_{k-1}.$$

Moreover $\bar{\alpha}_k = \alpha_k - (\tau_k - \tau_{k-1})$. The Kahan–Pal–Walker version of this algorithm squares some equations to update the squares of most of the involved quantities; see [164]. The implementation given in Gragg and Harrod (algorithm RKPW) is the following to add the data (λ, δ):

$\gamma_0^2 = 1$, $\beta_n^2 = \sigma_0^2 = \tau_0 = 0$, $\alpha_{n+1} = \lambda$, $\pi_0^2 = \delta^2$

for $k = 1, \ldots, n+1$

$\quad \rho_k^2 = \beta_{k-1}^2 + \pi_{k-1}^2$, $\bar{\beta}_{k-1}^2 = \gamma_{k-1}^2 \rho_k^2$

$\quad$ if $\rho_k^2 = 0$ then $\gamma_k^2 = 1$, $\sigma_k^2 = 0$

$\quad$ else $\gamma_k^2 = \beta_{k-1}^2 / \rho_k^2$, $\sigma_k^2 = \pi_{k-1}^2 / \rho_k^2$

$\quad \tau_k = \sigma_k^2 (\alpha_k - \lambda) - \gamma_k^2 \tau_{k-1}$

$\quad \bar{\alpha}_k = \alpha_k - (\tau_k - \tau_{k-1})$

$\quad$ if $\sigma_k^2 = 0$ then $\pi_k^2 = \sigma_{k-1}^2 \beta_{k-1}^2$

$\quad$ else $\pi_k^2 = \tau_k^2 / \sigma_k^2$

end

Note that if $\xi_1 = \alpha_1 - \lambda$ and

$$\xi_k = \alpha_k - \lambda - \frac{\beta_k^2}{\xi_{k-1}}$$

(which are the diagonal elements of the Cholesky-like factorization) then $\tau_k = \sigma_k^2 \xi_k$ and $\pi_k^2 = \tau_k \xi_k$. The solution is then obtained incrementally by adding one node and one weight after the other. Note that the order in which we introduce the new elements may have an efect on the results.

In [278], [279] Reichel considered a method very similar to the Gragg and Harrod method, although the implementation was slightly different.

5.6.4 Solution Using the QD Algorithm

First, for the direct problem of computing the nodes and weights from the Jacobi matrix, Laurie [220] proposed to use variants of the QL algorithm instead of the QR algorithm to compute the nodes and weights from the Jacobi matrix and the first moment. The main interest of this approach is that this algorithm can be "reversed" to give an algorithm named convqr to compute the Jacobi matrix from the nodes and weights. It turns out that this algorithm is very close to the Gragg and Harrod algorithm. But the Jacobi matrix is computed from bottom to top by adding a first row and a first column at each step. This corresponds to adding one node at a time going from the last one t_m to the first one t_1. However, these algorithms using orthogonal transformations are not too sensitive to the order in which the nodes are added.

In [220] Laurie proposed to use variants of the QD algorithm to recover the Jacobi matrix from nodes and (positive) weights. He used the stationary QD and the shifted progressive QD algorithms. The development of the algorithm pftoqd is formulated using partial fraction expansion and Stieltjes and Jacobi continued fractions but the basis is really the factorization of tridiagonal matrices. Laurie considered

$$r(\lambda) = \sum_{j=1}^{n} \frac{w_j}{\lambda - t_j},$$

where $t_j = \theta_j^{(n)}$. He used a Stieltjes continued fraction (or S-fraction) which is

$$r(\lambda) = \frac{e_0 \mid}{\mid \lambda - 1} - \frac{q_1 \mid}{\mid \lambda - 1} - \cdots - \frac{e_{n-1} \mid}{\mid \lambda - q_n},$$

where the coefficients are alternatively e's and q's. The coefficients of the S-fraction are coefficients for a two-term recursion of orthogonal polynomials

$$u_{k+1}(\lambda) = \lambda p_k(\lambda) - e_k u_k(\lambda),$$

$$p_{k+1}(\lambda) = u_{k+1}(\lambda) - q_{k+1} p_k(\lambda),$$

with initial conditions $u_0(\lambda) \equiv 0$, $p_0(\lambda) \equiv 1$. The coefficients q_k and e_k are elements of the LU factorization of J_n. They can be computed by QD algorithms. The stationary QD algorithm (dstqd) computes the S-fraction of $E_\sigma r$ of r for a shifted argument

$$E_\sigma(\lambda) r = r(\lambda + \sigma).$$

Let Zr be defined as

$$(Zr)(\lambda) = \lambda r(\lambda) - \lim_{\lambda \to \infty} \lambda r(\lambda).$$

The progressive QD algorithm (dqds) computes the S-fraction of $E_\sigma Zr$ from that of r. Laurie's algorithm pftoqd derivation starts by showing how to compute the S-fraction of $r^{(0)} + w/\lambda$ given the S-fraction of $r^{(0)}$. Let

$$(e_0, q_1, e_1, q_2, \ldots, e_{n-1}, q_n)$$

be an augmented QD row and $Q^{(k)}$ be the augmented QD row of $r^{(k)}$ with elements $e_j^{(k)}$ and $q_j^{(k)}$. Rutishauser's algorithm is the following.

1) Prepend the pair $(1, w)$ to $Q^{(0)}$ to form $Q^{(1)}$; the corresponding rational function is

$$r^{(1)}(\lambda) = \cfrac{1}{\lambda - \cfrac{w}{1 - r^{(0)}(\lambda)}}.$$

2) Apply the progressive QD algorithm with a zero shift to $Q^{(1)}$ to form $Q^{(2)}$; the corresponding rational function is

$$r^{(2)}(\lambda) = Zr^{(1)}(\lambda) = \lambda r^{(1)}(\lambda) - 1.$$

3) Discard the first element of $Q^{(2)}$ and append a zero to form $Q^{(3)}$; the corresponding rational function is

$$r^{(3)}(\lambda) = 1 - \frac{w}{\lambda r^{(2)}(\lambda)}.$$

Therefore,

$$r^{(3)}(\lambda) = \frac{w}{\lambda} + r^{(0)}(\lambda).$$

Laurie assumed that the nodes are sorted as $t_1 < t_2 < \cdots < t_m$ and denoted the differences as $\sigma_j = t_j - t_{j-1}$. The following algorithm computes the sum

$$r_1(\lambda) = \sum_{j=1}^{m} \frac{w_j}{\lambda - t_j}.$$

$$r_m(\lambda) = \frac{w_m}{\lambda - \sigma_m}$$

for $j = m - 1, \ldots, 1$

$$\tilde{r}_j(\lambda) = r_{j+1}(\lambda) + \frac{w_j}{\lambda}$$

$$r_j(\lambda) = \tilde{r}_j(\lambda - \sigma_j)$$

end

The same algorithm can be expressed using QD rows. This is Laurie's pftoqd algorithm:

$$Q = (\, w_1 \quad w_2 \quad \cdots \quad w_m \quad 0 \quad \cdots \quad 0 \,)$$

$$k = m + 1$$

$$Q(k) = \sigma_m$$

for $j = m - 1, \ldots, 1$

   apply dqd to $Q(j : k)$

   apply dstqd$(-\sigma_j)$ to $Q(j + 1 : k + 1)$

   $k = k + 1$

end

The number of floating point operations of this algorithm is $(9/2)n^2$, to be compared to $(11/2)n^2$ operations for convqr. When the QD row has been computed, the Jacobi matrix can be recovered from

$$\alpha_{k+1} = q_{k+1} + e_k, \quad \beta_k = q_k e_k.$$

The previous algorithms for computing the solution of the inverse problem will be compared on several examples in the second part of this book. For computations using also the QD algorithm see Cuyt [72].

5.7 Modifications of Weight Functions

In this section we consider measures α defined through a weight function w by $d\alpha(\lambda) = w(\lambda) \, d\lambda$. A problem that has been considered in many papers during the last 40 years is how to obtain the coefficients of the three-term recurrences of orthogonal polynomials related to a weight function $r(\lambda)w(\lambda)$ when the coefficients of the orthogonal polynomials related to w are known and r is a polynomial or a rational function. This problem was first studied by Christofell in 1858 when $r(\lambda) = \lambda - \beta$. The general solution was given by V. B. Uvarov [331], [332]. However, the solution is given using determinants and this is difficult to use for computation, although it has been used by Gautschi [131]. Another related problem is consideration of a weight function given as the sum of two weight functions with known orthogonal polynomials.

The general problem when r is the ratio of two polynomials can be broken into easier problems by writing

$$\int_a^b r(\lambda)w(\lambda) \, d\lambda) = \int_a^b \left(q(\lambda) + \sum_i \frac{a_i}{\lambda - t_i} + \sum_j \frac{b_j \lambda + c_j}{(\lambda - x_j)^2 + y_j^2} \right) w(\lambda) \, d\lambda,$$

where q is a real polynomial, t_i, $i = 1, 2, \ldots$ and $z_j = x_j \pm \imath y_j$, $\imath = \sqrt{-1}$, $j = 1, 2, \ldots$ are the real and complex roots of the denominator of r. Hence, it is enough to consider multiplication by polynomials and division by linear and quadratic factors if we know the roots of the denominator of the rational function. We have also to use an algorithm for sums of weight functions.

Let us consider some of the algorithms that have been proposed for solving these problems; see [202], [110], [142], [101], [99], [131], [307].

5.7.1 Sum of Weights

The problem considered in Fischer and Golub [110] is the following. Let $[l_j, u_j]$, $j = 1, \ldots, N$ with $l_1 \leq l_2 \leq \cdots \leq l_N$ be N intervals that can be disjoint or not and let w_j be a nonnegative weight function defined on $[l_j, u_j]$. There are orthogonal polynomials $p_k^{(j)}$ associated with every w_j satisfying three-term recurrences,

$$\lambda p_k^{(j)}(\lambda) = \beta_{k+1}^{(j)} p_{k+1}^{(j)}(\lambda) + \alpha_{k+1}^{(j)} p_k^{(j)}(\lambda) + \gamma_k^{(j)} p_{k-1}^{(j)}(\lambda), \qquad (5.10)$$

$$p_{-1}^{(j)}(\lambda) \equiv 0, \quad p_0^{(j)}(\lambda) \equiv 1.$$

Let $l = l_1$ and $u = \max_j u_j$ and

$$w(\lambda) = \sum_{j=1}^{N} \epsilon_j \chi_{[l_j, u_j]} w_j(\lambda),$$

where $|\epsilon_j| \leq 1$ and $\chi_{[l_j, u_j]}$ is the characteristic function of the interval $[l_j, u_j]$ with values 0 outside the interval and 1 inside the interval. The problem is to generate the coefficients $\beta_k, \alpha_k, \gamma_k$ of the orthogonal polynomial associated with w given the coefficients $\beta_k^{(j)}, \alpha_k^{(j)}, \gamma_k^{(j)}$ and the zero-order moments,

$$\mu_0^{(j)} = \int_{l_j}^{u_j} w_j(\lambda) \, d\lambda.$$

The inner product associated with w is

$$\langle f, g \rangle = \sum_{j=1}^{N} \epsilon_j \int_{l_j}^{u_j} f(\lambda) g(\lambda) w_j(\lambda) \, d\lambda.$$

All the algorithms considered in [110] need to compute $\langle p, 1 \rangle$ where p is a polynomial of degree less than or equal to $2n$ for a given n. This can be done by using Gauss quadrature with n nodes for every w_j.

The tridiagonal matrix $T_n^{(j)}$ defined by equation (5.10) and associated with the function w_j is not symmetric. However, as we have seen, it can be symmetrized using a diagonal matrix $D_n^{(j)}$. Then, $J_n^{(j)} = (D_n^{(j)})^{-1} T_n^{(j)} D_n^{(j)}$ is the Jacobi matrix associated with w_j. We will see in chapter 6 that the nodes of the Gauss quadrature are the eigenvalues of $J_n^{(j)}$ and the weights are the squares of the first elements of the eigenvectors. Then $\langle p, 1 \rangle$ can be computed as

$$\langle p, 1 \rangle = \sum_{j=1}^{N} \epsilon_j \mu_0^{(j)} (e^1)^T p(J_n^{(j)}) e^1.$$

The algorithm that worked the best for the examples considered in [110] is simply the Stieltjes procedure. For the sake of simplicity, let us take the number of intervals $N = 2$. Let ψ_k be the monic polynomial ($\beta_k = 1$) associated with w and let $\hat{\psi}_k$ be the corresponding orthonormal polynomial which satisfies

$$\lambda \hat{\psi}_k(\lambda) = \hat{\gamma}_{k+1} \hat{\psi}_{k+1}(\lambda) + \hat{\alpha}_{k+1} \hat{\psi}_k(\lambda) + \hat{\gamma}_k \hat{\psi}_{k-1}(\lambda),$$

$$\hat{\psi}_{-1}(\lambda) \equiv 0, \quad \hat{\psi}_0(\lambda) \equiv (\epsilon_1 \mu_0^{(1)} + \epsilon_2 \mu_0^{(2)})^{-1/2}.$$

Therefore $\hat{\psi}_k(\lambda) = \langle \psi_k, \psi_k \rangle^{-1/2} \psi_k(\lambda)$. The coefficients are given by

$$\hat{\alpha}_{k+1} = \alpha_{k+1} = \frac{\langle \lambda \psi_k, \psi_k \rangle}{\langle \psi_k, \psi_k \rangle},$$

$$\hat{\gamma}_k = \sqrt{\gamma_k} = \left(\frac{\langle \psi_k, \psi_k \rangle}{\langle \psi_{k-1}, \psi_{k-1} \rangle} \right)^{1/2}.$$

These coefficients can be computed in the following way. Let n be given and for $k < n$, $z_k^{(j)} = \psi_k(J_n^{(j)})e^1$, $j = 1, 2$, then we have the three-term recurrence

$$z_{k+1}^{(j)} = (J_n^{(j)} - \alpha_k I) z_k^{(j)} - \gamma_k z_{k-1}^{(j)}.$$

The starting vectors are $z_0^{(j)} = e^1$ and

$$z_1^{(j)} = \left(J_n^{(j)} - \frac{\epsilon_1 \mu_0^{(1)} (J_n^{(1)})_{1,1} + \epsilon_2 \mu_0^{(2)} (J_n^{(2)})_{1,1}}{\epsilon_1 \mu_0^{(1)} + \epsilon_2 \mu_0^{(2)}} \right) z_0^{(j)}.$$

Therefore, by using Gauss quadrature,

$$\langle \psi_k, \psi_k \rangle = \epsilon_1 \mu_0^{(1)} (z_k^{(1)})^T z_k^{(1)} + \epsilon_2 \mu_0^{(2)} (z_k^{(2)})^T z_k^{(2)},$$

and

$$\alpha_{k+1} = \frac{\epsilon_1 \mu_0^{(1)} (z_k^{(1)})^T J_n^{(1)} z_k^{(1)} + \epsilon_2 \mu_0^{(2)} (z_k^{(2)})^T J_n^{(2)} z_k^{(2)}}{\epsilon_1 \mu_0^{(1)} (z_k^{(1)})^T z_k^{(1)} + \epsilon_2 \mu_0^{(2)} (z_k^{(2)})^T z_k^{(2)}},$$

$$\gamma_k = \frac{\epsilon_1 \mu_0^{(1)} (z_k^{(1)})^T z_k^{(1)} + \epsilon_2 \mu_0^{(2)} (z_k^{(2)})^T z_k^{(2)}}{\epsilon_1 \mu_0^{(1)} (z_{k-1}^{(1)})^T z_{k-1}^{(1)} + \epsilon_2 \mu_0^{(2)} (z_{k-1}^{(2)})^T z_{k-1}^{(2)}}.$$

Two other algorithms proposed in [110] used modified moments

$$\mu_k = \langle q_k, 1 \rangle = \int_l^u q_k(\lambda) w(\lambda) \, d\lambda,$$

where q_k, $k = 1, 2, \ldots$ is a given suitable set of auxiliary polynomials and the modified Chebyshev algorithm. However, since the numerical results given in [110] are not better than with the Stieltjes method we do not report on them here.

In [101] Elhay, Golub and Kautsky considered the problem from another perspective. With two weight functions w_1 and w_2 they set up the following problem: Given the Jacobi matrices J_1 and J_2 of dimensions n_1 and n_2 corresponding to w_1 and w_2, find a Jacobi matrix J for the weight function $w = w_1 \pm w_2$. If it exists what will be its dimension?

Let $\mathcal{W}(J, n)$ denote the set of normalized weight functions corresponding to J of dimension n and thus to the first $2n$ moments.

THEOREM 5.9 *Given the Jacobi matrices J_1 and J_2 of dimensions n_1 and n_2 and $\epsilon_1 + \epsilon_2 = 1$, $\epsilon_1, \epsilon_2 \geq 0$, let $n = \min\{n_1, n_2\}$. Then there exists J of dimension n such that $w = \epsilon_1 w_1 + \epsilon_2 w_2$ belongs to $\mathcal{W}(J, n)$ for any $w_1 \in \mathcal{W}(J_1, n_1)$ and any $w_2 \in \mathcal{W}(J_2, n_2)$.*

Proof. See [101]. Clearly, the moments of a sum of weight functions are the sums of the corresponding moments. This gives the solution of the problem. $\quad\Box$

Three different algorithms are proposed in [101] to compute J from J_1 and J_2. Let us consider the one that seems the most robust. It is denoted as JJR and uses orthogonal matrices. Let

$$J_3 = \begin{pmatrix} J_1 & 0 \\ 0 & J_2 \end{pmatrix},$$

and some J_4 of which the leading $n \times n$ submatrix will be the required J. How to obtain J_4 is characterized in the following theorem.

THEOREM 5.10 *Given J_1 and J_2, the orthogonal matrix Q such that $J_4 = Q J_3 Q^T$ must satisfy*

$$Q^T e^1 = \begin{pmatrix} \sqrt{\epsilon_1} e^1 \\ \sqrt{\epsilon_2} e^1 \end{pmatrix}. \tag{5.11}$$

Proof. Let $J_i = Q_i \Lambda_i Q_i^T$, $i = 1, 2, 4$ be the respective spectral decompositions with Q_i orthogonal and Λ_i orthogonal. The weights of the corresponding Gauss quadratures are the squares of the first components of the eigenvectors $w^i = Q_i^T e^1$; see chapter 6. To merge the Gauss quadratures of J_1 and J_2, the requirement is

$$\Lambda_4 = \begin{pmatrix} \Lambda_1 & 0 \\ 0 & \Lambda_2 \end{pmatrix}, \quad w^4 = \begin{pmatrix} \sqrt{\epsilon_1} w^1 \\ \sqrt{\epsilon_2} w^2 \end{pmatrix}.$$

Then,

$$J_4 = Q_4 \begin{pmatrix} Q_1^T J_1 Q_1 & 0 \\ 0 & Q_2^T J_2 Q_2 \end{pmatrix} Q_4^T = Q J_3 Q^T,$$

where

$$Q = Q_4 \begin{pmatrix} Q_1^T & 0 \\ 0 & Q_2^T \end{pmatrix}.$$

Hence, we have

$$Q^T e^1 = \begin{pmatrix} Q_1 & 0 \\ 0 & Q_2 \end{pmatrix} w^4 = \begin{pmatrix} \sqrt{\epsilon_1} Q_1 w^1 \\ \sqrt{\epsilon_2} Q_2 w^2 \end{pmatrix} = \begin{pmatrix} \sqrt{\epsilon_1} e^1 \\ \sqrt{\epsilon_2} e^1 \end{pmatrix},$$

as required. $\quad\Box$

Therefore, Q must be constructed to satisfy equation (5.11) and such that J_4 is tridiagonal. The matrix Q can be built as the product of orthogonal matrices. The algorithm proposed in [101] to minimize the complexity is the following. First, a permutation similarity P is applied to J_3 and $Q^T e^1$. P selects the rows in the order $1, n+1, 2, n+2, \ldots, n, 2n$. The resulting permuted matrix $P J_3 P^T$ is a five-diagonal matrix (which can also be considered as a block tridiagonal matrix or a banded matrix) with a checkerboard-like structure

$$
\begin{array}{cccccc}
x & 0 & x & 0 & & \\
0 & x & 0 & x & 0 & \\
x & 0 & x & 0 & x & 0 \\
0 & x & 0 & x & 0 & x \\
& & \ddots & & \ddots & & \ddots \\
\end{array}
$$

Let R be a rotation matrix combining the first and second rows such that $R(\sqrt{\epsilon_1}e^1 + \sqrt{\epsilon_2}e^2) = e^1$. The matrix $\hat{J} = RPJ_3P^T R^T$ has the following structure

$$
\begin{array}{cccccc}
x & x & x & x & & \\
x & x & x & x & 0 & \\
x & x & x & 0 & x & 0 \\
x & x & 0 & x & 0 & x \\
& & \ddots & & \ddots & & \ddots \\
\end{array}
$$

Now, we can apply a series of rotations (not involving the first row) whose product is $\hat{R}$ such that $\hat{R}\hat{J}\hat{R}^T$ is tridiagonal. The matrix Q in theorem 5.10 is $Q = \hat{R}RP$. This is not expensive since $\hat{J}$ can first be turned into a five-diagonal matrix $\tilde{J}$ (zeroing the outer diagonals) in $O(n)$ operations. Then this last matrix can be reduced to tridiagonal structure in $O(n^2)$ operations. We need only the leading $n \times n$ submatrix of the result, which is J. The complexity is about $14n^2 + O(n)$. The corresponding code is provided in [101].

5.7.2 Multiplication by a Polynomial

Let r be a real polynomial of degree m strictly positive on $[a, b]$. We are interested in computing the Jacobi matrix corresponding to $r(\lambda)w(\lambda)$. The solution proposed by Fischer and Golub [142] is to use the modified Chebyshev algorithm since the modified moments can be easily computed.

Another algorithm is given in Gautschi [131] for multiplication by a linear factor which relies on Uvarov's results [332]. For a polynomial of degree larger than 1 this can be applied repeatedly if we know the roots of the polynomial. Generalizations of these problems to general linear functionals are studied in Bueno and Marcellán [43].

Other algorithms were proposed earlier by Kautsky and Golub [202] for general polynomials of which the roots are not necessarily known. However, we assume that the roots of r are outside of $[a, b]$ and that $r(\lambda)$ is positive on $[a, b]$. These algorithms are based on transformations of (not necessarily orthogonal) polynomial bases. Let J be a lower Hessenberg matrix of order k with nonzero superdiagonal elements $\beta_1, \ldots, \beta_{k-1}$ and $\beta_k \neq 0, p_0 \neq 0$; there exist polynomials p_j of exact degree j such that if $P = (p_0 \quad p_1 \quad \ldots \quad p_{k-1})^T$ we have the relation

$$\lambda P(\lambda) = JP(\lambda) + \beta_k p_k(\lambda)e^k. \tag{5.12}$$

Now, we are interested in the relations between two Hessenberg matrices. This is given in the following lemma.

LEMMA 5.11 *Let J and $\tilde{J}$ be two lower Hessenberg matrices of order k (with nonzero superdiagonal elements). There exists a unique (up to a scalar factor) nonsingular lower triangular matrix L and a vector c such that*

$$L\tilde{J} = JL + e^k c^T. \tag{5.13}$$

Moreover, if P, p_k and $\tilde{P}, \tilde{p}_k$ are the polynomial bases corresponding to J and $\tilde{J}$, then

$$P = L\tilde{P},$$

$$\beta_k p_k - \tilde{\beta}_k \tilde{p}_k (e^k)^T L e^k = c^T \tilde{P}.$$

Proof. Since P and $\tilde{P}$ are polynomial bases of exact degree, it is obvious that there exists a triangular matrix L such that $P = L\tilde{P}$. Substituting this relation into equation (5.12) and subtracting the same identity with tildes, we have

$$(JL - L\tilde{J})\tilde{P} + (\beta_k p_k - \tilde{\beta}_k \tilde{p}_k (e^k)^T L e^k)e^k = 0.$$

The scalar in the parenthesis in the left-hand side has to be a polynomial of degree less than k that can be written as $c^T \tilde{P}$ for some vector c. Since this is true for all $\tilde{P}$, it gives equation (5.13). □

Conversely, given J and a vector c but not knowing L, the matrix $\tilde{J}$ is determined by equation (5.13). When J and $\tilde{J}$ are symmetric (and therefore tridiagonal) this relation is similar to that of the Lanczos algorithm of chapter 4 if we know the first column $u = Le^1$ except that the matrix L is triangular and not orthogonal. Then Kautsky and Golub [202] were interested in the dependence of the result upon the vector c. This needs the introduction of a new definition. The mth perdiagonal of a matrix A is the set of elements $a_{i,j}$ such that $i + j = m + 1$ (so to speak, perdiagonals are "orthogonal" to the usual diagonals).

THEOREM 5.12 *Let J be a lower Hessenberg matrix (with nonzero superdiagonal elements) and c and u be given vectors. Then there exist a symmetric tridiagonal matrix $\tilde{J}$ and a lower triangular matrix L satisfying equation (5.13) and such that $Le^1 = u$. Moreover, if J is tridiagonal, then for $i \geq 0$ the first $k+i$ perdiagonals of L and the first $k+i-1$ perdiagonals of $\tilde{J}$ are independent of c_j, $j = i+1, \ldots, k$.*

Proof. See [202]. □

If the polynomial basis is orthonormal on an interval $[a, b]$ for a weight function w, the matrix J is tridiagonal and we have the following properties.

LEMMA 5.13

$$\int_a^b p_k P w \, d\lambda = 0, \qquad \int_a^b p_k^2 w \, d\lambda = 1,$$

$$\int_a^b P P^T w \, d\lambda = I, \qquad \int_a^b \lambda P P^T w \, d\lambda = J.$$

Proof. See Golub and Kautsky [145]. The results are obtained using the orthonormality of the polynomials.   □

If the basis defined by $\tilde{P}, \tilde{p}_k$ is also orthonormal on $[a, b]$ but for a weight function $\tilde{w}$, the following result gives an expression for $\tilde{J}$.

THEOREM 5.14 *Let*

$$M = \int_a^b PP^T \tilde{w} \, d\lambda, \quad M_1 = \int_a^b \lambda PP^T \tilde{w} \, d\lambda.$$

Then

$$M = LL^T,$$

with $P = L\tilde{P}$, L being lower triangular. The Jacobi matrix $\tilde{J}$ related to $\tilde{w}$ is

$$\tilde{J} = L^{-1} M_1 L^{-T}.$$

Proof. This is obtained by substituting $P = L\tilde{P}$ in the relations of lemma 5.13. □

In the simple case of a polynomial of degree 1, $r(\lambda) = \lambda - \beta$, we have the relations

$$\int_a^b \tilde{P}\tilde{P}^T (\lambda - \beta) \, d\lambda = I, \quad \int_a^b \lambda \tilde{P}\tilde{P}^T \, d\lambda = L^{-1} J L^{-T}.$$

Since $\tilde{P} = L^{-1} P$, this gives $L^{-1}(J - \beta I)L^{-T} = I$. Therefore,

$$J - \beta I = LL^T.$$

The lower triangular matrix L is the Cholesky factor of $J - \beta I$ (which is positive or negative definite) and $J_1 = L^T L + \beta I$ is, but for the last row and column, the Jacobi matrix for rw. Such a modification is called a Christoffel transformation, see Galant [118], [119]. However, in numerical computations, depending on the value of the shift β, the factorization of $J - \beta I$ may not always be fully accurate. Simple modifications of the LU factorization algorithm have been suggested in Bueno and Dopico [42] to overcome this problem.

Note that, since the polynomials p_k are orthonormal relative to w, the matrices M and M_1 contain modified moments. Now, let us come back to the situation where $\tilde{w} = rw$, r being a polynomial of degree m and we assume that we do not know the roots of r.

THEOREM 5.15 *Let $k > m$; then the first $k - m$ elements of c defined in equation (5.13) vanish.*

Proof. We have

$$c^T = \int_a^b (\beta_k p_k - \tilde{\beta}_k \tilde{p}_k (e^k)^T Le^k) \tilde{P}^T \tilde{w} \, d\lambda = \beta_k \int_a^b p_k P^T rw L^{-T} \, d\lambda.$$

The result follows because of the orthogonality of p_k and the fact that L^{-T} is upper triangular.   □

Let $\mathcal{Z}_{k,m}$ (or $\mathcal{Z}_m$ when the value of k is not specifically needed) be the set of matrices of order k whose first $2k-1-m$ perdiagonals vanish. The characterization of the dependence on c is given in the following results. The next lemma gives properties of matrices with vanishing perdiagonals; see [202].

LEMMA 5.16 *If J is lower Hessenberg, L lower triangular and $Z \in \mathcal{Z}_m$ then $Z^T, LZ, ZL^T \in \mathcal{Z}_m$ and $JZ \in \mathcal{Z}_{m+1}$. Moreover, if $\tilde{L}$ is a triangular matrix such that $LL^T - \tilde{L}\tilde{L}^T \in \mathcal{Z}_m$, then $L - \tilde{L} \in \mathcal{Z}_m$.*

The next theorem is a generalization of an identity of lemma 5.13.

THEOREM 5.17 *Let r be a polynomial of degree m. Then*

$$\int_a^b rPP^T w \, d\lambda - r(J) \in \mathcal{Z}_{m-1}, \ j \leq m.$$

Proof. Since r is a polynomial, it is sufficient to show that

$$\int_a^b \lambda^j PP^T w \, d\lambda - J^j \in \mathcal{Z}_{m-1}.$$

This is done by induction on j. □

The main result of Kautsky and Golub [202] is the following theorem.

THEOREM 5.18 *Let r be a polynomial of degree $m < k$ such that the Cholesky factorization $r(J) = L_1 L_1^T$ exists where L_1 is lower triangular. Let $J_1 = L_1^{-1} J L_1$ and $B = J[r(J)]^{-1}$. Then*

1) *B is symmetric and $B = [r(J)]^{-1}J$,*

2) *$J_1 = L_1^T B L_1$,*

3) *J_1 is a symmetric tridiagonal matrix,*

4) *$\tilde{J} - J_1 \in \mathcal{Z}_m$ where $\tilde{J}$ is the Jacobi matrix corresponding to $\tilde{w} = rw$.*

Proof. See [202]. The proof is obtained by using the spectral factorization of J. □

Given a shift β and a scalar $\sigma \neq 0$, consider a decomposition $\sigma(J - \beta I) = XR$ with X nonsingular and R upper triangular. The matrix

$$\hat{J} = \frac{1}{\sigma} RX + \beta I$$

is similar to J. Since J is also similar to $J_1 = L_1^T J L_1^{-T}$, we are interested in the conditions on X to have $\hat{J} = J_1$.

PROPOSITION 5.19 *If*

$$XX^T = \sigma^2 (J - \beta I)[r(J)]^{-1}(J - \beta I),$$

then $\hat{J} = J_1$.

Proof. We have

$$XX^T = XR[r(J)]^{-1} R^T X^T,$$

and $R = L_1^T$ follows from the uniqueness of the Cholesky factorization of $r(J)$. □

If r is a polynomial of degree 1 written as $r = \sigma(\lambda - \beta)$, the relation in proposition 5.19 is $XX^T = \sigma(J - \beta I)$ and (as we have already seen) we can choose

$X = R^T = L_1$. For general polynomials, one can proceed in the following way. Let $J = Z\Theta Z^T$ be the spectral decomposition of J with Z orthogonal and Θ diagonal and $D^2 = r(\Theta)$, $\Theta_\beta = \Theta - \beta I$. Then we must have

$$Z^T X X^T Z = \sigma^2 \Theta_\beta D^{-2} \Theta_\beta.$$

This implies that the matrix $U = (1/\sigma)D\Theta_\beta^{-1} Z^T X$ must be orthogonal. Since $Z^T X R = \sigma \Theta_\beta Z^T$, we have

$$UR = DZ^T. \tag{5.14}$$

Hence U and R can be obtained by the QR factorization of DZ^T. Since $X = \sigma Z\Theta_\beta D^{-1} U$,

$$J_1 = \hat{J} = X^{-1} JX = U^T \Theta U.$$

This gives us the theoretical solution of the problem. Using the spectral decomposition of J we have to compute the QR factorization of DZ^T in equation (5.14) and then $J_1 = U^T \Theta U$. From J_1 we can obtain $\tilde{J}$ of order $k - \lceil m/2 \rceil - 1$ corresponding to the weight $\tilde{w} = rw$. These theoretical results can be turned into several practical numerical methods to compute $\tilde{J}$; see Kautsky and Golub [202].

Choosing $c = 0$, one can use the relation $L_1 J_1 = JL_1$ to compute J_1 as long as $u = L_1 e^1$ is known. But, if $d = r(J)e^1$, then $u = \sqrt{d_1}d$ and d is obtained by $d = Zr(\Theta)Z^T e^1$. Only the $m+1$ first rows of Z are needed. This is a Lanczos-like algorithm.

The second algorithm is the polynomial shift implicit QR (PSI QR). It performs an implicit QR factorization of the matrix DZ^T. The matrix U is sought in the form $U = H_1 \cdots H_{k-1}$ where H_j is a symmetric Householder transformation with the requirement that the matrix

$$H_{k-1} \cdots H_1 \Theta H_1 \cdots H_{k-1}$$

should be tridiagonal. See [202] for details.

5.7.3 Weight Functions Not of One Sign

Kautsky and Elhay [201] generalized Jacobi matrices to arbitrary weight functions w. Let p_k, $k = 1, 2, \ldots$ be a set of monic polynomials and

$$P = (p_0(\lambda),\ p_1(\lambda), \ldots, p_{k-1}(\lambda))^T;$$

then there exists a lower Hessenberg matrix K with 1's on the superdiagonal such that

$$\lambda P(\lambda) = KP(\lambda) + p_k(\lambda)e^k.$$

From this relation we have that any zero of p_k is an eigenvalue of K. Let w be a weight function on $[a, b]$. We denote by M the symmetric Gramian matrix

$$M = \int_a^b PP^T w\, d\lambda.$$

Then

$$KM = \int_a^b \lambda PP^T w\, d\lambda - e^k c^T,$$

where

$$c = \int_a^b p_k P w \, d\lambda.$$

The matrix KM is a symmetric matrix plus a rank-one correction to its last row. The $k - 1$ first elements of c are determined by K and M since

$$c - (c^T e^k)e^k = (KM - MK^T)e^k.$$

The polynomial p_k is orthogonal to the previous polynomials p_j, $j = 0, \ldots, k - 1$ if and only if $c = 0$. When the polynomials of degrees from 0 to k are all mutually orthogonal, then M is diagonal and K is tridiagonal. If the polynomials are normalized M is the identity matrix. In the case of a sign changing weight function M may no longer be diagonal.

A particular polynomial p_{j-1} is orthogonal to all polynomial of smaller degrees if and only if the last row and column of the $j \times j$ principal submatrix of M can have a nonzero element only on the diagonal. Such a matrix M is called j-diagonal. The eigenvalues of the $(j - 1) \times (j - 1)$ principal submatrix are the zeros of p_{j-1}.

Elhay and Kautsky derived an algorithm which, given K and M for w, computes $\hat{K}$ and $\hat{M}$ for $\hat{w}(\lambda) = (\lambda - \beta)w(\lambda)$. Let

$$(K - \beta I)M = \begin{pmatrix} Y_j & y^j & \vdots \\ (y^j)^T & a_j & \cdots \\ \cdots & \vdots & \ddots \end{pmatrix},$$

$$L = \begin{pmatrix} L_j & 0 & \vdots \\ (l^j)^T & 1 & \cdots \\ \cdots & \vdots & \ddots \end{pmatrix}, \quad \hat{M} = \begin{pmatrix} \hat{M}_j & \hat{m}^j & \vdots \\ (\hat{m}^j)^T & \hat{d}_j & \cdots \\ \cdots & \vdots & \ddots \end{pmatrix}.$$

With these notations we obtain

$$Y_j = L_j \hat{M}_j L_j^T,$$

$$L_j(\hat{M}_j l^j + \hat{m}^j) = y^j,$$

$$a_j + \delta_{j,k} c^T e^k = (l^j)^T(\hat{M} l^j + 2\hat{m}^j) + \hat{d}_j.$$

The matrices L and $\hat{M}$ can be built one row and column at a time. The aim of Elhay and Kautsky was to construct an $\hat{M}$ which is i-diagonal for as many $i = 1, 2, \ldots, k$ as possible. We can set $\hat{m}^j = 0$ if $\hat{M}_j$ is nonsingular and then l^j is the solution of

$$\hat{M}_j l^j = z^j = L_j^{-1} y^j.$$

For a singular $\hat{M}_j$ we can choose either $l^j = 0$, $\hat{m}^j = z^j$ or a least squares solution of $\hat{M}_j l^j \simeq L_j^{-1} y^j$ and set $\hat{m}^j = z^j - \hat{M}_j l^j$. On this topic and for general polynomials, see also [98].

5.7.4 Division by a Polynomial

We are interested in cases for which (a) $r(\lambda) = 1/(\lambda - \beta)$ or (b) $r(\lambda) = 1/((\lambda - \beta)^2 + \gamma^2)$ and we assume that β is not in the interval $[a, b]$. This problem has been considered by Golub and Fischer [142] who used the modified Chebyshev algorithm and an algorithm due to Gautschi to compute the modified moments. The problem was also solved by Gautschi [131] using Uvarov's results. See also Paszkowski [271]. In the following we will describe the solution given by Elhay and Kautsky [99].

Many methods for a linear divisor as $\lambda - \beta$ are based on a so-called "inversion" of methods for multiplication by a linear factor. If $\tilde{w} = w/r$, then

$$\int_a^b f(\lambda)w(\lambda)\,d\lambda = \int_a^b f(\lambda)r(\lambda)\tilde{w}(\lambda)\,d\lambda.$$

Therefore, if we know how to compute J from $\tilde{J}$ by a multiplication algorithm, we might expect to compute $\tilde{J}$ from J by "inverting" this algorithm.

The work of Elhay and Kautsky is based on the following result; see [99].

THEOREM 5.20 *Let* $P(\lambda) = [p_0(\lambda), p_1(\lambda), \ldots, p_{k-1}(\lambda)]^T$ *and let* β_i *be the elements on the subdiagonal of* J. *If* r *is any analytic function, we have*

$$(J - \lambda I)P(\lambda)r(\lambda) = (J - \lambda I)r(J)P(\lambda) + \beta_k p_k(\lambda)(r(J) - r(\lambda)I)e^k.$$

If λ *is not an eigenvalue of* J,

$$P(\lambda)r(\lambda) = r(J)P(\lambda) + \beta_k p_k(\lambda)(J - \lambda I)^{-1}(r(J) - r(\lambda)I)e^k. \tag{5.15}$$

If the shift β *is a root of* p_k, *then* $r(\beta)$ *is an eigenvalue of* $r(J)$ *and* $P(\lambda)$ *is an eigenvector.*

Proof. See [99]. The proof is based on the fact that, if r is analytic, then J and $r(J)$ commute. □

Then, if $P(\lambda) = L\tilde{P}(\lambda)$ with L lower triangular, multiplying equation (5.15) on the right by $P(\lambda)^T/r(\lambda)$ and integrating, we obtain

$$I = r(J)LL^T$$
$$+ \beta_k \int_a^b (c^T\tilde{P}(\lambda) + \gamma_k\tilde{p}_k(\lambda))(J - \lambda I)^{-1}(r(J) - r(\lambda)I)e^k\tilde{P}(\lambda)^T\,d\tilde{\alpha}L^T,$$

where c is the vector of coefficients of $\tilde{P}$ in the development of p_k in terms of the $\tilde{p}_j$'s, $p_k(\lambda) = c^T\tilde{P} + \gamma_k\tilde{p}_k(\lambda)$. If we consider case (a) with $r(\lambda) = \lambda - \beta$, we have a much simpler relation since some terms cancel by orthogonality,

$$I = (J - \beta I)LL^T + \beta_k e^k c^T L^T.$$

Note that if we compute a UL Cholesky-like factorization $L_J^T L_J$ of $J - \beta I$ and take $L = L_J^{-1}$, then the previous relation is satisfied up to the last row. The solution chosen by Elhay and Kautsky is to compute the elements $l_{i,j}$ and d_i of a solution of the equation

$$I = \tilde{J}LL^T + e^k d^T,$$

for a given $l_{1,1} \neq 0$. Let α_j and η_j be the coefficients of the tridiagonal matrix $\tilde{J} = J - \beta I$ of order k. Let $l_{i,j}$ be the elements of L and L^T be the transpose of L. The first element of the first column of L is known. By identification we have $l_{2,1} = (1/l_{1,1} - \alpha_1 l_{1,1})/\eta_1$. The other elements of the first column are given by

$$l_{i+1,1} = -(\eta_{i-1} l_{i-1,1} + \alpha_i l_{i,1})/\eta_i, \ i = 2, \dots, k-1.$$

Then, for $j = 2, \dots, k$ we compute the diagonal elements using the equation

$$\eta_{j-1} l_{j,j}^2 = -\eta_{j-2} L_{j-2,1:j-2} L_{1:j-2,j}^T - \alpha_{j-1} L_{j-1,1:j-1} L_{1:j-1,j}^T$$
$$-\eta_{j-1} L_{j,1:j-1} L_{1:j-1,j}^T.$$

For $j = 2$ the first term in the right-hand side does not exist. We have used a Matlab-like notation for the rows and columns of L and L^T. The other elements of the column j are computed by

$$\eta_i l_{i+1,j} l_{j,j} = \delta_{i,j} - \eta_{i-1} L_{i-1,1:\min(i-1,j)} L_{1:\min(i-1,j),j}^T$$
$$-\alpha_i L_{i,1:\min(i,j)} L_{1:\min(i,j),j}^T$$
$$-\eta_i L_{i+1,1:j-1} L_{1:j-1,j}^T, \ i = j, \dots, k-1.$$

If one wants to compute also the vector d, the first element is $d_1 = -(\eta_{k-1} l_{k-1,1} + \alpha_k l_{k,1}) l_{1,1}$. The other elements can be computed at the end of the loop on j by

$$d_j = -(\eta_{k-1} L_{k-1,1:\min(k-1,j)} L_{1:\min(k-1,j),j}^T + \alpha_k L_{k,1:j} L_{1:j,j}^T).$$

Note that since $L^{-T} L^{-1} - \tilde{J} = e^k d^T L^{-T} L^{-1}$ and the matrix on the left-hand side is symmetric the row vector $d^T L^{-T} L^{-1}$ must have all its components equal to zero except for the last one. This can be considered as a check of the accuracy of the computation. This algorithm is known as the inverse Cholesky algorithm.

For the initial value we have $l_{1,1} = \sqrt{\tilde{\mu}_0/\mu_0}$ with

$$\tilde{\mu}_0 = \int_a^b \frac{1}{\lambda - \beta} \, d\alpha.$$

This Cauchy integral is known for classical measures or can be computed by an algorithm due to Gautschi [131]. For instance, for the Legendre measure, we have

$$\tilde{\mu}_0 = \log \left(\frac{\beta - 1}{\beta + 1} \right),$$

and for the Chebyshev case (first kind) we have $\tilde{\mu}_0 = \pi/\sqrt{\beta^2 - 1}$.

For case (b) one has to consider an equation of the type

$$I = \tilde{J}^2 L L^T + e^k d^T + e^{k-1} f^T,$$

where d and f are vectors. An algorithm for its solution is given in [99]. It requires the principal 2×2 matrix of L as input.

Chapter Six

Gauss Quadrature

6.1 Quadrature Rules

Given a measure α on the interval $[a, b]$ and a function f (such that its Riemann–Stieltjes integral and all the moments exist), a quadrature rule is a relation

$$\int_a^b f(\lambda)\,d\alpha = \sum_{j=1}^N w_j f(t_j) + R[f]. \tag{6.1}$$

The sum in the right-hand side is the approximation of the integral on the left-hand side and $R[f]$ is the remainder, which is usually not known exactly. The real numbers t_j are the nodes and w_j the weights of the quadrature rule. The rule is said to be of exact degree d if $R[p] = 0$ for all polynomials p of degree d and there are some polynomials q of degree $d + 1$ for which $R[q] \neq 0$.

Quadrature rules of degree $N - 1$ can be obtained by interpolation. The function f is approximated by an interpolation polynomial. For instance, we may use Lagrange interpolation which, given the nodes t_j, is written as

$$f(\lambda) \approx \sum_{i=1}^N f(t_i) l_i(\lambda),$$

where $l_i(\lambda)$ is a Lagrange polynomial

$$l_i(\lambda) = \prod_{\substack{j=1 \\ j \neq i}}^N \frac{\lambda - t_j}{t_i - t_j}.$$

The corresponding quadrature formula is obtained by integrating the interpolation formula. The nodes are the given interpolation points t_j and the weights are

$$w_i = \int_a^b l_i(\lambda)\,d\alpha(\lambda).$$

Such quadrature rules are called interpolatory. Newton–Cotes formulas are defined by taking the nodes to be equally spaced. When this is not the case, a popular choice for the nodes is the zeros of the Chebyshev polynomial of degree N. This is called the Fejér quadrature rule; see Gautschi [131] and Weideman and Trefethen [347]. Another interesting choice is the set of extrema of the Chebyshev polynomial of degree $N - 1$. This gives the Clenshaw–Curtis quadrature rule; see Clenshaw and Curtis [67], Trefethen [327] and the references therein. In both cases, the weights can be computed analytically. For relations of interpolatory quadrature with matrices, see Kautsky [198] and Kautsky and Elhay [200].

A way to obtain quadrature rules of higher degrees is to consider the nodes t_j to be unknowns. These quadrature rules are linked to moments and orthogonal polynomials; see Gautschi [129], [130]. To introduce the relation of quadrature with orthogonal polynomials we quote the following result from Gautschi [131].

THEOREM 6.1 *Let k be an integer, $0 \leq k \leq N$. The quadrature rule (6.1) has degree $d = N - 1 + k$ if and only if it is interpolatory and*

$$\int_a^b \prod_{j=1}^N (\lambda - t_j) p(x)\, d\alpha = 0, \quad \forall p \text{ polynomial of degree } \leq k - 1. \quad (6.2)$$

If the measure is positive $k = N$ is maximal for interpolatory quadrature since if $k = N + 1$ the condition in the last theorem would give that the polynomial

$$\prod_{j=1}^N (\lambda - t_j)$$

is orthogonal to itself, which is impossible when the measure is positive. The optimal quadrature rule of degree $2N - 1$ is called a Gauss quadrature rule [124] since it was introduced by C. F. Gauss at the beginning of the nineteenth century. Of course, it remains to explain how to compute the nodes and the weights.

In this chapter we consider the approximation of a Riemann–Stieltjes integral by Gauss quadrature. The general formula we will use is

$$I[f] = \int_a^b f(\lambda)\, d\alpha(\lambda) = \sum_{j=1}^N w_j f(t_j) + \sum_{k=1}^M v_k f(z_k) + R[f], \quad (6.3)$$

where the weights $[w_j]_{j=1}^N, [v_k]_{k=1}^M$ and the nodes $[t_j]_{j=1}^N$ are unknowns and the nodes $[z_k]_{k=1}^M$ are prescribed; see Davis and Rabinowitz [78], Gautschi [125], [128] and Golub and Welsch [160]. If $M = 0$, this leads to the Gauss rule with no prescribed nodes [124]. If $M = 1$ and $z_1 = a$ or $z_1 = b$ we have the Gauss–Radau rule [276]. If $M = 2$ and $z_1 = a, z_2 = b$, this is the Gauss–Lobatto rule [228]. Note that equation (6.3) implies that the Gauss rule will integrate exactly a polynomial of degree $2N - 1$ by evaluating the polynomial at N points.

The term $R[f]$ is the remainder which generally cannot be explicitly computed. If the measure α is a positive nondecreasing function and if f is smooth enough, it is known (see for instance Stoer and Bulirsch [314]) that

$$R[f] = \frac{f^{(2N+M)}(\eta)}{(2N + M)!} \int_a^b \prod_{k=1}^M (\lambda - z_k) \left[\prod_{j=1}^N (\lambda - t_j) \right]^2 d\alpha(\lambda), \quad a < \eta < b.$$
$$(6.4)$$

Note that for the Gauss rule, the remainder $R[f]$ has the sign of $f^{(2N)}(\eta)$. On the remainder of quadrature rules for analytic functions, see Gautschi and Varga [134]. For the sensitivity of Gauss quadrature to perturbations of the measure, see O'Leary, Strakoš and Tichý [252].

6.2 The Gauss Quadrature Rules

6.2.1 The Gauss Rule

One way to compute the nodes and weights is to use monomials λ^i, $i = 1, \ldots, 2N-1$ as functions and to solve by brute force (for instance, the Newton method) the nonlinear equations expressing the fact that the quadrature rule is exact. In this section we recall how the nodes and weights of the Gauss rule can be more easily obtained and show the connection of Gauss quadrature with orthogonal polynomials. From the theorems of chapter 2 for the measure α, there is a sequence of polynomials $p_0(\lambda), p_1(\lambda), \ldots$ which are orthonormal with respect to α:

$$\int_a^b p_i(\lambda)p_j(\lambda)\, d\alpha(\lambda) = \begin{cases} 1, & \text{if } i = j, \\ 0, & \text{otherwise,} \end{cases}$$

and the polynomial p_k is of exact degree k. Moreover, the roots of p_k are distinct, real and lie in the interval $[a, b]$. This set of orthonormal polynomials satisfies a three term recurrence relationship:

$$\gamma_j p_j(\lambda) = (\lambda - \omega_j)p_{j-1}(\lambda) - \gamma_{j-1}p_{j-2}(\lambda), \quad j = 1, 2, \ldots, N \qquad (6.5)$$

$$p_{-1}(\lambda) \equiv 0, \quad p_0(\lambda) \equiv 1,$$

assuming that $\int d\alpha = 1$. Let us assume that we know the coefficients ω_j and γ_j. As we have seen in previous chapters, the three-term recurrence can be written in matrix form as

$$\lambda P(\lambda) = J_N P(\lambda) + \gamma_N p_N(\lambda)e^N,$$

where

$$P(\lambda) = [p_0(\lambda)\, p_1(\lambda) \cdots p_{N-1}(\lambda)]^T, \qquad (6.6)$$

$$e^N = (0\, 0\, \cdots\, 0\, 1)^T,$$

and

$$J_N = \begin{pmatrix} \omega_1 & \gamma_1 & & & \\ \gamma_1 & \omega_2 & \gamma_2 & & \\ & \ddots & \ddots & \ddots & \\ & & \gamma_{N-2} & \omega_{N-1} & \gamma_{N-1} \\ & & & \gamma_{N-1} & \omega_N \end{pmatrix} \qquad (6.7)$$

is a Jacobi matrix. We note that all the eigenvalues of J_N are real and simple since $\gamma_i \neq 0$, $i = 1, \ldots, N-1$.

THEOREM 6.2 *The eigenvalues of J_N (the so-called Ritz values $\theta_j^{(N)}$ which are also the zeros of p_N) are the nodes t_j of the Gauss quadrature rule (i.e., $M = 0$). The weights w_j are the squares of the first elements of the normalized eigenvectors of J_N.*

Proof. This is shown in Wilf [349] and Golub and Welsch [160]. By theorem 6.1, the monic polynomial

$$\prod_{j=1}^{N}(\lambda - t_j)$$

is orthogonal to all polynomials of degree less than or equal to $N - 1$. Therefore, (up to a multiplicative constant) it is the orthogonal polynomial associated to α and the nodes of the quadrature rule are the zeros of the orthogonal polynomial, that is, the eigenvalues of J_N.

The vector $P(t_j)$ is an unnormalized eigenvector of J_N corresponding to the eigenvalue t_j. If q is an eigenvector with norm 1, we have $P(t_j) = \omega q$ with a scalar ω. As a consequence of the Christoffel–Darboux relation (see theorem 2.11), we have

$$w_j P(t_j)^T P(t_j) = 1, \, j = 1, \dots, N.$$

Then,

$$w_j P(t_j)^T P(t_j) = w_j \omega^2 \|q\|^2 = w_j \omega^2 = 1.$$

Hence, $w_j = 1/\omega^2$. To find ω we can pick any component of the eigenvector q, for instance, the first one that is different from zero. This gives $\omega = p_0(t_j)/q_1 = 1/q_1$. Then the weight is given by

$$w_j = q_1^2.$$

If the integral of the measure is not 1, we obtain

$$w_j = q_1^2 \mu_0 = q_1^2 \int_a^b d\alpha(\lambda).$$

$\square$

Therefore, the knowledge of the Jacobi matrix (and eventually of the first moment) allows us to compute the nodes and weights of the Gauss quadrature rule. It is shown in Golub and Welsch [160] how the squares of the first components of the eigenvectors can be computed without having to compute the other components with a QR-like method. On the QR algorithm applied to tridiagonal matrices, see also Gates and Gragg [123]. Expressions for the first element of an eigenvector are also given in chapter 3, but we will see that in many cases we do not have to compute the nodes and weights to be able to use the quadrature rule.

For the Gauss quadrature rule (renaming the weights and nodes w_j^G and t_j^G), we have

$$I[f] = \int_a^b f(\lambda) \, d\alpha(\lambda) = \sum_{j=1}^{N} w_j^G f(t_j^G) + R_G[f],$$

with

$$R_G[f] = \frac{f^{(2N)}(\eta)}{(2N)!} \int_a^b \left[\prod_{j=1}^{N} (\lambda - t_j^G) \right]^2 d\alpha(\lambda).$$

The monic polynomial

$$\prod_{j=1}^{N}(t_j^G - \lambda),$$

which is the determinant χ_N of $J_N - \lambda I$, can be written as $\gamma_1 \cdots \gamma_{N-1} p_N(\lambda)$. This is seen by comparing the three-term recurrence for the polynomials and the recurrence for the determinant of J_N, see chapter 3. But p_N is an orthonormal polynomial related to the measure α. Then

$$\int_a^b \left[\prod_{j=1}^{N}(\lambda - t_j^G)\right]^2 d\alpha(\lambda) = (\gamma_1 \cdots \gamma_{N-1})^2,$$

and the next theorem follows.

THEOREM 6.3 *Suppose f is such that $f^{(2n)}(\xi) > 0$, $\forall n$, $\forall \xi$, $a < \xi < b$, and let*

$$L_G[f] = \sum_{j=1}^{N} w_j^G f(t_j^G).$$

The Gauss rule is exact for polynomials of degree less than or equal to $2N - 1$ and we have

$$L_G[f] \leq I[f].$$

Moreover $\forall N$, $\exists \eta \in [a, b]$ such that

$$I[f] - L_G[f] = (\gamma_1 \cdots \gamma_{N-1})^2 \frac{f^{(2N)}(\eta)}{(2N)!}.$$

Proof. The main idea of the proof is to use a Hermite interpolatory polynomial of degree $2N - 1$ on the N nodes, which allows us to express the remainder as an integral of the difference between the function and its interpolatory polynomial and to apply the mean value theorem (since the measure is positive and increasing). As we know the sign of the remainder, we easily obtain bounds. $\quad\Box$

We remark that if we know bounds of $f^{(2N)}$, we can bound the absolute value of the error of the Gauss quadrature rule. For examples, see Calvetti, Golub and Reichel [52].

An inverse problem of reconstruction of a weight function given the Gauss quadrature nodes was considered by Kautsky [199].

To summarize, we have seen that, if we know the coefficients of the three-term recurrence for the orthonormal polynomials associated with the measure α and the first moment, then we can compute the nodes and weights of the Gauss quadrature rule.

6.2.2 The Gauss–Radau Rule

To obtain the Gauss–Radau quadrature rule ($M = 1$ in equations (6.3) and (6.4)), we have to extend the matrix J_N in equation (6.7) in such a way that it has one prescribed eigenvalue; see Golub [139].

If we assume that the prescribed node is the left end of the integration interval $z_1 = a$, we wish to construct p_{N+1} such that $p_{N+1}(a) = 0$. From the recurrence relation (6.5), we have

$$0 = \gamma_{N+1} p_{N+1}(a) = (a - \omega_{N+1}) p_N(a) - \gamma_N p_{N-1}(a).$$

This gives

$$\omega_{N+1} = a - \gamma_N \frac{p_{N-1}(a)}{p_N(a)}.$$

Therefore we have to compute the ratio $p_{N-1}(a)/p_N(a)$ without using the three-term recurrence. We note that we have

$$(J_N - aI)P(a) = -\gamma_N p_N(a) e^N.$$

Let us denote $\delta(a) = [\delta_1(a), \ldots, \delta_N(a)]^T$ with

$$\delta_l(a) = -\gamma_N \frac{p_{l-1}(a)}{p_N(a)}, \quad l = 1, \ldots, N.$$

This gives $\omega_{N+1} = a + \delta_N(a)$ and $\delta(a)$ satisfies

$$(J_N - aI)\delta(a) = \gamma_N^2 e^N. \tag{6.8}$$

From these relations we have the solution of the problem by performing the following steps:

1) we generate γ_N;

2) we solve the tridiagonal system (6.8) for $\delta(a)$; this gives $\delta_N(a)$;

3) we compute $\omega_{N+1} = a + \delta_N(a)$.

Then the tridiagonal matrix $\hat{J}_{N+1}$ defined as

$$\hat{J}_{N+1} = \begin{pmatrix} J_N & \gamma_N e^N \\ \gamma_N (e^N)^T & \omega_{N+1} \end{pmatrix} \tag{6.9}$$

has the prescribed node a as an eigenvalue and gives the nodes and the weights of the corresponding quadrature rule we were looking for. As for the Gauss rule, the nodes are the eigenvalues and the weights are the squares of the first components of the eigenvectors. Something similar is done if $z_1 = b$. Therefore, the algorithm is to compute as for the Gauss quadrature rule and to modify the last element to obtain the prescribed node.

For the Gauss–Radau rule (see Stoer and Bulirsch [314]) the remainder R_{GR} is

$$R_{GR}[f] = \frac{f^{(2N+1)}(\eta)}{(2N+1)!} \int_a^b (\lambda - z_1) \left[\prod_{j=1}^N (\lambda - t_j) \right]^2 d\alpha(\lambda). \tag{6.10}$$

This is proved by constructing an interpolatory polynomial for the function and its derivative on the t_j's and for the function on z_1. Therefore, if we know the sign of the derivatives of f, we can bound the remainder. This is stated in the following theorem.

THEOREM 6.4 *Suppose f is such that $f^{(2n+1)}(\xi) < 0, \forall n, \forall \xi, a < \xi < b$. Let U_{GR} be defined as*

$$U_{GR}[f] = \sum_{j=1}^{N} w_j^a f(t_j^a) + v_1^a f(a),$$

w_j^a, v_1^a, t_j^a *being the weights and nodes computed with $z_1 = a$, and let L_{GR} be defined as*

$$L_{GR}[f] = \sum_{j=1}^{N} w_j^b f(t_j^b) + v_1^b f(b),$$

w_j^b, v_1^b, t_j^b *being the weights and nodes computed with $z_1 = b$. The Gauss–Radau rule is exact for polynomials of degree less than or equal to $2N$ and we have*

$$L_{GR}[f] \le I[f] \le U_{GR}[f].$$

Moreover $\forall N \exists \eta_U, \eta_L \in [a, b]$ such that

$$I[f] - U_{GR}[f] = \frac{f^{(2N+1)}(\eta_U)}{(2N+1)!} \int_a^b (\lambda - a) \left[\prod_{j=1}^{N} (\lambda - t_j^a) \right]^2 d\alpha(\lambda),$$

$$I[f] - L_{GR}[f] = \frac{f^{(2N+1)}(\eta_L)}{(2N+1)!} \int_a^b (\lambda - b) \left[\prod_{j=1}^{N} (\lambda - t_j^b) \right]^2 d\alpha(\lambda).$$

Proof. With our hypothesis the sign of the remainder is easily obtained. It is negative if we choose $z_1 = a$, positive if we choose $z_1 = b$. □

Remarks:

1) if the sign of the derivatives of f is positive, the bounds are reversed;

2) it is enough to suppose that there exists an n_0 such that $f^{(2n_0+1)}(\eta) < 0$ but then $N = n_0$ is fixed.

6.2.3 The Gauss–Lobatto Rule

In this section we consider the Gauss–Lobatto quadrature rule ($M = 2$ in equations (6.3) and (6.4)), with the ends of the integration interval $z_1 = a$ and $z_2 = b$ as prescribed nodes. As in the Gauss–Radau rule, we should modify the matrix of the Gauss quadrature rule; see Golub [139]. Here, we would like to have

$$p_{N+1}(a) = p_{N+1}(b) = 0.$$

Using the recurrence relation (6.5) for the polynomials, this leads to a linear system of order 2 for the unknowns ω_{N+1} and γ_N:

$$\begin{pmatrix} p_N(a) & p_{N-1}(a) \\ p_N(b) & p_{N-1}(b) \end{pmatrix} \begin{pmatrix} \omega_{N+1} \\ \gamma_N \end{pmatrix} = \begin{pmatrix} a\, p_N(a) \\ b\, p_N(b) \end{pmatrix}. \tag{6.11}$$

Let δ and μ be defined as vectors with components

$$\delta_l = -\frac{p_{l-1}(a)}{\gamma_N p_N(a)}, \quad \mu_l = -\frac{p_{l-1}(b)}{\gamma_N p_N(b)}, \; l = 1, \ldots, N;$$

then

$$(J_N - aI)\delta = e^N, \quad (J_N - bI)\mu = e^N,$$

and the linear system (6.11) can be written as

$$\begin{pmatrix} 1 & -\delta_N \\ 1 & -\mu_N \end{pmatrix} \begin{pmatrix} \omega_{N+1} \\ \gamma_N^2 \end{pmatrix} = \begin{pmatrix} a \\ b \end{pmatrix},$$

whose solution gives the unknowns we need. The tridiagonal matrix $\hat{J}_{N+1}$ is then defined as in the Gauss–Radau rule in equation (6.9).

Having computed the nodes and weights (eigenvalues and squares of the first components of the eigenvectors), we have

$$\int_a^b f(\lambda) d\alpha(\lambda) = \sum_{j=1}^N w_j^{GL} f(t_j^{GL}) + v_1^{GL} f(a) + v_2^{GL} f(b) + R_{GL}[f].$$

This gives the following result.

THEOREM 6.5 *Suppose f is such that $f^{(2n)}(\xi) > 0$, $\forall n$, $\forall \xi$, $a < \xi < b$ and let*

$$U_{GL}[f] = \sum_{j=1}^N w_j^{GL} f(t_j^{GL}) + v_1^{GL} f(a) + v_2^{GL} f(b),$$

t_j^{GL}, w_j^{GL}, v_1^{GL} and v_2^{GL} *being the nodes and weights computed with a and b as prescribed nodes. The Gauss–Lobatto quadrature rule is exact for polynomials of degree less than or equal to $2N + 1$ and we have*

$$I[f] \leq U_{GL}[f].$$

Moreover $\forall N \; \exists \eta \in [a, b]$ such that

$$I[f] - U_{GL}[f] = \frac{f^{(2N+2)}(\eta)}{(2N+2)!} \int_a^b (\lambda - a)(\lambda - b) \left[\prod_{j=1}^N (\lambda - t_j^{GL}) \right]^2 d\alpha(\lambda).$$

In [145] Golub and Kautsky studied quadrature rules of the form

$$\sum_{j=1}^N \sum_{i=1}^{n_i} w_{i,j} f^{(i-1)}(t_j) + \sum_{j=1}^M \sum_{i=1}^{m_i} v_{i,j} f^{(i-1)}(z_j),$$

where $f^{(i)}(x)$ denotes the value of the ith derivative of f at x. There are N free nodes t_j of multiplicities n_i and M fixed nodes z_j of multiplicities m_i. It is proved in [145] that the approach we used to obtain the Gauss–Radau and Gauss–Lobatto rules by modifying one or two elements in the last row of the Jacobi matrix cannot be extended to a double prescribed node or to the case where the two prescribed nodes are on the same side of the integration interval. However, it works when the two nodes are on opposite sides of the interval and not only when they are the end points.

6.2.4 Computation of the Gauss Rules

The nodes and weights can be computed by the Golub and Welsch QR algorithm [160]. However, we do not always need to compute the eigenvalues and eigenvectors of the tridiagonal matrix J_N. Let Z_N be the orthogonal matrix of the eigenvectors of J_N (or $\hat{J}_N$) whose columns we denote by z^i and Θ_N be the diagonal matrix of the eigenvalues $t_i = \theta_i^{(N)}$ which gives the nodes of the Gauss quadrature rule. We have seen that the weights w_i are given by the squares of the first components of the eigenvectors

$$w_i = (z_1^i)^2 = ((e^1)^T z^i)^2.$$

Then we can express the quadrature rule as a function of the Jacobi matrix J_N.

THEOREM 6.6

$$\sum_{l=1}^N w_l f(t_l) = (e^1)^T f(J_N) e^1.$$

Proof. Since the weights are the squares of the first components of the eigenvectors, we have

$$\sum_{l=1}^N w_l f(t_l) = \sum_{l=1}^N (e^1)^T z^l f(t_l)(z^l)^T e^1$$

$$= (e^1)^T \left(\sum_{l=1}^N z^l f(t_l)(z^l)^T \right) e^1$$

$$= (e^1)^T Z_N f(\Theta_N) Z_N^T e^1$$

$$= (e^1)^T f(J_N) e^1.$$

This concludes the proof.   □

The same statement is true for the Gauss–Radau and Gauss–Lobatto rules replacing J_N by the appropriate modified Jacobi matrix. Therefore, in some cases when the $(1, 1)$ element of the matrix $f(J_N)$ (or its modified version) is easily computable (for instance, if $f(\lambda) = 1/\lambda$; see chapter 3), we do not need to compute the eigenvalues and the first components of the eigenvectors of J_N.

On the computation of Gauss quadrature rules see also Beckermann [23].

6.3 The Anti-Gauss Quadrature Rule

Anti-Gauss quadrature rules were introduced by Laurie in [218]; see also [222]. The idea is to construct a quadrature rule whose error is equal but of opposite sign to the error of the Gauss rule. This was motivated by the need to estimate the error term of the Gauss rule. Even though we have an analytic expression for $R_G[f]$, it is not easy to find an accurate estimate of this term since it involves an unknown point η in the interval of integration. Let

$$L_G^N[f] = \sum_{j=1}^{N} w_j^G f(t_j^G)$$

be the Gauss rule approximation with N nodes. It is exact for polynomials of degree up to $2N - 1$, that is,

$$L_G^N[p] = I[p] \quad \text{for all polynomials of degree } 2N - 1.$$

The usual way of obtaining an estimate of $I[f] - L_G^N[f]$ is to use another quadrature rule $Q[f]$ of degree greater than $2N - 1$ and to estimate the error as $Q[f] - L_G^N[f]$. There are several possible ways to do this. One can for instance use $L_G^{N+1}[f]$, but this may not be very precise and requires the recalculation of a new Gauss quadrature. We will see in the next section that for certain measures it is possible to find a $(2N + 1)$-point rule containing the original N nodes of the Gauss rule. This is known as a Kronrod rule. However, Gauss–Kronrod rules do not always exist. The idea of Laurie is to construct a quadrature rule with $N + 1$ nodes called an anti-Gauss rule,

$$H^{N+1}[f] = \sum_{j=1}^{N+1} \varpi_j f(\vartheta_j),$$

such that

$$I[p] - H^{N+1}[p] = -(I[p] - L_G^N[p]), \tag{6.12}$$

for all polynomials of degree $2N + 1$. Then, the error of the Gauss rule can be estimated as

$$\frac{1}{2}(H^{N+1}[f] - L_G^N[f]).$$

Using this anti-Gauss rule, the integral $I[f]$ can also be approximated by

$$\frac{1}{2}(H^{N+1}[f] + L_G^N[f]).$$

From equation (6.12) we have

$$H^{N+1}[p] = 2I[p] - L_G^N[p],$$

for all polynomials p of degree $2N + 1$. Hence, H^{N+1} is a Gauss rule with $N + 1$ nodes for the functional $\mathcal{I}(\cdot) = 2I[\cdot] - L_G^N[\cdot]$. Associated with this functional is a sequence of orthonormal polynomials $\tilde{p}_j$, $0, \ldots, N + 1$ and a tridiagonal matrix $\tilde{J}_{N+1}$. We have

$$I[pq] = \mathcal{I}(pq)$$

for p a polynomial of degree $N - 1$ and q a polynomial of degree N. Using the Stieltjes procedure of chapter 5 to obtain an expression of the three-term recurrence coefficients we see (Laurie [218] or Calvetti, Reichel and Sgallari [54]) that $\tilde{p}_j = p_j$, $j = 0, \ldots, N$ and the first coefficients of the tridiagonal matrices J_N for the Gauss rule and $\tilde{J}_{N+1}$ for the anti-Gauss rule are the same, $\tilde{\omega}_j = \omega_j$, $j = 1, \ldots, N$

and $\tilde{\gamma}_j = \gamma_j$, $j = 1, \ldots, N-1$. We also have $L_G^N[\tilde{p}_N^2] = L_G^N[p_N^2] = 0$ because the nodes are the zeros of p_N. This implies that

$$\mathcal{I}(\tilde{p}_N^2) = 2I(\tilde{p}_N^2),$$

and therefore $\tilde{\gamma}_N^2 = 2\gamma_N^2$. Moreover, $\tilde{\omega}_{N+1} = \omega_{N+1}$. The tridiagonal matrix $\tilde{J}_{N+1}$ is

$$\tilde{J}_{N+1} = \begin{pmatrix} \omega_1 & \gamma_1 & & & & \\ \gamma_1 & \omega_2 & \gamma_2 & & & \\ & \ddots & \ddots & \ddots & & \\ & & \gamma_{N-2} & \omega_{N-1} & \gamma_{N-1} & \\ & & & \gamma_{N-1} & \omega_N & \sqrt{2}\gamma_N \\ & & & & \sqrt{2}\gamma_N & \omega_{N+1} \end{pmatrix}. \quad (6.13)$$

The $N+1$ nodes of the anti-Gauss rule are the eigenvalues of $\tilde{J}_{N+1}$ and the weights are the squares of the first components of the eigenvectors. Note that $\tilde{J}_{N+1}$ is a low-rank modification of J_{N+1}. As for the Gauss rule we have

$$H^{N+1}[f] = (e^1)^T f(\tilde{J}_{N+1})e^1.$$

We have that the weights are strictly positive and the anti-Gauss nodes interlace the Gauss nodes because of the Cauchy interlace theorem. This implies that the anti-Gauss nodes ϑ_j, $j = 2, \ldots, N$ are inside the integration interval, see Laurie [218]. However, the first and the last nodes can eventually be outside of the integration interval. Actually, in some cases, the matrix $\tilde{J}_{N+1}$ can be indefinite even if J_N is positive definite.

There is nothing magical in asking for an error that is the opposite of the Gauss rule error. Therefore, the work of Laurie has been generalized (see Patterson [273], [274], Calvetti and Reichel [49], Ehrich [96], Spalević [309]) to a quadrature rule $S^{N+1}[f]$ such that

$$I[p] - S^{N+1}[p] = -\gamma(I[p] - L_G^N[p])$$

for all polynomials of degree $2N + 1$. The parameter γ is positive and less than or equal to 1; for $\gamma = 1$ we recover Laurie's method. Then the error of the Gauss rule can be estimated as

$$\frac{1}{1+\gamma}(S^{N+1}[f] - L_G^N[f]).$$

The integral $I[f]$ can also be approximated by

$$\frac{1}{1+\gamma}(H^{N+1}[f] + \gamma L_G^N[f]).$$

The $N + 1$ nodes of the anti-Gauss rule are the eigenvalues and the weights are the squares of the first components of the eigenvectors of the matrix

$$\tilde{J}_{N+1} = \begin{pmatrix} \omega_1 & \gamma_1 & & & & \\ \gamma_1 & \omega_2 & \gamma_2 & & & \\ & \ddots & \ddots & \ddots & & \\ & & \gamma_{N-2} & \omega_{N-1} & \gamma_{N-1} & \\ & & & \gamma_{N-1} & \omega_N & \gamma_N\sqrt{1+\gamma} \\ & & & & \gamma_N\sqrt{1+\gamma} & \omega_{N+1} \end{pmatrix}. \quad (6.14)$$

Eventually, γ can be chosen such that $\tilde{J}_{N+1}$ is positive definite. This can be seen by computing the diagonal elements of the Cholesky-like factorization of $\tilde{J}_{N+1}$,

$$\tilde{\delta}_i = \delta_i, \ i = 1, \dots, N, \quad \tilde{\delta}_{N+1} = \omega_{N+1} - (1 + \gamma)\frac{\gamma_N^2}{\delta_N}.$$

If $\tilde{\delta}_{N+1} \leq 0$, then we have to decrease the value of γ.

6.4 The Gauss–Kronrod Quadrature Rule

As noted in the previous section, a Gauss–Kronrod rule is a formula

$$K^{2N+1}[f] = \sum_{j=1}^{2N+1} w_j^K f(t_j^K), \tag{6.15}$$

such that N of the nodes t_j^K coincide with the nodes t_j^G of the N-point Gauss rule and the rule K^{2N+1} is exact for polynomials of degree less than or equal to $3N+1$. This was introduced by A. S. Kronrod [209], [210] for the purpose of estimating the error in the Gauss formula. The advantage is to be able to reuse the N function values already computed for the Gauss rule. In a short note [133] Gautschi pointed out that this idea was already proposed in 1894 by R. Skutsch [308].

Here, we describe the properties of the Jacobi matrices given by Laurie in [219] and the algorithm proposed by Calvetti, Golub, Gragg and Reichel [59]. For the computation of the Gauss–Kronrod rule, see also Xu [353], Boutry [33] and Ammar, Calvetti and Reichel [6] and also the review paper by Monegato [243]. We are interested in rules where the nodes and weights are real and the weights are positive. From a result in Gautschi [127] the Jacobi–Kronrod matrix exists and is real if and only if the Kronrod rule exists and is real and positive.

Let $\hat{\omega}_j$ and $\hat{\gamma}_j$ be the nonzero coefficients of the symmetric Jacobi–Kronrod matrix $\hat{J}_{2N+1}$. The fact that K^{2N+1} is exact for polynomials of degree less than or equal to $3N + 1$ implies that the first $3N + 1$ coefficients in the sequence $\hat{\omega}_1, \hat{\gamma}_1, \hat{\omega}_2, \hat{\gamma}_2, \dots$ equal the corresponding coefficients in the sequence for J_N. Therefore, if N is odd

$$\hat{\omega}_j = \omega_j, \quad \hat{\gamma}_j = \gamma_j, \ j = 1, \dots, \frac{3N + 1}{2},$$

and if N is even

$$\hat{\omega}_j = \omega_j, \ j = 1, \dots, \frac{3N}{2} + 1, \quad \hat{\gamma}_j = \gamma_j, \ j = 1, \dots, \frac{3N}{2}.$$

Hence, the number of matrix elements we have to determine depends on the parity of N. Then the important result which leads to an algorithm to compute the Kronrod rule is the following lemma due to Laurie [219].

LEMMA 6.7 *The characteristic polynomial of the trailing principal $N \times N$ submatrix of $\hat{J}_{2N+1}$ is the same as that of its leading principal $N \times N$ matrix. In other words, these matrices have the same eigenvalues.*

Proof. See Laurie [219]. The proof considers ϕ_k and ψ_k which are respectively the characteristic polynomials of the leading and trailing matrices of order k. ϕ_{2N+1} is obtained by expanding the determinant along the $(N+1)$st row of the matrix. This expansion shows that any common zero of ϕ_k and ψ_k is a zero of ϕ_{2N+1}. Conversely, if ϕ_N is a factor of ϕ_{2N+1}, then $\phi_{N-1}\psi_N$ must be divisible by ϕ_N. The polynomials ϕ_{N-1} and ϕ_N being mutually prime, ψ_N is divisible by ϕ_N. These polynomials have the same leading coefficient and are therefore identical.   □

The algorithm given by Laurie in [219] relies on mixed moments and computes the entries of $\hat{J}_{2N+1}$ by a modified Chebyshev algorithm, see chapter 5. It is also described in Gautschi's book [131]. Here we describe the approach proposed in [59]. Let

$$\hat{J}_{2N+1} = \begin{pmatrix} J_N & \gamma_N e^N & 0 \\ \gamma_N (e^N)^T & \omega_{N+1} & \gamma_{N+1}(e^1)^T \\ 0 & \gamma_{N+1}e^1 & \check{J}_N \end{pmatrix}.$$

Lemma 6.7 says that $\check{J}_N$ has the same eigenvalues as J_N. Moreover, some elements of the matrix $\check{J}_N$ are known. When N is odd, $\check{J}_N$ can be written as

$$\check{J}_N = \begin{pmatrix} J_{N+2:\frac{3N+1}{2}} & \gamma_{\frac{3N+1}{2}} e^{\frac{N-1}{2}} \\ \gamma_{\frac{3N+1}{2}}(e^{\frac{N-1}{2}})^T & \hat{J}_N^* \end{pmatrix},$$

where $J_{N+2:\frac{3N+1}{2}}$ is a principal block of J_N going from row $N+2$ to row $\frac{3N+1}{2}$ and $\hat{J}_N^*$ is an unknown tridiagonal matrix of order $(N+1)/2$. Similarly, when N is even we have

$$\check{J}_N = \begin{pmatrix} J_{N+2:\frac{3N}{2}+1} & \hat{\gamma}_{\frac{3N}{2}+1} e^{\frac{N}{2}} \\ \hat{\gamma}_{\frac{3N}{2}+1}(e^{\frac{N}{2}})^T & \hat{J}_N^* \end{pmatrix},$$

where $\hat{J}_N^*$ is an unknown tridiagonal matrix of order $N/2$. Moreover, the entry $\hat{\gamma}_{\frac{3N}{2}+1}$ is unknown. There is a Kronrod rule with real nodes and positive weights if there exists a real matrix $\check{J}_N$ which has the same eigenvalues as J_N.

The algorithm in [59] first determines the eigenvalues as well as the first and last components of the eigenvectors of J_N. Of course, this gives the Gauss rule L_G^N. Expressions for the first and last components of a tridiagonal matrix are given in chapter 3, proposition 3.4, or can be computed with the QR algorithm.

The second step of the algorithm is to compute the first components of the normalized eigenvectors of $\check{J}_N$ for reasons we will see soon. Remember that we do not know all the coefficients of this matrix. To be able to obtain this result, we use a method due to Boley and Golub [31] when N is even. With the matrix $\check{J}_N$ we can associate an unknown measure $\check{\alpha}$ and a Gauss rule whose nodes $\theta_j^{(N)}$ we know because of lemma 6.7 (they are the same as those of J_N) and weights $\check{w}_j$ which are unknown since they are the squares of the first components of the eigenvectors we seek. For the first N moments $\check{\mu}_k$ associated with $\check{\alpha}$ we have

$$\check{\mu}_k = \sum_{j=1}^{N} (\theta_j^{(N)})^k \check{w}_j, \quad k = 0, \dots, N-1.$$

The largest principal leading submatrix of $\breve{J}_N$ for which all coefficients are known is $\breve{J}_{N/2} = J_{N+2:\frac{3N}{2}+1}$ of order $N/2$. We can compute its eigenvalues θ_j^* and the squares of the first components of the eigenvectors w_j^*, $j = 1, \ldots, N/2$. There is also a Gauss rule associated with this matrix. It is exact for polynomials of degree less than or equal to $2(N/2) - 1 = N - 1$, in particular for the N monomials λ^j, $j = 0, \ldots, N - 1$. Both quadrature rules can be regarded as discretizations of $\breve{\alpha}$ and therefore

$$\breve{\mu}^k = \sum_{j=1}^{N/2} (\theta_j^*)^k w_j^* = \sum_{j=1}^{N} (\theta_j^{(N)})^k \breve{w}_j, \ k = 0, \ldots, N - 1.$$

This gives a linear system of N equations in N unknowns which can be solved for the $\breve{w}_j$. However, the matrix of this linear system is a Vandermonde matrix and therefore often badly conditioned. It was proposed in [59] to use a Lagrange interpolation polynomial

$$l_k(\theta) = \prod_{\substack{m=1 \\ m \neq k}}^{N} \frac{\theta - \theta_m^{(N)}}{\theta_k^{(N)} - \theta_m^{(N)}}.$$

Then the solution is written as

$$\breve{w}_k = \sum_{j=1}^{N/2} l_k(\theta_j^*) w_j^*, \ k = 1, \ldots, N.$$

The Kronrod quadrature rule fails to exist if one of the components of the solution is negative. The same method cannot be used when N is odd because the largest principal leading submatrix whose coefficients are known is $J_{N+2:\frac{3N+1}{2}}$ of order $(N - 1)/2$. The associated Gauss rule is exact only for polynomials of degree less than or equal to $N - 2$. This is not sufficient to match the N moments. Before proceeding as before, the unknown coefficient $\omega_{(3N+3)/2}$ must be determined. This can be done but it is rather technical and we refer to [59] for details.

When the eigenvalues and the last components of eigenvectors of J_N and the first components of eigenvectors of $\breve{J}_N$ are computed, one must compute the eigenvalues and first components of eigenvectors of $\hat{J}_{2N+1}$. This is similar to what is done in the divide and conquer algorithm for computing eigenvalues of tridiagonal matrices; see Cuppen [71] and Dongarra and Sorensen [89]. We can use the spectral decomposition of J_N and $\breve{J}_N$,

$$J_N = Z_N \Theta_N Z_N^T, \quad \breve{J}_N = \breve{Z}_N \Theta_N \breve{Z}_N^T,$$

where Z_N and $\breve{Z}_N$ are the matrices whose columns are the normalized eigenvectors. The matrix $\breve{Z}_N$ is unknown but it turns out that we just need the first elements of its columns which we know. Then $\hat{J}_{2N+1}$ is similar to

$$\begin{pmatrix} \Theta_N & \gamma_N Z_N^T e^N & 0 \\ \gamma_N (e^N)^T Z_N & \omega_{N+1} & \gamma_{N+1} (e^1)^T \breve{Z}_N \\ 0 & \gamma_{N+1} \breve{Z}_N^T e^1 & \Theta_N \end{pmatrix}. \tag{6.16}$$

Remark that $(e^N)^T Z_N$ (resp. $(e^1)^T \breve{Z}_N$) are the last (resp. first) components of the eigenvectors of J_N (resp. $\breve{J}_N$). By applying a symmetric permutation the matrix (6.16) is similar to an arrowhead matrix

$$
\begin{pmatrix}
\Theta_N & & \gamma_N Z_N^T e^N \\
& \Theta_N & \gamma_{N+1} \breve{Z}_N^T e^1 \\
\gamma_N (e^N)^T Z_N & \gamma_{N+1}(e^1)^T \breve{Z}_N & \omega_{N+1}
\end{pmatrix}.
$$

Then, to isolate the eigenvalues of Θ_N which are eigenvalues of $\hat{J}_{2N+1}$ we apply rotations to annihilate the N first terms in row $2N+1$ and column $2N+1$. Let the matrix G represents the product of these N rotations, then the matrix is similar to

$$
\begin{pmatrix}
\Theta_N & & \\
& \Theta_N & c \\
& c^T & \omega_{N+1}
\end{pmatrix},
$$

where $c = (\xi_1, \ldots, \xi_N)^T$ and the ξ_j are the last N components of

$$
[G^T (\gamma_N (e^N)^T Z_N, \gamma_{N+1}(e^1)^T \breve{Z}_N)]^T.
$$

Therefore, we now look for the eigenvalues of the arrowhead matrix

$$
\Upsilon_N = \begin{pmatrix}
\Theta_N & c \\
c^T & \omega_{N+1}
\end{pmatrix}.
$$

By looking at the Schur complement of ω_{N+1} the eigenvalues of Υ_N are seen to be solutions of the secular equation

$$
g(\theta) = \theta - \omega_{N+1} - \sum_{j=1}^{N} \frac{\xi_j^2}{\theta_j^{(N)} - \theta} = 0.
$$

The $N + 1$ new eigenvalues interlace the nodes of the Gauss rule $\theta_j^{(N)}$, which are the poles of $g(\theta)$. For algorithms to compute solutions of secular equations, see chapter 9. The first components of the eigenvectors of $\hat{J}_{2N+1}$ are computed using an approach suggested by Gu and Eisenstat [167], see also Boutry [33].

Patterson [272] has extended the Kronrod rules by constructing quadrature rules which have the Kronrod nodes prescribed. Elhay and Kautsky [98] have investigated generalized Kronrod–Patterson embedded quadrature rules. More precisely, they look for sequences $Q_i[f]$, $i = 1, \ldots, n$ of quadrature rules such that $Q_1[f]$ has nodes $\{v_j^{(1)}\}_{j=1}^{k_1}$ of order $2k_1$, $Q_2[f]$ has nodes $\{v_j^{(1)}\}_{j=1}^{k_1} \cup \{v_j^{(2)}\}_{j=1}^{k_2}$ with order $2k_2 + k_1$ and more generally $Q_n[f]$ has nodes $\cup_{i=1}^{n}\{v_j^{(i)}\}_{j=1}^{k_i}$ and order $k_n + \sum_{i=1}^{n} k_i$. Certain of these rules may fail to exist because no orthogonal polynomial of the appropriate degree exists, since the measures are not positive, the orthogonal polynomial may have complex roots, the real roots may be outside of the interval of integration or some weights may be negative.

6.5 The Nonsymmetric Gauss Quadrature Rules

6.5.1 The Gauss Rule

In this section we consider the case where the measure α can be written as

$$\alpha(\lambda) = \sum_{k=1}^{l} \alpha_k \delta_k, \quad \lambda_l \leq \lambda < \lambda_{l+1}, \, l = 1, \ldots, N - 1,$$

where $\alpha_k \neq \delta_k$ and $\alpha_k \delta_k \geq 0$ with jumps at the values λ_l, $l = 1, \ldots, N-1$. In this case α is still a positive increasing function. For variable-signed weight functions; see Struble [321] and chapter 2.

We assume that there exist two sequences of mutually orthogonal (sometimes called bi-orthogonal) polynomials p and q such that

$$\gamma_j p_j(\lambda) = (\lambda - \omega_j)p_{j-1}(\lambda) - \beta_{j-1}p_{j-2}(\lambda), \quad p_{-1}(\lambda) \equiv 0, \quad p_0(\lambda) \equiv 1,$$
$$\beta_j q_j(\lambda) = (\lambda - \omega_j)q_{j-1}(\lambda) - \gamma_{j-1}q_{j-2}(\lambda), \quad q_{-1}(\lambda) \equiv 0, \quad q_0(\lambda) \equiv 1,$$

with $\langle p_i, q_j \rangle = 0$, $i \neq j$. Let

$$P(\lambda)^T = [p_0(\lambda) \, p_1(\lambda) \cdots p_{N-1}(\lambda)],$$

$$Q(\lambda)^T = [q_0(\lambda) \, q_1(\lambda) \cdots q_{N-1}(\lambda)],$$

and

$$J_N = \begin{pmatrix} \omega_1 & \gamma_1 & & & \\ \beta_1 & \omega_2 & \gamma_2 & & \\ & \ddots & \ddots & \ddots & \\ & & \beta_{N-2} & \omega_{N-1} & \gamma_{N-1} \\ & & & \beta_{N-1} & \omega_N \end{pmatrix}.$$

Then, in matrix form, we can write

$$\lambda P(\lambda) = J_N P(\lambda) + \gamma_N p_N(\lambda)e^N,$$
$$\lambda Q(\lambda) = J_N^T Q(\lambda) + \beta_N q_N(\lambda)e^N.$$

The two sets of polynomials differ only by multiplicative functions.

PROPOSITION 6.8

$$p_j(\lambda) = \frac{\beta_j \cdots \beta_1}{\gamma_j \cdots \gamma_1} q_j(\lambda).$$

Proof. The result is proved by induction. We have

$$\gamma_1 p_1(\lambda) = \lambda - \omega_1, \quad \beta_1 q_1(\lambda) = \lambda - \omega_1.$$

Therefore

$$p_1(\lambda) = \frac{\beta_1}{\gamma_1} q_1(\lambda).$$

Now, assume that

$$p_{j-1}(\lambda) = \frac{\beta_{j-1} \cdots \beta_1}{\gamma_{j-1} \cdots \gamma_1} q_{j-1}(\lambda).$$

We have

$$\gamma_j p_j(\lambda) = (\lambda - \omega_j)p_{j-1}(\lambda) - \beta_{j-1}p_{j-2}(\lambda)$$

$$= (\lambda - \omega_j)\frac{\beta_{j-1}\cdots\beta_1}{\gamma_{j-1}\cdots\gamma_1}q_{j-1}(\lambda) - \beta_{j-1}\frac{\beta_{j-2}\cdots\beta_1}{\gamma_{j-2}\cdots\gamma_1}q_{j-2}(\lambda).$$

Multiplying by $(\gamma_{j-1}\cdots\gamma_1)/(\beta_{j-1}\cdots\beta_1)$ we obtain the result. Hence, q_N is a multiple of p_N and the polynomials have the same roots which are also the common real eigenvalues of J_N and J_N^T. $\square$

In the applications we have in mind, it is possible to choose γ_j and β_j such that

$$\gamma_j = \pm\beta_j,$$

with, for instance, $\gamma_j \geq 0$. Then, we have

$$p_j(\lambda) = \pm q_j(\lambda).$$

We define the nonsymmetric quadrature rule as

$$\int_a^b f(\lambda)\,d\alpha(\lambda) = \sum_{j=1}^N f(\theta_j)s_j t_j + R[f], \tag{6.17}$$

where θ_j is an eigenvalue of J_N, s_j is the first component of the eigenvector u_j of J_N corresponding to θ_j and t_j is the first component of the eigenvector v_j of J_N^T corresponding to the same eigenvalue, normalized such that $v_j^T u_j = 1$.

We have the following results from [149].

PROPOSITION 6.9 *Assume that $\gamma_j\beta_j \neq 0$; then the nonsymmetric Gauss quadrature rule (6.17) is exact for polynomials of degree less than or equal to $N - 1$.*

Proof. Assuming f is a polynomial of degree $N - 1$, it can be written as

$$f(\lambda) = \sum_{k=0}^{N-1} c_k p_k(\lambda),$$

and because of the orthonormality properties

$$\langle p_j, 1\rangle = \int_a^b p_j\,d\alpha = 0, \ \forall j \neq 1,$$

then

$$\int_a^b f(\lambda)\,d\alpha(\lambda) = c_0.$$

For the quadrature rule, we have

$$\sum_{j=1}^N f(\theta_j)s_j t_j q_l(\theta_j) = \sum_{j=1}^N \left(\sum_{k=0}^{N-1} c_k p_k(\theta_j)\right) s_j t_j q_l(\theta_j)$$

$$= \sum_{k=0}^{N-1} c_k \sum_{j=1}^N p_k(\theta_j)s_j t_j q_l(\theta_j).$$

But $p_k(\theta_j)s_j$ and $q_l(\theta_j)t_j$ are, respectively, the components of the eigenvectors of J_N and J_N^T corresponding to θ_j. Therefore they are orthonormal with the normalization that we chose. Hence,

$$\sum_{j=1}^{N} f(\theta_j)s_j t_j q_l(\theta_j) = c_l, \ l = 0, \ldots, N-1,$$

and consequently

$$\sum_{j=1}^{N} f(\theta_j)s_j t_j = c_0,$$

which proves the result.   □

Now, we extend the exactness result to polynomials of higher degree.

THEOREM 6.10 *Assume that $\gamma_j \beta_j \neq 0$; then the nonsymmetric Gauss quadrature rule (6.17) is exact for polynomials of degree less than or equal to $2N - 1$.*

Proof. Suppose f is a polynomial of degree $2N - 1$. Then f can be written as

$$f(\lambda) = p_N(\lambda)s(\lambda) + r(\lambda),$$

where s and r are polynomials of degree less than or equal to $N - 1$. Then,

$$\int_a^b f(\lambda)\, d\alpha(\lambda) = \int_a^b p_N(\lambda)s(\lambda)\, d\alpha(\lambda) + \int_a^b r(\lambda)\, d\alpha(\lambda) = \int_a^b r(\lambda)\, d\alpha(\lambda),$$

since p_N is orthogonal to any polynomial of degree less than or equal to $N - 1$ because of the orthogonality property of the p's and q's. For the quadrature rule applied to the function f, we obtain

$$\sum_{j=1}^{N} p_N(\theta_j)s(\theta_j)s_j t_j + \sum_{j=1}^{N} r(\theta_j)s_j t_j.$$

Since θ_j is an eigenvalue of J_N, it is a root of p_N and the first sum in the quadrature rule vanishes,

$$\sum_{j=1}^{N} p_N(\theta_j)s(\theta_j)s_j t_j = 0.$$

On the other hand the polynomial r is of degree less than $N - 1$ and the quadrature rule has been proven to be exact for polynomials of degree less than $N-1$; therefore

$$\int_a^b r(\lambda)\, d\alpha(\lambda) = \sum_{j=1}^{N} r(\theta_j)s_j t_j,$$

which proves the result.   □

Regarding expressions for the remainder, we can do exactly the same as for the Gauss rule. We can write

$$R[f] = \frac{f^{(2N)}(\eta)}{(2N)!} \int_a^b p_N(\lambda)^2\, d\alpha(\lambda).$$

6.5.2 The Gauss–Radau and Gauss–Lobatto Rules

Now, we extend the Gauss–Radau and Gauss–Lobatto rules to the nonsymmetric case. This is almost identical to the symmetric case. For the Gauss–Radau rule, assume that the prescribed node is a, the left end of the interval. Then, we would like to have $p_{N+1}(a) = q_{N+1}(a) = 0$. This gives

$$(a - \omega_{N+1})p_N(a) - \beta_N p_{N-1}(a) = 0.$$

If we denote $\delta(a) = [\delta_1(a), \ldots, \delta_N(a)]^T$, with

$$\delta_l(a) = -\beta_N \frac{p_{l-1}(a)}{p_N(a)},$$

we have

$$\omega_{N+1} = a + \delta_N(a),$$

where

$$(J_N - aI)\delta(a) = \gamma_N \beta_N e^N.$$

Therefore, the algorithm is essentially the same as previously discussed for the Gauss rule.

For the Gauss–Lobatto rule, the algorithm is also almost the same as for the symmetric case. We would like to compute p_{N+1} and q_{N+1} such that

$$p_{N+1}(a) = p_{N+1}(b) = 0, \quad q_{N+1}(a) = q_{N+1}(b) = 0.$$

This leads to solve the linear system

$$\begin{pmatrix} p_N(a) & p_{N-1}(a) \\ p_N(b) & p_{N-1}(b) \end{pmatrix} \begin{pmatrix} \omega_{N+1} \\ \beta_N \end{pmatrix} = \begin{pmatrix} a p_N(a) \\ b p_N(b) \end{pmatrix}.$$

The linear system for the q's whose solution is $(\omega_{N+1}, \gamma_N)^T$ can be shown to have the same solution for ω_{N+1} and $\gamma_N = \pm\beta_N$ depending on the sign relations between the p's and the q's.

Let $\delta(a)$ and $\mu(b)$ be the solutions of

$$(J_N - aI)\delta(a) = e^N, \quad (J_N - bI)\mu(b) = e^N.$$

Then we have

$$\begin{pmatrix} 1 & -\delta(a)_N \\ 1 & -\mu(b)_N \end{pmatrix} \begin{pmatrix} \omega_{N+1} \\ \beta_N^2 \end{pmatrix} = \begin{pmatrix} a \\ b \end{pmatrix}.$$

When we have the solution of this system, we choose $\gamma_N = \pm\beta_N$ and $\gamma_N \geq 0$.

As in the symmetric case, we do not always need to compute the eigenvalues and eigenvectors of J_N (or its modifications) but only the $(1, 1)$ element of $f(J_N)$.

6.6 The Block Gauss Quadrature Rules

6.6.1 The Block Gauss Rule

In this section we consider the block case using matrix polynomials and quadrature rules for matrix measures. We use the results of Golub and Meurant [149]. For other related approaches see Sinap and Van Assche [305], Sinap [304] and Dette and Studden [82]. The problem is to find a quadrature rule for a symmetric matrix measure. The integral $\int_a^b f(\lambda)d\alpha(\lambda)$ is now a 2×2 symmetric matrix. The most general quadrature formula is of the form

$$\int_a^b f(\lambda)d\alpha(\lambda) = \sum_{j=1}^N W_j f(T_j) W_j + R[f],$$

where W_j and T_j are symmetric 2×2 matrices. They are the equivalent of the weights and the nodes in the scalar case. In this rule, we have $6N$ unknowns. It can be simplified using the spectral decomposition of T_j, $T_j = Q_j \Lambda_j Q_j^T$, where Q_j is the orthonormal matrix of the eigenvectors, and Λ_j the diagonal matrix of the eigenvalues of T_j. This gives

$$\sum_{j=1}^N W_j Q_j f(\Lambda_j) Q_j^T W_j.$$

But $W_j Q_j f(\Lambda_j) Q_j^T W_j$ can be written as $f(\lambda_1)z^1(z^1)^T + f(\lambda_2)z^2(z^2)^T$, where the vector z^i has two components. Therefore, changing notations, we can write the quadrature rule as

$$\sum_{j=1}^{2N} f(t_j)w^j (w^j)^T,$$

where t_j is a scalar and w^j is a vector with two components. In this quadrature rule, there are also $6N$ unknowns, the nodes t_j and the two components of w_j, $j = 1, \ldots, 2N$. We have seen that there exist orthogonal matrix polynomials related to α such that

$$\lambda p_{j-1}(\lambda) = p_j(\lambda)\Gamma_j + p_{j-1}(\lambda)\Omega_j + p_{j-2}(\lambda)\Gamma_{j-1}^T, \qquad (6.18)$$

$$p_0(\lambda) \equiv I_2, \quad p_{-1}(\lambda) \equiv 0.$$

This can be written as

$$\lambda[p_0(\lambda), \ldots, p_{N-1}(\lambda)] = [p_0(\lambda), \ldots, p_{N-1}(\lambda)]J_N + [0, \ldots, 0, p_N(\lambda)\Gamma_N],$$

where

$$J_N = \begin{pmatrix} \Omega_1 & \Gamma_1^T & & & \\ \Gamma_1 & \Omega_2 & \Gamma_2^T & & \\ & \ddots & \ddots & \ddots & \\ & & \Gamma_{N-2} & \Omega_{N-1} & \Gamma_{N-1}^T \\ & & & \Gamma_{N-1} & \Omega_N \end{pmatrix} \qquad (6.19)$$

is a block tridiagonal matrix of order $2N$ and a banded matrix with at most five nonzero elements in a row.

Let us denote $P(\lambda) = [p_0(\lambda), \ldots, p_{N-1}(\lambda)]^T$. Since J_N is symmetric, we have in matrix form

$$J_N P(\lambda) = \lambda P(\lambda) - [0, \ldots, 0, p_N(\lambda)\Gamma_N]^T.$$

We note that, if λ is an eigenvalue, say θ_r, of J_N and if we choose $u = u_r$ to be a two-element vector whose components are the first two components of an eigenvector corresponding to θ_r, then $P(\theta_r)u$ is this eigenvector (because of the relations that are satisfied) and if Γ_N is nonsingular, $p_N^T(\theta_r)u = 0$. The main difference with the scalar case is that, although the eigenvalues of J_N are real, it might be that they are of multiplicity greater than 1.

The nodes of the quadrature rule are the zeros of the determinant of the matrix orthogonal polynomials that is the eigenvalues of J_N. Finally, we define the block quadrature rule as

$$\int_a^b f(\lambda)\, d\alpha(\lambda) = \sum_{i=1}^{2N} f(\theta_i) u_i u_i^T + R[f], \tag{6.20}$$

where $2N$ is the order of J_N, the eigenvalues θ_i are those of J_N and u_i is the vector consisting of the two first components of the corresponding eigenvector, normalized as before. In fact, if there are multiple eigenvalues, the quadrature rule should be written as follows. Let $\theta_i, i = 1, \ldots, l$ be the set of distinct eigenvalues and n_i their multiplicities. The quadrature rule is then

$$\sum_{i=1}^{l} \left(\sum_{j=1}^{n_i} (w_i^j)(w_i^j)^T \right) f(\theta_i). \tag{6.21}$$

Unfortunately, to prove that the block quadrature rule is exact for polynomials of degree up to $2N - 1$, we cannot use the same method as for the scalar case using a factorization of the given polynomial because of commutativity problems with matrix polynomials. Therefore, we use another (more involved) approach that has been proposed in a different setting by Basu and Bose [22].

We consider all the monomials $\lambda^k, k = 0, 1, \ldots$. Let M_k be the moment matrix, defined as

$$M_k = \int_a^b \lambda^k \, d\alpha(\lambda).$$

We write the (matrix) orthonormal polynomials p_j associated with the measure α as

$$p_j(\lambda) = \sum_{k=0}^{j} p_k^{(j)} \lambda^k,$$

$p_k^{(j)}$ being a matrix of order 2. Then we have

$$\int_a^b p_j^T(\lambda)\, d\alpha(\lambda) = \sum_{k=0}^{j} (p_k^{(j)})^T \int_a^b \lambda^k \, d\alpha(\lambda) = \sum_{k=0}^{j} (p_k^{(j)})^T M_k,$$

and more generally

$$\int_a^b p_j^T(\lambda)\lambda^q \, d\alpha(\lambda) = \sum_{k=0}^{j} (p_k^{(j)})^T M_{k+q}. \tag{6.22}$$

Let us write this equation for $j = N-1$. From the orthogonality of the polynomials, we have

$$\int_a^b p_{N-1}^T(\lambda)\lambda^q \, d\alpha(\lambda) = 0, \quad q = 0, \ldots, N-2. \tag{6.23}$$

Let H_N be the block Hankel matrix of order $2N$, defined as

$$H_N = \begin{pmatrix} M_0 & \cdots & M_{N-1} \\ \vdots & & \vdots \\ M_{N-1} & \cdots & M_{2N-2} \end{pmatrix}. \tag{6.24}$$

Then using equations (6.22) and (6.23) we have

$$H_N \begin{pmatrix} p_0^{(N-1)} \\ \vdots \\ p_{N-2}^{(N-1)} \\ p_{N-1}^{(N-1)} \end{pmatrix} = \begin{pmatrix} 0 \\ \vdots \\ 0 \\ \int_a^b p_{N-1}^T(\lambda)\lambda^{N-1} \, d\alpha(\lambda) \end{pmatrix}. \tag{6.25}$$

Let us introduce some additional notations. Let L_N be a block upper triangular matrix of order $2N$,

$$L_N = \begin{pmatrix} p_0^{(0)} & p_0^{(1)} & \cdots & p_0^{(N-1)} \\ & p_1^{(1)} & \cdots & p_1^{(N-1)} \\ & & \ddots & \vdots \\ & & & p_{N-1}^{(N-1)} \end{pmatrix}. \tag{6.26}$$

Let V_N be a $4N \times 2N$ matrix defined in block form as

$$V_N = \begin{pmatrix} B_1 \\ B_2 \\ \vdots \\ B_l \end{pmatrix}, \tag{6.27}$$

where B_j is a $2n_j \times 2N$ matrix defined as

$$B_j = \begin{pmatrix} I_2 & \theta_j I_2 & \cdots & \theta_j^{N-1} I_2 \\ \vdots & \vdots & \vdots & \vdots \\ I_2 & \theta_j I_2 & \cdots & \theta_j^{N-1} I_2 \end{pmatrix}, \tag{6.28}$$

where the values θ_j are the l distinct eigenvalues of J_N and n_j their multiplicities.

PROPOSITION 6.11 *Let L_N be defined by equation (6.26) and V_N by equations (6.27) and (6.28). Then*

$$V_N L_N = \begin{pmatrix} C_1 \\ C_2 \\ \vdots \\ C_l \end{pmatrix},$$

where C_j is a $2n_j \times 2N$ matrix,

$$C_j = \begin{pmatrix} p_0(\theta_j) & p_1(\theta_j) & \cdots & p_{N-1}(\theta_j) \\ \vdots & \vdots & \vdots & \vdots \\ p_0(\theta_j) & p_1(\theta_j) & \cdots & p_{N-1}(\theta_j) \end{pmatrix}.$$

Proof. This is straightforward by the definition of the polynomials $p_j(\lambda)$. $\qquad\square$

PROPOSITION 6.12 *Let L_N be defined by equation (6.26) and H_N by equation (6.24). We have*

$$L_N^T H_N L_N = I,$$

where I is the identity matrix.

Proof. The generic term of $H_N L_N$ is

$$(H_N L_N)_{ij} = \sum_{s=1}^{j} M_{s+i-2}\, p_{s-1}^{(j-1)},$$

and therefore the generic block term of $L_N^T H_N L_N$ is

$$(L_N^T H_N L_N)_{ij} = \sum_{r=1}^{i} \sum_{s=1}^{j} \int_a^b (p_{r-1}^{(i-1)})^T \lambda^{s+r-2}\, d\alpha(\lambda) p_{s-1}^{(j-1)}.$$

Splitting the power of λ in two parts and using

$$\sum_{s=1}^{j} \lambda^{s-1} p_{s-1}^{(j-1)} = p_{j-1}(\lambda),$$

we can see that we have

$$(L_N^T H_N L_N)_{ij} = \int_a^b p_{i-1}^T(\lambda)\, d\alpha(\lambda)\, p_{j-1}(\lambda).$$

From the orthonormality properties, we obtain

$$(L_N^T H_N L_N)_{ij} = \begin{cases} I_2, & \text{if } i = j, \\ 0, & \text{otherwise,} \end{cases}$$

which proves the result. $\qquad\square$

The last result implies that the inverse of H_N is $H_N^{-1} = L_N L_N^T$. Let K_m be defined as

$$K_m(\mu, \lambda) = \sum_{j=0}^{m} p_j(\mu) p_j^T(\lambda),$$

and let K_i^j be a $2n_i \times 2n_j$ matrix

$$K_i^j = \begin{pmatrix} K_{N-1}(\theta_i, \theta_j) & \cdots & K_{N-1}(\theta_i, \theta_j) \\ \vdots & \vdots & \vdots \\ K_{N-1}(\theta_i, \theta_j) & \cdots & K_{N-1}(\theta_i, \theta_j) \end{pmatrix},$$

and finally let

$$K = \begin{pmatrix} K_1^1 & K_1^2 & \dots & K_1^l \\ K_2^1 & \dots & \dots & K_2^l \\ \vdots & & & \vdots \\ K_l^1 & \dots & \dots & K_l^l \end{pmatrix}. \tag{6.29}$$

Then, we have the following factorization of the matrix K.

PROPOSITION 6.13 *Let L_N be defined by equation (6.26), V_N by equations (6.27) and (6.28) and K by equation (6.29). Then*

$$V_N L_N (V_N L_N)^T = K.$$

Proof. This is proved using the definition of K_i^j. $\qquad\square$

Now, we define a $2N \times 4N$ matrix W_N^T whose only nonzero components in row i are in position $(i, 2i-1)$ and $(i, 2i)$ and are successively the two components of $(w_1^1)^T, \ldots, (w_1^{n_1})^T, (w_2^1)^T, \ldots, (w_2^{n_2})^T, \ldots (w_l^1)^T, \ldots, (w_l^{n_l})^T$. We choose the weights w_j^i such that they are normalized as

$$(w_i^k)^T K_{N-1}(\theta_i, \theta_j) w_j^l = \delta_{k,l}.$$

PROPOSITION 6.14 *Let K be defined by equation (6.29). With the previous definition of W_N we have*

$$W_N^T K W_N = I.$$

Proof. This is obvious from the way the w_i^j's are constructed. $\qquad\square$

This leads to the following result.

PROPOSITION 6.15 *Let V_N be defined by equations (6.27) and W_N be defined as above. Then $W_N^T V_N$ is a nonsingular $2N \times 2N$ matrix.*

Proof.

$$W_N^T V_N H_N^{-1} V_N^T W_N = W_N^T V_N L_N L_N^T V_N^T W_N = W_N^T K W_N = I.$$

This shows that $W_N^T V_N$ is nonsingular. $\qquad\square$

We now give the main result concerning the exactness of the block quadrature rule.

THEOREM 6.16 *The quadrature rule (6.20) or (6.21) is exact for polynomials of order less than or equal to $2N - 1$.*

Proof. From the proof of proposition 6.15, we have

$$H_N^{-1} = (W_N^T V_N)^{-1} (V_N^T W_N)^{-1}.$$

Therefore,

$$H_N = (V_N^T W_N)(W_N^T V_N).$$

By identification of the block entries in the two matrices we have,

$$M_k = \sum_{i=1}^{l} \left(\sum_{j=1}^{n_i} (w_i^j)(w_i^j)^T \right) \theta_i^k, \quad k = 0, \ldots, 2N - 2,$$

which proves that the quadrature rule is exact up to degree $2N - 2$. It remains to prove that it is exact for $k = 2N - 1$. Writing equation (6.25) for $N + 1$, we have

$$H_{N+1} \begin{pmatrix} p_0^{(N)} \\ \vdots \\ p_{N-1}^{(N)} \\ p_N^{(N)} \end{pmatrix} = \begin{pmatrix} 0 \\ \vdots \\ 0 \\ \int_a^b p_N^T(\lambda) \lambda^N \, d\alpha(\lambda) \end{pmatrix}.$$

Writing the $(N-1)$st block row of this equality, we obtain

$$M_{2N-1} \, p_N^{(N)} = - \sum_{r=0}^{N-1} M_{N+r-1} \, p_r^{(N)}. \tag{6.30}$$

We have seen before that

$$M_{N+r-1} = \sum_{i=1}^{l} \left(\sum_{j=1}^{n_i} w_i^j (w_i^j)^T \right) \theta_i^{N+r-1}.$$

By substitution into equation (6.30), we have

$$M_{2N-1} \, p_N^{(N)} = - \sum_{r=0}^{N-1} \sum_{i=1}^{l} \sum_{j=1}^{n_i} w_i^j (w_i^j)^T \theta_i^{N+r-1} p_r^{(N)}.$$

We put the sum over r on the last two terms and use the fact that

$$(w_i^j)^T \sum_{r=0}^{N-1} \theta_i^r p_r^{(N)} = (w_i^j)^T p_N(\theta_i) - (w_i^j)^T \theta_i^N p_N^{(N)},$$

because of

$$\sum_{r=0}^{N-1} \theta_i^r p_r^{(N)} = \sum_{r=0}^{N} \theta_i^r p_r^{(N)} - \theta_i^N p_N^{(N)},$$

and

$$(w_i^j)^T p_N(\theta_i) = 0,$$

because since w_i^j is proportional to the two first components of an eigenvector we have $p_N(\theta_i)^T w_i^j = 0$. This shows that

$$M_{2N-1} p_N^{(N)} = \sum_{i=1}^{l} \sum_{j=1}^{n_i} (w_i^j)(w_i^j)^T \theta_i^{2N-1} p_N^{(N)}.$$

Since $p_N^{(N)}$ is nonsingular, this proves the desired result.   □

   To obtain expressions for the remainder, we would like to use a similar approach as for the scalar case. However, there are some differences, since in the block case the quadrature rule is exact for polynomials of order $2N-1$ and since we have $2N$ nodes, we cannot interpolate with a Hermite polynomial and we have to use a Lagrange polynomial. By theorems 2.1.1.1 and 2.1.4.1 of Stoer and Bulirsch [314], there exists a polynomial q of degree $2N-1$ such that

$$q(\theta_j) = f(\theta_j), \quad j = 1, \dots, 2N$$

and

$$f(x) - q(x) = \frac{s(x) f^{(2N)}(\xi(x))}{(2N)!},$$

where

$$s(x) = (x - \theta_1) \cdots (x - \theta_{2N}).$$

Applying the mean value theorem, the remainder $R[f]$, which is a 2×2 matrix, can be written as

$$R[f] = \frac{f^{(2N)}(\eta)}{(2N)!} \int_a^b s(\lambda) \, d\alpha(\lambda).$$

Unfortunately, the elements of s do not have a constant sign over the interval $[a, b]$. Therefore this representation formula for the remainder is of little practical use, except eventually to obtain bounds of the norm of the remainder.

6.6.2 The Block Gauss–Radau Rule

We now extend the process described for scalar polynomials to the block analog of the Gauss–Radau quadrature rule. We would like a to be a double eigenvalue of J_{N+1}. We have

$$J_{N+1}P(a) = aP(a) - [0, \ldots, 0, p_{N+1}(a)\Gamma_{N+1}]^T.$$

Then, we require $p_{N+1}(a) \equiv 0$. From the block three-term recurrence this translates into

$$ap_N(a) - p_N(a)\Omega_{N+1} - p_{N-1}(a)\Gamma_N^T = 0.$$

Therefore, if $p_N(a)$ is nonsingular, we have

$$\Omega_{N+1} = aI_2 - p_N(a)^{-1}p_{N-1}(a)\Gamma_N^T.$$

We must compute the right-hand side. This can be done by remarking that

$$J_N \begin{pmatrix} p_0(a)^T \\ \vdots \\ p_{N-1}(a)^T \end{pmatrix} = a \begin{pmatrix} p_0(a)^T \\ \vdots \\ p_{N-1}(a)^T \end{pmatrix} - \begin{pmatrix} 0 \\ \vdots \\ \Gamma_N^T p_N(a)^T \end{pmatrix}.$$

Multiplying on the right by $p_N(a)^{-T}$, we obtain the matrix equation

$$(J_N - aI) \begin{pmatrix} -p_0(a)^T p_N(a)^{-T} \\ \vdots \\ -p_{N-1}(a)^T p_N(a)^{-T} \end{pmatrix} = \begin{pmatrix} 0 \\ \vdots \\ \Gamma_N^T \end{pmatrix}.$$

Thus, we first solve

$$(J_N - aI) \begin{pmatrix} \delta_0(a) \\ \vdots \\ \delta_{N-1}(a) \end{pmatrix} = \begin{pmatrix} 0 \\ \vdots \\ \Gamma_N^T \end{pmatrix}.$$

This is a block tridiagonal linear system. Then we have,

$$\Omega_{N+1} = aI_2 + \delta_{N-1}(a)^T \Gamma_N^T, \tag{6.31}$$

which gives J_{N+1} with a double prescribed eigenvalue. The block Gauss–Radau rule is exact for polynomials of degree $2N$.

6.6.3 The Block Gauss–Lobatto Rule

The generalization of the Gauss–Lobatto construction to the block case is a little
more difficult. We would like to have a and b as double eigenvalues of the matrix
J_{N+1}. This leads to satisfying the two following matrix equations

$$ap_N(a) - p_N(a)\Omega_{N+1} - p_{N-1}(a)\Gamma_N^T = 0,$$

$$bp_N(b) - p_N(b)\Omega_{N+1} - p_{N-1}(b)\Gamma_N^T = 0.$$

This can be written as a linear system

$$\begin{pmatrix} I_2 & p_N^{-1}(a)p_{N-1}(a) \\ I_2 & p_N^{-1}(b)p_{N-1}(b) \end{pmatrix} \begin{pmatrix} \Omega_{N+1} \\ \Gamma_N^T \end{pmatrix} = \begin{pmatrix} aI_2 \\ bI_2 \end{pmatrix}. \tag{6.32}$$

We now consider the problem of computing $p_N^{-1}(\lambda)p_{N-1}(\lambda)$. Let $\delta(\lambda)$ be the
solution of

$$(J_N - \lambda I)\delta(\lambda) = (0 \ldots 0 \ I_2)^T.$$

Then, as before,

$$\delta_{N-1}(\lambda) = -p_{N-1}(\lambda)^T p_N(\lambda)^{-T}\Gamma_N^{-T}.$$

We can show that $\delta_{N-1}(\lambda)$ is symmetric. We consider solving a 2×2 block linear
system of the form

$$\begin{pmatrix} I & X \\ I & Y \end{pmatrix} \begin{pmatrix} U \\ V \end{pmatrix} = \begin{pmatrix} aI \\ bI \end{pmatrix}.$$

We use a block factorization

$$\begin{pmatrix} I & X \\ I & Y \end{pmatrix} = \begin{pmatrix} I & 0 \\ I & W \end{pmatrix} \begin{pmatrix} I & X \\ 0 & Z \end{pmatrix} = \begin{pmatrix} I & X \\ I & X+WZ \end{pmatrix};$$

thus $WZ = Y - X$.

The solution of the system for the forward step

$$\begin{pmatrix} I & 0 \\ I & W \end{pmatrix} \begin{pmatrix} U_1 \\ V_1 \end{pmatrix} = \begin{pmatrix} aI \\ bI \end{pmatrix}$$

gives

$$U_1 = aI, \quad WV_1 = (b - a)I.$$

The backward step is

$$\begin{pmatrix} I & X \\ 0 & Z \end{pmatrix} \begin{pmatrix} U \\ V \end{pmatrix} = \begin{pmatrix} U_1 \\ V_1 \end{pmatrix},$$

and we obtain

$$ZV = V_1 = W^{-1}(b - a)I,$$

or

$$(WZ)V = (b - a)I.$$

Therefore
$$V = (b - a)(Y - X)^{-1}.$$
Using this result for the linear system in equation (6.32), we have
$$Y - X = p_N^{-1}(b)p_{N-1}(b) - p_N^{-1}(a)p_{N-1}(a) = \Gamma_N(\delta_{N-1}(a) - \delta_{N-1}(b)).$$
This means that
$$\Gamma_N^T = (b - a)(\delta_{N-1}(a) - \delta_{N-1}(b))^{-1}\Gamma_N^{-1},$$
or
$$\Gamma_N^T\Gamma_N = (b - a)(\delta_{N-1}(a) - \delta_{N-1}(b))^{-1}.$$
Thus, Γ_N is given as a Cholesky factorization of the right-hand side matrix. This matrix is positive definite because $\delta_{N-1}(a)$ is a diagonal block of the inverse of $(J_N - aI)^{-1}$, which is positive definite because the eigenvalues of J_N are larger than a, and $-\delta_{N-1}(b)$ is the negative of a diagonal block of $(J_N - bI)^{-1}$, which is negative definite because the eigenvalues of J_N are smaller than b.

From Γ_N, we can compute Ω_{N+1}:
$$\Omega_{N+1} = aI_2 + \Gamma_N\delta_{N-1}(a)\Gamma_N^T.$$
The block Gauss–Lobatto rule is exact for polynomials of degree $2N + 1$.

6.6.4 Computation of the Block Gauss Rules

As for the scalar case, it is not always necessary to compute the nodes and the weights for the block quadrature rules.

THEOREM 6.17 *We have*
$$\sum_{i=1}^{2N} f(\theta_i)u_iu_i^T = e^T f(J_N)e, \tag{6.33}$$
where $e^T = (I_2\ 0\dots0)$.

Proof. The quadrature rule is
$$\sum_{i=1}^{2N} u_if(\theta_i)u_i^T.$$
If z^i are the eigenvectors of J_N then $u_i = e^Tz^i$ and
$$\sum_{i=1}^{2N} u_if(\theta_i)u_i^T = \sum_{i=1}^{2N} e^Tz^i f(\theta_i)(z^i)^Te$$
$$= e^T\left(\sum_{i=1}^{2N} z^i f(\theta_i)(z^i)^T\right)e$$
$$= e^T Z_N f(\Theta_N)Z_N^Te$$
$$= e^T f(J_N)e,$$
where Z_N is the matrix of the eigenvectors and Θ_N the diagonal matrix of the eigenvalues of J_N. However, since J_N is a block tridiagonal matrix, it may not be easy to compute elements of $f(J_N)$. Nevertheless, we will see that it can be done if $f(\lambda) = 1/\lambda$. $\square$

Chapter Seven

Bounds for Bilinear Forms $u^T f(A)v$

7.1 Introduction

As we said in chapter 1, we are interested in computing bounds or approximations for bilinear forms

$$u^T f(A)v, \tag{7.1}$$

where A is a symmetric square matrix of order n, u and v are given vectors and f is a smooth (possibly C^∞) function on a given interval of the real line. For the relation of this problem to matrix moments, see Golub [140], [141]. There are many different areas of scientific computing where such estimates are required, for instance, solid state physics, physics problems leading to ill-posed linear systems, computing error bounds for iterative methods and so on.

We will also consider a generalization of the form (7.1),

$$W^T f(A)W,$$

where W is an $n \times m$ matrix. For specificity, we will consider $m = 2$.

In this short chapter we summarize how the results and techniques developed in chapters 2 to 6 can be used to approximate the bilinear form (7.1). Assuming that the matrix A is symmetric, we can use the spectral decomposition of A written as

$$A = Q\Lambda Q^T,$$

where Q is the orthonormal matrix whose columns are the normalized eigenvectors of A and Λ is a diagonal matrix whose diagonal elements are the eigenvalues λ_i of A, which we order as

$$\lambda_1 \le \lambda_2 \le \cdots \le \lambda_n.$$

The definition of a function of a symmetric matrix is

$$f(A) = Qf(\Lambda)Q^T;$$

see, for instance, [189]. Therefore,

$$u^T f(A)v = u^T Qf(\Lambda)Q^T v,$$
$$= \gamma^T f(\Lambda)\beta,$$
$$= \sum_{i=1}^{n} f(\lambda_i)\gamma_i \beta_i.$$

This last sum can be considered as a Riemann–Stieltjes integral (see chapter 2),

$$I[f] = u^T f(A)v = \int_a^b f(\lambda) \, d\alpha(\lambda), \tag{7.2}$$

where the measure α is piecewise constant and defined by

$$\alpha(\lambda) = \begin{cases} 0, & \text{if } \lambda < a = \lambda_1, \\ \sum_{j=1}^{i} \gamma_j \beta_j, & \text{if } \lambda_i \leq \lambda < \lambda_{i+1}, \\ \sum_{j=1}^{n} \gamma_j \beta_j, & \text{if } b = \lambda_n \leq \lambda. \end{cases}$$

When $u = v$, we remark that α is an increasing positive function as well as when $\gamma_j \beta_j > 0$.

The block generalization is obtained in the following way. Let W be an $n \times 2$ matrix, $W = (w_1 \; w_2)$, then

$$W^T f(A) W = W^T Q f(\Lambda) Q^T W = \omega f(\Lambda) \omega^T,$$

where, of course, ω is a $2 \times n$ matrix such that

$$\omega = (\omega_1 \ldots \omega_n),$$

and ω_i is a vector with two components. With these notations, we have

$$W^T f(A) W = \sum_{i=1}^{n} f(\lambda_i) \omega_i \omega_i^T.$$

This can be written as a matrix Riemann–Stieltjes integral (see chapter 2),

$$I_B[f] = W^T f(A) W = \int_a^b f(\lambda) \, d\alpha(\lambda). \tag{7.3}$$

$I_B[f]$ is a 2×2 matrix where the entries of the (matrix) measure α are piecewise constant and defined by

$$\alpha(\lambda) = \sum_{k=1}^{i} \omega_k \omega_k^T, \quad \lambda_i \leq \lambda < \lambda_{i+1}.$$

As we have seen in chapter 6, a way to obtain bounds for the Riemann–Stieltjes integrals (7.2) or (7.3) is to use Gauss, Gauss–Radau and Gauss–Lobatto quadrature rules or their block equivalents. When $u = v$, the measure is a positive increasing function. If the given function f is such that its derivatives have a constant sign on the interval of integration, then we can obtain bounds for the quadratic form $u^T f(A) u$. If the signs are not constant, the quadrature rules give only approximations of the quadratic form. From chapter 2 we know that when the measure is positive, it is possible to define a sequence of polynomials $p_0(\lambda), p_1(\lambda), \ldots$ that are orthonormal with respect to the measure α. The nodes and weights of the quadrature rules are obtained from the Jacobi matrix (or some modifications of it) whose nonzero entries are the coefficients of the polynomials three-term recurrence; see chapter 6.

7.2 The Case $u = v$

How do we generate the Jacobi matrix corresponding to a measure α that is unknown, since we do not know the eigenvalues of A? When A is symmetric and

$u = v$, an elegant solution is to use the Lanczos algorithm with the first Lanczos vector chosen as $v^1 = u/\|u\|$. We have seen in theorem 4.2 that the measure α for which the Lanczos polynomials are orthonormal is defined by the eigenvalues of A and the components of the vector $Q^T v^1$. This is precisely what we need for the quadrature rules. Hence, the nodes and weights are given by the eigenvalues and first components of the eigenvectors of the tridiagonal Lanczos matrix which is a Jacobi matrix.

The algorithm is the following:

1) Normalize u if necessary to obtain v^1.

2) Run k iterations of the Lanczos algorithm with A starting from v^1 and compute the Jacobi matrix J_k.

3) If we use the Gauss–Radau or Gauss–Lobatto rules, modify J_k to $\tilde{J}_k$ accordingly. For the Gauss rule $\tilde{J}_k = J_k$.

4) If this is feasible, compute $(e^1)^T f(\tilde{J}_k) e^1$. Otherwise, compute the eigenvalues and the first components of the eigenvectors using the Golub and Welsch algorithm to obtain the approximations from the Gauss, Gauss–Radau and Gauss–Lobatto quadrature rules.

Let n be the order of the matrix A and V_k be the $n \times k$ matrix whose columns are the Lanczos vectors. Assume for the sake of simplicity that A has distinct eigenvalues. Then after n Lanczos iterations we have $AV_n = V_n J_n$. If Q (resp. Z) is the matrix of the eigenvectors of A (resp. J_n), we have the relation $V_n Z = Q$. Assuming u is of norm 1, we have $u = V_n e^1$; therefore

$$u^T f(A) u = (e^1)^T V_n^T Q f(\Lambda) Q^T V_n e^1 = (e^1)^T Z^T f(\Lambda) Z e^1.$$

But J_n has the same eigenvalues as A. Hence $u^T f(A) u = (e^1)^T f(J_n) e^1$. The remainder of the Gauss quadrature rule can thus be written as

$$R[f] = (e^1)^T f(J_n) e^1 - (e^1)^T f(J_k) e^1.$$

From this expression of the remainder we see that the convergence of the Gauss quadrature approximation to the integral depends on the convergence of the Ritz values (which are the eigenvalues of J_k and the nodes of the quadrature rule) to the eigenvalues of A.

7.3 The Case $u \neq v$

A first possibility is to use the identity

$$u^T f(A) v = [(u + v)^T f(A)(u + v) - (u - v)^T f(A)(u - v)]/4.$$

If the signs of the derivatives of the function f are constant we can obtain lower and upper bounds of the two terms on the right-hand side and combine them to obtain bounds of $u^T f(A) v$. However, this has the disadvantage that we have to run the Lanczos algorithm twice, one time with $u + v$ and one time with $u - v$ as the starting vector.

Another possibility is to apply the nonsymmetric Lanczos algorithm to the symmetric matrix A. The main difference with the symmetric Lanczos algorithm is

that the algorithm may break down. As we have seen in chapter 4, we can monitor
the signs of the computed vectors during the algorithm to know wether we have a
lower or an upper bound. Otherwise the framework of the algorithm is the same as
for the case $u = v$. Note that as in the first possibility we have two matrix-vector
multiplications per iteration.

A way to get around the breakdown problem is to introduce a parameter δ and
use $v^1 = u/\delta$ and $\tilde{v}^1 = \delta u + v$. This will give an estimate of $u^T f(A)v/\delta +$
$u^T f(A)u$. Using the bounds we can compute for $u^T f(A)u$, we can obtain bounds
for $u^T f(A)v$. Of course, the problem is to determine a good value of δ.

7.4 The Block Case

The general framework is the same as for the Lanczos algorithm. We have to deal
with a block tridiagonal matrix and we have to compute $e^T f(J_k)e$ where $e = (I_2\, 0\, \dots\, 0)^T$. The difficulty of such a computation depends on the function f.

For the generation of the matrix orthogonal polynomials we use the block Lanczos algorithm. However, we have seen that we have to start the algorithm from an
$n \times 2$ matrix X_0 such that $X_0^T X_0 = I_2$. Considering the bilinear form $u^T f(A)v$
we would like to use $X_0 = [u\, v]$ but this does not fulfill the condition on the starting matrix. Therefore, we have to orthogonalize the pair $[u\, v]$ before starting the
algorithm. Let u and v be independent vectors and $n_u = \|u\|$; then we compute

$$\tilde{u} = \frac{u}{n_u}, \quad \bar{v} = v - \frac{u^T v}{n_u^2}u, \quad n_v = \|\bar{v}\|, \quad \tilde{v} = \frac{\bar{v}}{n_v},$$

and we set $X_0 = [\tilde{u}\, \tilde{v}]$. Of course, this does not directly compute $u^T f(A)v$. Let J^1
be the leading 2×2 submatrix of the matrix $f(J_k)$ where J_k is the block tridiagonal
matrix constructed by the block Lanczos algorithm. Then, an approximation of
$u^T f(A)v$ is given by

$$u^T f(A)v \approx (u^T v)J^1_{1,1} + n_u n_v J^1_{1,2}.$$

Note that when u and v are orthogonal $\bar{v} = v$, we just have to normalize both
vectors and the approximation is $n_u n_v J^1_{1,2}$. At the same time we have estimates of
$u^T f(A)u$ and $v^T f(A)v$ given by

$$u^T f(A)u \approx n_u^2 J^1_{1,1}, \quad v^T f(A)v \approx n_v^2 J^1_{2,2} + 2(u^T v)\frac{n_u}{n_v} J^1_{1,2} + \frac{(u^T v)^2}{n_u^2} J^1_{1,1}.$$

Therefore, in only one run with the block Lanczos algorithm we can obtain three
estimates.

7.5 Other Algorithms for $u \neq v$

A possibility is to approximate $f(A)v$ by $\|v\|V_k f(J_k)e^1$ (see Druskin and Knizhn-
erman [91], [92], [93], [95]) and then to compute the bilinear form as $u^T \|v\|V_k f(J_k)e^1$.
Note that we can compute $u^T V_k$ during the Lanczos iterations without storing the

Lanczos vectors. For computing $f(A)v$ see also Eiermann and Ernst [97], Afanas-jew, Eiermann, Ernst and Güttel [2], [3], Frommer and Simoncini [117], Hochbruck and Lubich [192] and chapter 13 of Higham [189].

A variant of the ideas described above has been proposed by Sidje, Burrage and Philippe [300]. This paper uses an augmented Lanczos algorithm to handle the case $u \neq v$. Assume the Lanczos algorithm is started from $v^1 = v/\|v\|$. After k iterations we obtain a matrix V_k whose columns are the Lanczos vectors and a tridiagonal matrix J_k. The matrix V_k is augmented to

$$\hat{V}_{k+1} = (V_k \quad \hat{v}^{k+1}),$$

the vector $\hat{v}^{k+1}$ being defined using the vector u as

$$\hat{v}^{k+1} = \frac{(I - P_k)u}{\|(I - P_k)u\|},$$

where $P_k = V_k V_k^T$ is the orthogonal projector on the Krylov subspace spanned by the columns of V_k. The approximation of $u^T f(A)v$ is taken as

$$u^T \|v\| \hat{V}_{k+1} f(\hat{V}_{k+1}^T A \hat{V}_{k+1}) e^1.$$

It remains to see what is the structure of $\hat{V}_{k+1}^T A \hat{V}_{k+1}$. Clearly, we have

$$\hat{J}_{k+1} = \hat{V}_{k+1}^T A \hat{V}_{k+1} = \begin{pmatrix} J_k & V_k^T A \hat{v}^{k+1} \\ (\hat{v}^{k+1})^T A V_k & (\hat{v}^{k+1})^T A \hat{v}^{k+1} \end{pmatrix}.$$

Using the results from the k first Lanczos steps and orthogonality, the off-diagonal part of the last row is

$$(\hat{v}^{k+1})^T A V_k = (\hat{v}^{k+1})^T (V_k J_k + \eta_k v^{k+1} (e^k)^T) = \eta_k (\hat{v}^{k+1})^T v^{k+1} (e^k)^T.$$

Hence, the matrix $\hat{J}_{k+1}$ is tridiagonal, and denoting $\hat{\eta}_k = \eta_k (\hat{v}^{k+1})^T v^{k+1}$ and $\hat{\alpha}_{k+1} = (\hat{v}^{k+1})^T A \hat{v}^{k+1}$, we write it as

$$\hat{J}_{k+1} = \begin{pmatrix} J_k & \hat{\eta}_k e^k \\ \hat{\eta}_k (e^k)^T & \hat{\alpha}_{k+1} \end{pmatrix}.$$

How can we compute $\hat{\eta}_k$ and $\hat{\alpha}_{k+1}$? First we need $\hat{v}^{k+1}$. We remark that

$$(I - P_k)u = (I - v^k (v^k)^T) \cdots (I - v^1 (v^1)^T)u.$$

Therefore, the vector $(I - P_k)u$ can be computed incrementally as a new Lanczos vector becomes available. The norm is equal to

$$\|(I - P_k)u\|^2 = \|(I - P_{k-1})u\|^2 + [(v^k)^T u]^2.$$

When we have the vector $\hat{v}^{k+1}$ we can compute $\hat{\eta}_k$. The value $\hat{\alpha}_{k+1}$ can be computed directly but this needs an extra matrix-vector multiplication. Sidje, Burrage and Philippe developed a recurrence for calculating $\hat{\alpha}_{k+1}$,

$$\hat{\alpha}_{k+1} \|(I - P_k)u\|^2 = \hat{\alpha}_k \|(I - P_{k-1})u\|^2 - 2\eta_k (v^{k+1})^T u (v^k)^T u - \alpha_k [(v^k)^T u]^2.$$

The stability of this recurrence has not been investigated so far. The advantage of the Sidje, Burrage and Philippe algorithm is that only one matrix-vector multiplication is needed per iteration and it is not necessary to store all the Lanczos vectors.

Chapter Eight

Extensions to Nonsymmetric Matrices

When the matrix A is nonsymmetric (and not diagonalizable), there are several equivalent ways to define $f(A)$; see Higham [189] and also Frommer and Simoncini [117]. Some particular functions can be defined from their power series. A general way of defining $f(A)$ is through the Jordan canonical form of A even though this is not a practical means of computing the matrix function. Another definition uses a Hermite interpolation polynomial with interpolation conditions on the derivatives of f at the eigenvalues of A. Finally, when the function f is analytic on and inside a closed contour Γ that encloses the spectrum of A, the matrix function can be defined as a Cauchy integral

$$f(A) = \frac{1}{2\pi i} \int_\Gamma f(z)(zI - A)^{-1} \, dz.$$

For nonsymmetric matrices A most of the research has focused on computing approximations of $f(A)v$ where v is a given vector, see Eiermann and Ernst [97], Knizhnerman [205], Hochbruck and Hochstenbach [191], Druskin and Knizhnerman [93], [94] and many others. However, there are a few papers in the literature dealing with estimating $u^T A^{-1} v$ or its equivalent for complex matrices and vectors. They used the nonsymmetric Lanczos algorithm or the Arnoldi algorithm. Let us mention Freund and Hochbruck [114], Saylor and Smolarski [293], [294] and Calvetti, Kim and Reichel [58]. More recently this problem has also been considered by Golub, Stoll and Wathen [163] using generalizations of the LSQR algorithm of Paige and Saunders [256]. This generalization of the LSQR algorithm was introduced in the past by Saunders, Simon and Yip [292] for the sake of solving nonsymmetric linear systems. Generalizations of the Vorobyev moment problem [338] to the nonsymmetric case were studied by Strakoš [316] and Strakoš and Tichý [319]. These last papers justify the recourse to the nonsymmetric Lanczos and Arnoldi algorithms to compute estimates of the bilinear form without using complex quadrature rules. Strakoš proved that k iterations of the nonsymmetric Lanczos algorithm matches the first $2k$ moments of A, that is,

$$u^H A^j v = (e^1)^T J_k^j e^1, \ j = 0, 1, \dots, 2k - 1,$$

where J_k is the nonsymmetric tridiagonal matrix of the nonsymmetric Lanczos algorithm. The Hessenberg matrix computed by the Arnoldi algorithm matches only k moments after k iterations.

In this chapter we briefly review some of these approaches for computing estimates of bilinear forms.

8.1 Rules Based on the Nonsymmetric Lanczos Algorithm

In the paper [293] Saylor and Smolarski described Gauss quadrature rules in the complex plane. Their motivation was to compute estimates of $u^T A^{-1} v$. In electromagnetics, one problem is to consider waves impinging on an obstacle and to analyze the scattered wave received by an antenna. The solution of a linear system $Ax = v$ gives the field x from the signal v and the signal received by the antenna is represented by $u^T x$. It is known as the scattering amplitude. The scattering cross section of the obstacle is $|u^T A^{-1} v|^2$. The same problem arises also in other areas of physics. For complex vectors u and v the usual Euclidean inner product is

$$\langle u, v \rangle = \sum_{i=1}^{n} u_i \bar{v}_i.$$

Saylor and Smolarski considered bilinear forms

$$(u, v)_w = \sum_{i=1}^{n} u_i v_i w_i,$$

where the values w_i are weights. The vectors u and v are said to be formally orthogonal if $(u, v)_w = 0$.

In the complex plane, a weighted inner product can be naturally defined as a contour integral

$$((f, g))_w = \int_{\gamma} f(\zeta) g(\zeta) w(\zeta) \, d\zeta,$$

where γ is an arc. The main problem with this definition is that, in general, $((\zeta g, g))_w \neq ((f, \zeta g))_w$. This means that a sequence of orthogonal polynomials does not satisfy a three-term recurrence. Saylor and Smolarski used a line integral

$$\langle f, g \rangle_w = \int_{\gamma} f(\zeta) \overline{g(\zeta)} w(\zeta) \, |d\zeta|,$$

where $|d\zeta|$ is the arc length. The corresponding normalized formally (or formal) orthogonal polynomials satisfy a three-term recurrence but they may fail to exist at some stages. A Gauss quadrature formula is then

$$\int_{\gamma} f(\zeta) w(\zeta) \, |d\zeta| = \sum_{i=1}^{k} \omega_i f(\zeta_i) + R_k.$$

Besides the fact that the formally orthogonal polynomials ϕ_i may fail to exist, the derivation is then more or less the same as in the real case. If we assume that the polynomials exist, the nodes are the eigenvalues of the tridiagonal matrix and the weights are the squares of the first components of the eigenvectors divided by $\phi_i^2(\zeta_j)$. The nodes and weights may be complex numbers.

So far, this does not solve the problem of computing the scattering amplitude. Saylor and Smolarski used the biconjugate gradient method (denoted as BCG or BiCG) which can be derived from the nonsymmetric Lanczos algorithm, see Fletcher

[112]. In BiCG there are two sets of residual vectors r^k which belong to $\mathcal{K}(A, r^0)$ and $\tilde{r}^k$ which belong to $\mathcal{K}(A^H, \tilde{r}^0)$. The vectors $r^k = p_k(A)r^0$ and $\tilde{r}^k = \tilde{p}_k(A)\tilde{r}^0$ satisfy three-term recurrences. Moreover, $p_k(\zeta) = \tilde{p}_k(\zeta)$.

The inner product of r^k and $\tilde{r}^k$ can be written as a line integral (where γ is an arc connecting the eigenvalues of A) and the polynomials p_k are formally orthogonal relative to this bilinear form. The tridiagonal matrix can be obtained from the BiCG coefficients in the same spirit that the Lanczos coefficients can be derived from those of CG (and reciprocally) in the symmetric (or Hermitian) case. Details are given in [294]. When the nodes and weights are computed we have that the approximation of the scattering amplitude is

$$u^T A^{-1} v \approx \sum_{i=1}^{k} \frac{\omega_i}{\zeta_i}.$$

On this topic, see also the addendum [294] and the reports of Warnick [344], [345], [346].

Interesting numerical experiments comparing different methods are given in Strakoš and Tichý [319].

8.2 Rules Based on the Arnoldi Algorithm

The paper [58] by Calvetti, Kim and Reichel considers complex nonsymmetric matrices and apply the Arnoldi algorithm we have described at the beginning of chapter 4, which in matrix form is written as

$$AV_k = V_k H_k + h_{k+1,k} v^{k+1} (e^k)^H,$$

where the upper index H denotes the conjugate transpose since we are dealing with complex matrices. The matrix H_k is upper Hessenberg with elements $h_{i,j}$. Let

$$\langle f, g \rangle = v^H f(A)^H g(A) v, \tag{8.1}$$

be a quadratic form where the functions f and g are assumed to be analytic in a neighborhood of the eigenvalues of A. The quadratic form is represented as an integral

$$\langle f, g \rangle = \frac{1}{4\pi^2} \int_\gamma \int_\gamma \overline{f(z)} g(w) v^H (\bar{z}I - A^H)^{-1} (wI - A)^{-1} v \, \overline{dz} \, dw,$$

where the contour γ contains the spectrum of A. Then, the quadratic form (8.1) is approximated by

$$\langle f, g \rangle_k = \|v\|^2 (e^1)^H f(H_k)^H g(H_k) e^1,$$

where the functions f and g must be also analytic in a neighborhood of the spectrum of H_k which gives approximations to the spectrum of A.

The Arnoldi vectors v^j are given by a polynomial $\hat{p}_j$ of degree j applied to the first vector v, $v^j = \hat{p}_{j-1}(A)v$. These polynomials are orthonormal with respect to the quadratic form (8.1). The monic orthogonal polynomials p_j associated with the polynomials $\hat{p}_j$ satisfy a recurrence relation

$$p_j(\lambda) = (\lambda - c_{j,j}) p_{j-1}(\lambda) - \sum_{k=1}^{j-1} c_{k,j} p_{k-1}(\lambda),$$

with $p_0(\lambda) \equiv 1$ and

$$c_{k,j} = \frac{\langle p_{k-1}, \lambda p_{k-1} \rangle}{\langle p_{k-1}, p_{k-1} \rangle}.$$

The nonzero entries of H_k are given in terms of the monic polynomials as

$$h_{j+1,j} = \frac{\langle p_j, p_j \rangle^{1/2}}{\langle p_{j-1}, p_{j-1} \rangle^{1/2}}, \quad h_{k,j} = \frac{\langle p_{k-1}, \lambda p_{j-1} \rangle}{\langle p_{k-1}, p_{k-1} \rangle^{1/2} \langle p_{j-1}, p_{j-1} \rangle^{1/2}}, \quad j < k.$$

Note that the coefficients $h_{j+1,j}$ are positive. Hence, the polynomials are well defined up to $k = n(A)$, the grade of A with respect to v.

Let $W_k = (P_k \oplus P_{k+1}) \cup (P_{k+1} \oplus P_k)$ and P_k denote the set of polynomials of degree at most k. Then,

$$\langle f, g \rangle_k = \langle f, g \rangle, \quad \forall f, g \in W_{k-1}.$$

This result is due to Freund and Hochbruck [114] and a proof is also given in [58]. Calvetti, Kim and Reichel introduced the bilinear forms

$$\langle f, g \rangle^{(r,s)} = \|v\|^2 (v^r)^H f(A)^H g(A) v^s, \tag{8.2}$$

where v^r and v^s are Arnoldi vectors, and an approximation

$$\langle f, g \rangle_k^{(r,s)} = \|v\|^2 (e^r)^H f(H_k)^H g(H_k) e^s.$$

Then,

$$\langle f, g \rangle^{(r,s)} = \langle f, g \rangle_k^{(r,s)},$$

for all integers r and s smaller than k and all polynomials f and g such that $f \in P_{k-r+1}$ and $g \in P_{k-s}$ or $f \in P_{k-r}$ and $g \in P_{k-s+1}$.

Calvetti, Kim and Reichel [58] introduced also an anti-Arnoldi quadrature rule denoted by $[f, g]_{k+1}$ for which

$$\langle f, g \rangle - \langle f, g \rangle_k = -(\langle f, g \rangle - [f, g]_{k+1}), \quad \forall f, g \in W_k.$$

The derivation is similar to what has been done by Laurie [218] for the symmetric case. The Hessenberg matrix $\tilde{H}_{k+1}$ for the anti-Arnoldi rule is obtained from H_{k+1} by multiplying the elements of the last row and the last column by $\sqrt{2}$ except for the diagonal element $h_{k+1,k+1}$.

If the functions f and g are such that we have expansions

$$f(A)v = \sum_{i=0}^{m_f} \eta_i \hat{p}_i(A)v, \quad g(A)v = \sum_{i=0}^{m_g} \xi_i \hat{p}_i(A)v,$$

then, if the coefficients η_i and ξ_i are sufficiently small for $i \geq k+1$, $\mathrm{Re}(\langle f, g \rangle_k)$ and $\mathrm{Re}([f, g]_{k+1})$ (resp. $\mathrm{Im}(\langle f, g \rangle_k)$ and $\mathrm{Im}([f, g]_{k+1})$) can be shown to give lower or upper bounds of $\mathrm{Re}(\langle f, g \rangle)$ (resp. $\mathrm{Im}(\langle f, g \rangle)$). The conditions given in [56] are difficult to check but this result shows that there exist cases for which the Arnoldi and anti-Arnoldi quadrature rules do give bounds.

If a vector u belongs to the Krylov subspace span by the Arnoldi vectors, we have $u = \sum_{r=1}^{l} \beta_r v^r$, then

$$u^H g(A) v = \frac{1}{\|v\|} \sum_{r=1}^{l} \beta_r \langle 1, g \rangle^{(r,1)}.$$

In [56], it is proposed to use

$$\frac{1}{\|v\|} \sum_{r=1}^{l} \max\{\beta_r \langle 1, g \rangle_k^{(r,1)}, \beta_r [1, g]_{k+1}^{(r,1)}\} \text{ and } \frac{1}{\|v\|} \sum_{r=1}^{l} \min\{\beta_r \langle 1, g \rangle_k^{(r,1)}, \beta_r [1, g]_{k+1}^{(r,1)}\},$$

as estimates of upper and lower bounds of $u^H g(A) v$.

Chapter Nine

Solving Secular Equations

9.1 Examples of Secular Equations

What are secular equations? The term "secular" comes from the latin "saecularis" which is related to "saeculum", which means "century". So secular refers to something that is done or happens every century. It is also used to refer to something that is several centuries old. It appeared in mathematics to denote equations related to the motion of planets and celestial mechanics. For instance, it appears in the title of a 1829 paper of A. L. Cauchy (1789–1857) *"Sur l'équation à l'aide de laquelle on détermine les inégalités séculaires des mouvements des planètes"* (Oeuvres Complètes (IIème Série), v 9 (1891), pp 174–195). There is also a paper by J. J. Sylvester (1814–1897) whose title is *"On the equation to the secular inequalities in the planetary theory"* (Phil. Mag., v 5 n 16 (1883), pp 267-269).

In modern applied mathematics, the term "secular" is used to refer to equations that involve matrices like the inverse of $A - \lambda I$ or powers of the inverse where λ is a real number. Let us now consider a few examples.

9.1.1 Eigenvalues of a Tridiagonal Matrix

Several methods reduce the problem of computing the eigenvalues of a symmetric matrix A to the simpler problem of computing the eigenvalues of a (sequence of) symmetric tridiagonal matrices J_k. A well-known example is the Lanczos algorithm, which generates such a matrix at each iteration. In fact, a new row and a new column are appended at each iteration to the previous matrix. We can consider what happens when we go from the step k (with a matrix J_k whose eigenvalues we assume we know) to step $k + 1$. We look for an eigenvalue λ and an eigenvector $x = \begin{pmatrix} y & \zeta \end{pmatrix}^T$ of J_{k+1} where y is a vector of dimension k and ζ is a real number. This gives the two equations

$$J_k y + \eta_k \zeta e^k = \lambda y,$$

$$\eta_k y_k + \alpha_{k+1} \zeta = \lambda \zeta,$$

where y_k is the last component of y, α_j, $j = 1, \ldots, k + 1$ are the diagonal entries of J_{k+1} and η_j, $j = 1, \ldots, k$ are the entries on the subdiagonal. By eliminating the vector y from these two equations we have

$$(\alpha_{k+1} - \eta_k^2 ((e^k)^T (J_k - \lambda I)^{-1} e^k) \zeta = \lambda \zeta.$$

This equation shows why solving secular equations is related to the main topic of this book, since $(e^k)^T (J_k - \lambda I)^{-1} e^k$ is a quadratic form. We can divide by

ζ if it is nonzero. Otherwise, λ is an eigenvalue of J_k, but this is impossible if J_{k+1} is unreduced. By using the spectral decomposition of J_k we obtain that the eigenvalues of J_{k+1} are solutions of the following "secular equation" for λ:

$$\alpha_{k+1} - \eta_k^2 \sum_{j=1}^{k} \frac{(\xi_j)^2}{\theta_j - \lambda} = \lambda,$$

where $\xi_j = z_k^j$ is the kth (i.e., last) component of the jth eigenvector of J_k and the θ_j's are the eigenvalues of J_k, which are called the Ritz values. Therefore, to obtain the eigenvalues of J_{k+1} from those of J_k we have to solve

$$f(\lambda) = \lambda - \alpha_{k+1} + \eta_k^2 \sum_{j=1}^{k} \frac{\xi_j^2}{\theta_j - \lambda} = 0. \tag{9.1}$$

The secular function f has poles at the eigenvalues (Ritz values) of J_k for $\lambda = \theta_j = \theta_j^{(k)}, j = 1 \dots, k$. We easily see that f is a strictly increasing function between two consecutive poles. There is only one zero of f in each interval between poles. An example with four poles ($k = 4$) is displayed in figure 9.1. In this small example, to obtain the eigenvalues of J_{k+1} we have to compute the five zeros of f in equation (9.1). The figure illustrates also the interlacing property of the eigenvalues of J_k and J_{k+1} which is known as the Cauchy interlacing theorem; see chapter 3. The zeros that we wish to compute will be the poles of the secular function for the next Lanczos iteration. This illustrates the convergence of the Ritz values toward the eigenvalues of A; see [239].

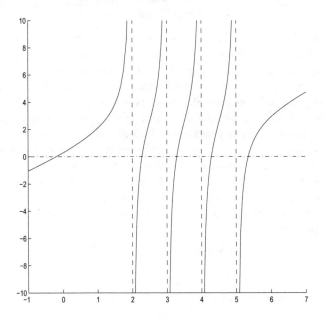

Figure 9.1 Example of secular function (9.1) with $k = 4$

9.1.2 Modification by a Rank-One Matrix

Assume that we know the eigenvalues of a matrix A and we would like to compute the eigenvalues of a rank-one modification of A. Therefore, we have

$$Ax = \lambda x,$$

where we assume that we know the eigenvalues λ and we want to compute μ such that

$$(A + cc^T)y = \mu y,$$

where c is a given vector (not orthogonal to an eigenvector of A). Clearly μ is not an eigenvalue of A. Therefore $A - \mu I$ is nonsingular and we can obtain an equation for μ by writing

$$y = -(A - \mu I)^{-1}cc^T y.$$

Multiplying by c^T to the left, we have

$$c^T y = -c^T (A - \mu I)^{-1}cc^T y.$$

The vector c is not orthogonal to the eigenvector y since otherwise μ is an eigenvalue of A and y is an eigenvector, but this is impossible with our hypothesis on c. Thus we can divide by $c^T y \neq 0$ and we obtain the secular equation for μ,

$$1 + c^T (A - \mu I)^{-1}c = 0.$$

Using the spectral decomposition of $A = Q\Lambda Q^T$ with Q orthogonal and Λ diagonal and $z = Q^T c$, we have

$$1 + \sum_{j=1}^{n} \frac{(z_j)^2}{\lambda_j - \mu} = 0, \tag{9.2}$$

where λ_j are the eigenvalues of A. For eigenvalues after a rank-one modification of the matrix, see Bunch, Nielsen and Sorensen [45]. This situation arises, for instance, in the divide and conquer method for computing the eigenvalues of a tridiagonal matrix. This method was introduced by Cuppen [71], and through the work of Dongarra and Sorensen [89] found its way to being one of the tridiagonal eigensolvers of LAPACK. The matrix is split into two pieces. Knowing the eigenvalues of the two parts, the eigenvalues of the full matrix are recovered by solving a secular equation. This splitting is done recursively until the matrices have a size small enough to use efficiently another method (QR for instance). This method is especially useful on parallel computers. To obtain the solution at a given step, the matrix whose eigenvalues μ_i are sought is $D + \rho cc^T$, where D is a diagonal matrix with diagonal elements d_j and ρ is a real number. Then we have the following interlacing property:

$$d_1 \leq \mu_1 \leq d_2 \leq \cdots \leq d_n \leq \mu_n \text{ if } \rho > 0,$$

$$\mu_1 \leq d_1 \leq \mu_2 \leq \cdots \leq d_{n-1} \leq \mu_n \leq d_n \text{ if } \rho < 0.$$

The secular equation to be solved is

$$f(\mu) = 1 + \rho \sum_{j=1}^{n} \frac{(c_j)^2}{d_j - \mu} = 0. \tag{9.3}$$

For $\rho > 0$, the function f is increasing between the poles. An example of a secular function with $\rho = 1$, $d = [2\ 3\ 4\ 5]^T$ and $c = [1\ 1\ 1\ 1]^T$ is given in figure 9.2. With the same data for d and c but with $\rho = -1$ we have figure 9.3. The function f is then decreasing in each interval between poles. In this example, there are four zeros to be computed.

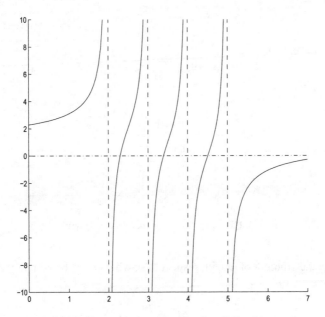

Figure 9.2 Example of secular function (9.3) with $\rho > 0$

When there is a rank-k change to A, we seek μ such that

$$(A + CC^T)y = \mu y,$$

where the matrix C is $n \times k$. The secular equation for μ is

$$\det(I + C^T(A - \mu I)^{-1}) = 0.$$

On this topic, see also Arbenz and Golub [8].

9.1.3 Constrained Eigenvalue Problem

We wish to find a vector x of norm one which is the solution of

$$\max_x x^T A x,$$

satisfying the constraint $c^T x = 0$ where c is a given vector. We introduce a functional φ with two Lagrange multipliers λ and μ corresponding to the two constraints,

$$\varphi(x, \lambda, \mu) = x^T A x - \lambda(x^T x - 1) + 2\mu x^T c.$$

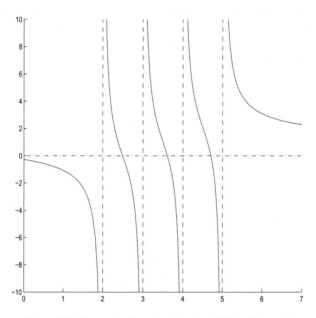

Figure 9.3 Example of secular function (9.3) with $\rho < 0$

Computing the gradient of φ with respect to x, which must be zero at the solution, we find the equation

$$Ax - \lambda x + \mu c = 0,$$

from which we have $x = -\mu(A - \lambda I)^{-1}c$. If λ is not an eigenvalue of A ($\mu \neq 0$) and using the constraint $c^T x = 0$ we have the secular equation

$$c^T(A - \lambda I)^{-1}c = 0. \tag{9.4}$$

Using the spectral decomposition of $A = Q\Lambda Q^T$ and $d = Q^T c$, the secular equation is

$$f(\lambda) = \sum_{j=1}^{n} \frac{d_j^2}{\lambda_j - \lambda} = 0. \tag{9.5}$$

The function f for this type of problem is shown in figure 9.4. There are $n - 1$ solutions to the secular equation. When we have the values of λ that are solutions of equation (9.5), we use the constraint $x^T x = 1$ to remark that

$$x^T x = \mu^2 c^T(A - \lambda I)^{-2}c = 1.$$

Therefore,

$$\mu^2 = \frac{1}{c^T(A - \lambda I)^{-2}c}$$

and

$$x = -\mu(A - \lambda I)^{-1}c.$$

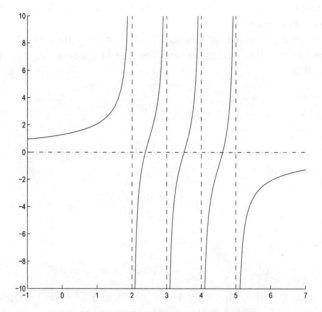

Figure 9.4 Example of secular function (9.5)

This problem can be generalized by replacing the only constraint $c^T x = 0$ with a set of m constraints $N^T x = t$ where N is a rectangular matrix of dimension $n \times m$ with $m < n$ and t is a given vector of dimension m. This has been considered by Gander, Golub and von Matt [121]. The matrix N is assumed to be of full rank, that is, m. The matrix N and the vector t must satisfied $\|(N^T)^\dagger t\| < 1$ where $(N^T)^\dagger$ is the pseudoinverse. Otherwise, the constraint of x being of norm one cannot be satisfied. If $\|(N^T)^\dagger t\| = 1$, the solution is $x = (N^T)^\dagger t$.

Gander, Golub and von Matt simplified the problem by using a QR factorization of the matrix N. Let P be an orthogonal matrix and R be an upper triangular matrix of order m such that

$$P^T N = \begin{pmatrix} R \\ 0 \end{pmatrix}.$$

Then, if we denote

$$P^T A P = \begin{pmatrix} B & \Gamma^T \\ \Gamma & C \end{pmatrix}, \quad P^T x = \begin{pmatrix} y \\ z \end{pmatrix},$$

we have

$$x^T A x = y^T B y + 2 z^T \Gamma y + z^T C z, \quad R^T y = t, \quad y^T y + z^T z = 1.$$

This implies $y = R^{-T} t$. Denoting $\alpha^2 = 1 - y^T y$ and $b = -\Gamma y$, the problem reduces to

$$\min_z z^T C z - 2 b^T z,$$

with the constraint $z^T z = \alpha^2$. This leads to an eigenvalue problem with a quadratic constraint (see next section).

Using the same technique for the simpler previous maximization problem with the constraint $c^T x = 0$ (c being a vector), one can find a Householder transformation H such that

$$
H^T c = \begin{pmatrix} r \\ 0 \\ \vdots \\ 0 \end{pmatrix}.
$$

Then, using the matrix

$$
H^T A H = \begin{pmatrix} \beta & g^T \\ g & C \end{pmatrix},
$$

and $H^T x = (y \quad z)^T$ where y is a scalar, we have

$$
c^T x = (r \quad 0 \quad \cdots \quad 0) H^T x = r y = 0.
$$

Therefore, $y = 0$ and we have $x^T A x = z^T C z$ and $z^T z = 1$. The constraint $c^T x = 0$ has been eliminated and the problem reduced to

$$
\max_z z^T C z,
$$

with $z^T z = 1$. The solution is, of course, that z is the eigenvector of C (which is symmetric) associated with the largest eigenvalue. The solution x is recovered by

$$
x = H \begin{pmatrix} 0 \\ z \end{pmatrix}.
$$

Looking back at the secular equation (9.4), we see that it can be written as

$$
c^T (A - \lambda I)^{-1} c = c^T H \left[\begin{pmatrix} \beta & g^T \\ g & C \end{pmatrix} - \lambda I \right]^{-1} H^T c = 0.
$$

This implies that the $(1, 1)$ entry of

$$
\begin{pmatrix} \beta - \lambda & g^T \\ g & C - \lambda I \end{pmatrix}^{-1}
$$

must be zero. Looking for the solution of

$$
\begin{pmatrix} \beta - \lambda & g^T \\ g & C - \lambda I \end{pmatrix} \begin{pmatrix} w \\ z \end{pmatrix} = e^1,
$$

with the constraint that $w = 0$, we find that

$$
(C - \lambda I) z = 0, \quad g^T z = 1.
$$

Therefore, (as we already know), λ is an eigenvalue of C.

9.1.4 Eigenvalue Problem with a Quadratic Constraint

We consider the problem

$$\min_x x^T A x - 2 c^T x,$$

with the constraint $x^T x = \alpha^2$. Introducing a Lagrange multiplier and using the stationary values of the Lagrange functional, we have

$$(A - \lambda I) x = c, \quad x^T x = \alpha^2. \tag{9.6}$$

Let $A = Q \Lambda Q^T$ be the spectral decomposition of A. The Lagrange equations (9.6) can be written as

$$\Lambda Q^T x - \lambda Q^T x = Q^T c, \quad x^T Q Q^T x = \alpha^2.$$

Introducing $y = Q^T x$ and $d = Q^T c$, we have

$$\Lambda y - \lambda y = d, \quad y^T y = \alpha^2.$$

Assume for the sake of simplicity that all the eigenvalues λ_i of A are simple. If λ is equal to one of the eigenvalues, say λ_j, we must have $d_j = 0$. For all $i \neq j$ we have

$$y_i = \frac{d_i}{\lambda_i - \lambda}.$$

Then, there is a solution or not, whether we have

$$\sum_{i \neq j} \left(\frac{d_i}{\lambda_i - \lambda_j} \right)^2 = \alpha^2,$$

or not. If λ is not an eigenvalue of A, the inverse of $A - \lambda I$ exists and we obtain the secular equation

$$c^T (A - \lambda I)^{-2} c = \alpha^2.$$

Using the spectral decomposition of A this is written as

$$f(\lambda) = \sum_{i=1}^{n} \left(\frac{d_i}{\lambda_i - \lambda} \right)^2 - \alpha^2 = 0. \tag{9.7}$$

An example of such a secular function is displayed in figure 9.5. We see that, contrary to the previous cases and depending on the data, we may have two zeros in an interval between poles. However, for the problem under consideration we are only interested in the smallest zero of f. It is located left of the first pole (as in the example) if $d_1 \neq 0$.

On problems with quadratic constraints, see Gander [120] and Golub and von Matt [157].

9.2 Secular Equation Solvers

We have seen in the previous section that there exist several types of secular equations. In some situations we need to compute all the zeros of the secular function,

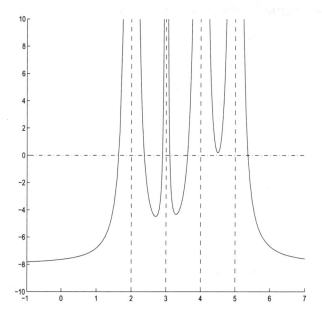

Figure 9.5 Example of secular function (9.7)

whereas in other cases we are interested in only one zero, generally the smallest or the largest.

The numerical solution of secular equations has been studied for a long time. A good summary of the techniques used to solve secular equations is given in Melman [233] where a numerical comparison of methods is provided; see also [232]. Melman considered solving the equation

$$1 + \rho \sum_{j=1}^{n} \frac{(c_j)^2}{d_j - \lambda} = 0, \tag{9.8}$$

with $\rho > 0$, corresponding to figure 9.2. When we look for the solution in the interval $]d_i, d_{i+1}[$, we make a change of variable $\lambda = d_i + \rho t$. Denoting $\delta_j = (d_j - d_i)/\rho$, the secular equation becomes

$$f(t) = 1 + \sum_{j=1}^{i} \frac{c_j^2}{\delta_j - t} + \sum_{j=i+1}^{n} \frac{c_j^2}{\delta_j - t} = 1 + \psi(t) + \phi(t) = 0. \tag{9.9}$$

Note that

$$\psi(t) = \sum_{j=1}^{i-1} \frac{c_j^2}{\delta_j - t} - \frac{c_i^2}{t},$$

since $\delta_i = 0$. The function ψ has poles $\delta_1, \ldots, \delta_{i-1}, 0$ with $\delta_j < 0$, $j = 1, \ldots, i - 1$. The solution is sought in the interval $]0, \delta_{i+1}[$ with $\delta_{i+1} > 0$. For the sake of simplicity, let us denote $\delta = \delta_{i+1}$. In $]0, \delta[$ we have $\psi(t) < 0$ and $\phi(t) > 0$.

There are two cases depending on whether or not we know all the poles of the secular function f. Let us assume that we are in the first case, so we are able to compute values of ψ and ϕ and their derivatives. It is generally agreed that the Newton method is not well suited to obtaining the solution of secular equations, although this conclusion depends on the choice of the starting point and the choice of the function to which we apply the algorithm; for instance see Reinsch [281], [282], who used the Newton method on $1/f$. Bunch, Nielsen and Sorensen [45] interpolated ψ to first order by a rational function $p/(q-t)$ and ϕ by $r + s/(\delta - t)$. This is called osculatory interpolation. The parameters p, q, r, s are determined by matching the exact values of the function and the first derivative of ψ or ϕ at some given point $\bar{t}$ (to the right of the exact solution) where f has a negative value. These parameters are given by

$$q = \bar{t} + \psi(\bar{t})/\psi'(\bar{t}),$$

$$p = \psi(\bar{t})^2/\psi'(\bar{t}),$$

$$r = \phi(\bar{t}) - (\delta - \bar{t})\phi'(\bar{t}),$$

$$s = (\delta - \bar{t})^2\phi'(\bar{t}).$$

For computing them we have

$$\psi'(t) = \sum_{j=1}^{i} \frac{c_j^2}{(\delta_j - t)^2}$$

and

$$\phi'(t) = \sum_{j=i+1}^{n} \frac{c_j^2}{(\delta_j - t)^2}.$$

Then the new iterate is obtained by solving the quadratic equation

$$1 + \frac{p}{q - t} + r + \frac{s}{\delta - t} = 0.$$

This equation has two roots, of which only one is of interest. This method is called BNS1 in Melman [45] and "approaching from the left" in Li [226]. Provided the initial guess is carefully chosen, it gives a monotonically increasing sequence which converges quadratically to the solution.

Let us consider the example of figure 9.2. It was obtained with $\rho = 1$, $d = [2, 3, 4, 5]$ and $c = [1, 1, 1, 1]$. We would like to compute the zero in the second interval $]2, 3[$. Therefore, after the change of variable, the interval of interest is $]0, 1[$. Figure 9.6 shows ψ (solid) and ϕ (dots) as a function of t. It is interesting to compare these functions with their interpolants. In this case, the interpolation of ψ is almost exact. Whatever is $\bar{t}$ in $]0, \delta[$ we find $q = 0$ up to machine precision and almost $p = c_1^2$. When the interpolation point is varied in $]0, \delta[$ (over 60 equidistant sample points) the minimum relative difference between the function ψ and its interpolant $\bar{\psi}$ (computed as $\|\psi - \bar{\psi}\|/\|\psi\|$) is 0 and the maximum is $1.11 \ 10^{-11}$. The results are less satisfactory for the function ϕ. When the interpolation point

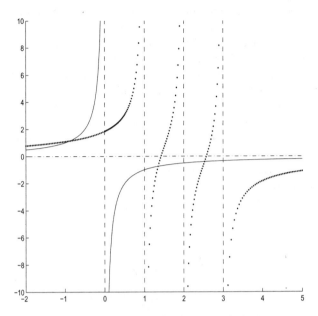

Figure 9.6 Functions ψ (solid) and ϕ (dots)

is varied, the minimum of the relative difference is 0.0296 and the maximum is 0.2237.

To obtain an efficient method it remains to find an initial guess and to solve the quadratic equation giving the next iterate. For the starting point, we follow Melman's strategy of looking for a zero of an interpolant of the function $tf(t)$. When seeking the ith root, this function is written as

$$tf(t) = t\left(1 - \frac{c_i^2}{t} + \frac{c_{i+1}^2}{\delta - t} + h(t)\right).$$

The function h is written in two parts $h = h_1 + h_2$,

$$h_1(t) = \sum_{j=1}^{i-1} \frac{c_j^2}{\delta_j - t}, \quad h_2(t) = \sum_{j=i+2}^{n} \frac{c_j^2}{\delta_j - t}.$$

The functions h_1 and h_2 do not have poles in the interval of concern. They are interpolated as in BNS1 by rational functions of the form $p/(q - t)$. The parameters are found by interpolating the function values at points 0 and δ. This gives a function $\bar{h} = \bar{h}_1 + \bar{h}_2$. The starting point is obtained by computing the zero (in the interval $]0, \delta[$) of the function

$$t\left(1 - \frac{c_i^2}{t} + \frac{c_{i+1}^2}{\delta - t} + \bar{h}(t)\right).$$

This can be done with a standard zero finder. In our example $\bar{h}_1 \equiv 0$. The computed starting point t_s (translated back to $]2, 3[$) is 2.2958817871588815. Denoting the iterates by t^ν, it takes three iterations to reach $|t^\nu - t^{\nu-1}|/|t^{\nu-1}| \leq 10^{-10}$. The computed zero is 2.2960896453121185 and the value of the function f is $-4.4409\ 10^{-16}$, which is generally the best we can hope for since this is the round-off level.

In fact, with such a good starting point the Newton iteration also converges in three iterations, the computed solution being the same. However, if we take as a starting point $2 + 0.1(t_s - 2)$, the Newton method takes nine iterations while BNS1 needs only four iterations. This shows that the general belief about the Newton method depends very much on the starting point. However, the methods using osculatory interpolation are much less sensitive to the starting point than the Newton method for the function f.

Interpolating ψ by $\bar{r} + \bar{s}/t$ and ϕ by $\bar{p}/(\bar{q} - t)$, one obtains another method called BNS2 or "approaching from the right". Then,

$$\bar{r} = \psi(\bar{t}) - (\delta - \bar{t})\psi'(\bar{t}),$$

$$\bar{s} = (\delta - \bar{t})^2\psi'(\bar{t}),$$

$$\bar{q} = \bar{t} + \phi(\bar{t})/\phi'(\bar{t}),$$

$$\bar{p} = \phi(\bar{t})^2/\phi'(\bar{t}).$$

For our small example, the interpolation of ψ is done relative to the pole 0. When the interpolation point is varied in $]0, \delta[$ (over 60 equidistant sample points) the minimum relative difference between the function ψ and its interpolant $\tilde{\psi}$ is 0 and the maximum is $2.4614\ 10^{-19}$. For the function ϕ the results are worse; the minimum of the relative difference is 0.08699 and the maximum is 0.51767.

It takes three iterations for BNS2 to reach the stopping criteria. The computed zero is 2.2960896453121186 (the last decimal digit is different from the BNS1 solution) and the value of the function f is $1.3323\ 10^{-15}$. In fact, the starting point is not really adapted to this method since it is smaller than the root. The first iterate jumps to the other side of the root and then the iterates decrease monotonically. This is also why the value of the function at the approximate root is positive. If we take as a starting point $2 + 1.2(t_s - 2)$ the Newton method takes five iterations while BNS2 needs four iterations.

R. C. Li [226] proposed to use $r + s/(\delta - t)$ to interpolate both functions ψ and ϕ. He called this method "the middle way" (MW). In fact, we interpolate ψ, which has a pole at 0 by $\bar{r} + \bar{s}/t$ and ϕ, which has a pole at δ by $r + s/(\delta - t)$. This method is not monotonic, but has quadratic convergence.

It takes three iterations for "the middle way" to reach the stopping criteria. The computed zero is 2.2960896453121185 (the same as the BNS1 solution) and the value of the function f is $-4.4409\ 10^{-16}$.

The fixed weight method FW1 fixes the coefficient for the pole at 0. Hence, f is interpolated to first order by

$$r - \xi_1^2/t + s/(\delta - t)$$

at a point $\bar{t}$; ξ_1 is the coefficient corresponding to the pole 0, which is c_2 in our example. Therefore,

$$r = f(\bar{t}) - (\delta - \bar{t})f'(\bar{t}) + \frac{\xi_1^2\delta}{\bar{t}^2},$$

$$s = (\delta - \bar{t})^2 \left[f'(\bar{t}) - \frac{\xi_1^2}{\bar{t}^2} \right].$$

The next iterate is found by computing the root of the interpolant located in $]0, \delta[$. The convergence is quadratic. It takes three iterations for the FW1 method to reach the stopping criteria. The computed zero is 2.2960896453121186 and the value of the function f is $1.3323 \, 10^{-15}$.

The method FW2 fixes the coefficient for the pole at δ. The interpolant of f is

$$\bar{r} + \bar{s}/t + \xi_2^2/(\delta - t),$$

where ξ_2 is the coefficient corresponding to the pole δ which is c_3 in our example. The parameters are given by

$$\bar{r} = f + tf' - \frac{\xi_2^2 \delta}{(\delta - \bar{t})^2},$$

$$\bar{s} = -t^2 \left[f' - \frac{\xi_2^2}{(\delta - \bar{t})^2} \right].$$

The convergence is quadratic. It takes three iterations for the FW2 method to reach the stopping criteria. The computed zero is 2.2960896453121185 and the value of the function f is (by chance) 0.

The Gragg method (GR) interpolates f at $\bar{t}$ to second order with the function

$$a + b/t + c/(\delta - t).$$

This gives

$$c = \frac{(\delta - \bar{t})^3}{\delta} f' + \frac{\bar{t}(\delta - \bar{t})^3}{2\delta} f'',$$

$$b = \frac{\bar{t}^3}{2\delta} [f''(\delta - \bar{t}) - 2f'],$$

$$a = f - \frac{f''\bar{t}}{2}(\delta - \bar{t}) + (2\bar{t} - \delta)f'.$$

Then we solve the quadratic equation

$$at^2 - t(a\delta - b + c) - b\delta = 0.$$

The convergence is cubic. It takes three iterations for the Gragg method to reach the stopping criteria. The computed zero is 2.2960896453121186 and the value of the function f is $1.7764 \, 10^{-15}$.

9.3 Numerical Experiments

Let us do more numerical experiments. We consider an example inspired by one used by Melman. The dimension is $n = 4$ and $d = [1, 1 + \beta, 3, 4]$. Defining $v^T = [\gamma, \omega, 1, 1]$, we have $c^T = v^T / \|v\|$. Let us choose $\beta = 1$, but a small weight $\gamma = 10^{-2}$ and $\omega = 1$. Results for the methods we have reviewed before are

Table 9.1 $\beta = 1, \gamma = 10^{-2}, \omega = 1$

Method	No. it.	Root−1	$f(\text{Root})$
BNS1	2	$2.068910657999406\ 10^{-5}$	$-4.2294\ 10^{-12}$
BNS2	2	$2.068910658015177\ 10^{-5}$	$8.0522\ 10^{-12}$
MW	2	$2.068910657999406\ 10^{-5}$	$-4.2294\ 10^{-12}$
FW1	2	$2.068910657989172\ 10^{-5}$	$-1.2199\ 10^{-11}$
FW2	2	$2.068910658007030\ 10^{-5}$	$1.7082\ 10^{-12}$
GR	2	$2.068910657999392\ 10^{-5}$	$-4.2398\ 10^{-12}$

Table 9.2 $\beta = 1, \gamma = 1, \omega = 10^{-2}$

Method	No. it.	Root−1	$f(\text{Root})$
BNS1	3	$2.539918603315181\ 10^{-1}$	$-3.3307\ 10^{-16}$
BNS2	2	$2.539918603315182\ 10^{-1}$	$-1.1102\ 10^{-16}$
MW	3	$2.539918603315181\ 10^{-1}$	$-3.3307\ 10^{-16}$
FW1	3	$2.539918603315182\ 10^{-1}$	$3.3307\ 10^{-16}$
FW2	3	$2.539918603315181\ 10^{-1}$	$-3.3307\ 10^{-16}$
GR	3	$2.539918603315181\ 10^{-1}$	$-3.3307\ 10^{-16}$

Table 9.3 $\beta = 10^{-2}, \gamma = 1, \omega = 1$

Method	No. it.	Root−1	$f(\text{Root})$
BNS1	2	$4.939569815595898\ 10^{-3}$	$7.8160\ 10^{-14}$
BNS2	2	$4.939569815595901\ 10^{-3}$	$1.4921\ 10^{-13}$
MW	2	$4.939569815595898\ 10^{-3}$	$7.8160\ 10^{-14}$
FW1	2	$4.939569815595903\ 10^{-3}$	$1.8474\ 10^{-13}$
FW2	2	$4.939569815595914\ 10^{-3}$	$4.0501\ 10^{-13}$
GR	2	$4.939569815595873\ 10^{-3}$	$-4.0501\ 10^{-13}$

given in table 9.1 for computing the root in $]1, 1 + \beta[$. Note that the root is close to 1 and the values of the function at the solution are much larger than in the previous example, even though they are still acceptable.

In the results of table 9.2 we use $\gamma = 1$ and a small weight $\omega = 10^{-2}$. The values of the function at the approximate solution are smaller. Then, for the results in table 9.3, we use a small interval with $\beta = 10^{-2}$ and unit weights $\gamma = 1, \omega = 1$.

For these examples, there is not much difference between the methods and the numbers of iterations are quite small. This is mainly due to the good choice of the starting point proposed by Melman.

PART 2
Applications

Chapter Ten

Examples of Gauss Quadrature Rules

Until the 1960s, quadrature rules were computed by hand or using desk calculators and published in the form of tables (see, for instance, the list given in [320]) giving the nodes and weights for a given degree of approximation. It was a great improvement when some software appeared allowing the computation of nodes and weights. This can be done by brute force solving systems of nonlinear equations; see, for instance, the Fortran codes and the tables published in the book of Stroud and Secrest [320]. However, the method of choice today is to compute the nodes and weights using the Jacobi matrix corresponding to the orthogonal polynomials associated with the given measure and interval of integration. In this chapter we give some examples of computation of Gauss quadrature rules. We compare numerically several methods to obtain the nodes and weights. We also give examples of computation of integrals as well as examples of the use of modification algorithms for the given measure.

10.1 The Golub and Welsch Approach

The result saying that the nodes are given by the eigenvalues of the Jacobi matrix and the weights are given by the squares of the first components of the normalized eigenvectors was already known at the beginning of the 1960s; see Goertzel, Waldinger and Agresta [137] or Wilf [349]. Golub and Welsch [160] used these results and devised an algorithm based on QR with a Wilkinson-like shift to compute the nodes and weights. It is constructed in such a way that only the first components of the eigenvectors are computed. Moreover, they published Algol procedures implementing their algorithm. We translated these procedures to Matlab and computed some n-point Gauss quadrature rules.

For classical orthogonal polynomials, the coefficients of the recurrences are explicitly known; see for instance Gautschi [131], Laurie [221] or chapter 2. Table 10.1 gives the results for the Legendre weight function equal to 1 on $[-1, 1]$ using the computed values of the coefficients of the Jacobi matrix, which are known explicitly; see chapter 2. Of course, today such a table has some interest only to compare different numerical methods or to look at the properties of the nodes and weights. It is more useful to have a good routine to compute the quadrature rule. We see that, up to the precision of the computation, the nodes and weights are symmetric around 0. Table 10.2 shows the nodes and weights for a 10-point Gauss rule using the Chebyshev weight function. Obviously, the (constant) weights are π/n which is $3.141592653589793 \ 10^{-1}$ when rounded to 16 decimal figures. The results are correct up to the last three decimal digits. The nodes are symmetric around 0 up to the last two decimal digits.

Table 10.1 Legendre weight function, $n = 10$, Golub and Welsch

Nodes	Weights
$-9.739065285171721 \ 10^{-1}$	$6.667134430868844 \ 10^{-2}$
$-8.650633666889848 \ 10^{-1}$	$1.494513491505808 \ 10^{-1}$
$-6.794095682990242 \ 10^{-1}$	$2.190863625159823 \ 10^{-1}$
$-4.333953941292464 \ 10^{-1}$	$2.692667193099961 \ 10^{-1}$
$-1.488743389816314 \ 10^{-1}$	$2.955242247147535 \ 10^{-1}$
$1.488743389816312 \ 10^{-1}$	$2.955242247147525 \ 10^{-1}$
$4.333953941292474 \ 10^{-1}$	$2.692667193099962 \ 10^{-1}$
$6.794095682990244 \ 10^{-1}$	$2.190863625159821 \ 10^{-1}$
$8.650633666889842 \ 10^{-1}$	$1.494513491505805 \ 10^{-1}$
$9.739065285171717 \ 10^{-1}$	$6.667134430868807 \ 10^{-2}$

Table 10.2 Chebyshev weight function of the first kind on $[-1, 1]$, $n = 10$, Golub and Welsch

Nodes	Weights
$-9.876883405951373 \ 10^{-1}$	$3.141592653589779 \ 10^{-1}$
$-8.910065241883682 \ 10^{-1}$	$3.141592653589815 \ 10^{-1}$
$-7.071067811865478 \ 10^{-1}$	$3.141592653589777 \ 10^{-1}$
$-4.539904997395468 \ 10^{-1}$	$3.141592653589801 \ 10^{-1}$
$-1.564344650402311 \ 10^{-1}$	$3.141592653589792 \ 10^{-1}$
$1.564344650402312 \ 10^{-1}$	$3.141592653589789 \ 10^{-1}$
$4.539904997395469 \ 10^{-1}$	$3.141592653589798 \ 10^{-1}$
$7.071067811865472 \ 10^{-1}$	$3.141592653589789 \ 10^{-1}$
$8.910065241883679 \ 10^{-1}$	$3.141592653589789 \ 10^{-1}$
$9.876883405951381 \ 10^{-1}$	$3.141592653589803 \ 10^{-1}$

10.2 Comparisons with Tables

In the Golub and Welsch paper [160], comparisons were made with tabulated results in Concus et al. [69]. Let us do the same comparisons with our Matlab implementation. They are done for the generalized Laguerre weight function $\lambda^\alpha e^{-\lambda}$ with $\alpha = -0.75$ on the interval $[0, \infty)$. Table 10.3 gives the results of the Golub and Welsch algorithm and table 10.4 shows the relative differences with the tabulated results as reported in [160]. We observe that the errors are slightly smaller than in the Golub and Welsch paper. This is due to the use of IEEE arithmetic. We have also implemented a variable-precision version of the Golub and Welsch algorithm. The computations in extended precision show that the Concus et al. results are accurate to the two last decimal digits; in fact, the nodes are more accurate than the weights. The relative errors of the double-precision Golub and Welsch algorithm with the variable-precision results using 32 decimal digits are in table 10.5. The error on the seventh weight is smaller than with the Concus et al. results. Otherwise, the errors are of the same magnitude.

Table 10.3 Generalized Laguerre weight function with $\alpha = -0.75$ on $[0, \infty)$, $n = 10$, Golub and Welsch

Nodes	Weights
$2.766655867079714 \ 10^{-2}$	2.566765557790772
$4.547844226059476 \ 10^{-1}$	$7.733479703443403 \ 10^{-1}$
1.382425761158596	$2.331328349732204 \ 10^{-1}$
2.833980012092694	$4.643674708956692 \ 10^{-2}$
4.850971448764913	$5.549123502036255 \ 10^{-3}$
7.500010942642828	$3.656466626776365 \ 10^{-4}$
$1.088840802383440 \ 10^{1}$	$1.186879857102432 \ 10^{-5}$
$1.519947804423760 \ 10^{1}$	$1.584410942056775 \ 10^{-7}$
$2.078921462107011 \ 10^{1}$	$6.193266726796800 \ 10^{-10}$
$2.857306016492211 \ 10^{1}$	$3.037759926517505 \ 10^{-13}$

Table 10.4 Generalized Laguerre weight function with $\alpha = -0.75$ on $[0, \infty)$, $n = 10$, Golub and Welsch, comparison with Concus et al. [69]

Relative error on nodes	Relative error on weights
$2.257239358109354 \ 10^{-15}$	$1.730151039708872 \ 10^{-16}$
$3.051509048681502 \ 10^{-15}$	$8.613636297234880 \ 10^{-16}$
$1.927434610931477 \ 10^{-15}$	$6.190847943742033 \ 10^{-15}$
$1.253612821417562 \ 10^{-15}$	$1.793121440790894 \ 10^{-15}$
$1.830928978001428 \ 10^{-16}$	$9.378364756201140 \ 10^{-16}$
$4.736944660441535 \ 10^{-16}$	$4.151229031776791 \ 10^{-15}$
$3.262840326174143 \ 10^{-16}$	$1.541513368114529 \ 10^{-14}$
$4.674783789891258 \ 10^{-16}$	$3.174213198624942 \ 10^{-15}$
0	$6.511112251549391 \ 10^{-15}$
$1.243378783474516 \ 10^{-16}$	$1.661984460767555 \ 10^{-15}$

Another possibility is to compute the moments which are known for the Laguerre polynomials and then compute the recurrence coefficients by the Gautschi algorithm described in chapter 5 based on the factorization of the Hankel matrix. The results are given in table 10.6. We see that the errors using the moments are much larger than when computing directly from the (known) recurrence coefficients.

10.3 Using the Full QR Algorithm

The Golub and Welsch paper [160] uses the QR algorithm to compute the eigenvalues of the Jacobi matrix. However, it computes only the first component of the eigenvectors to save computational time. Of course, we can use the standard Matlab eigenvalue solver, the price to pay being having to compute all the components

Table 10.5 Generalized Laguerre weight function with $\alpha = -0.75$ on $[0, \infty)$, $n = 10$, Golub and Welsch, comparison with variable precision (32 digits)

Relative error on nodes	Relative error on weights
$3.887467783410548 \ 10^{-15}$	$1.730151039708872 \ 10^{-16}$
$2.075026153103423 \ 10^{-15}$	$7.178030247695734 \ 10^{-16}$
$1.766815060020521 \ 10^{-15}$	$5.000300262253175 \ 10^{-15}$
$1.253612821417562 \ 10^{-15}$	$4.034523241779535 \ 10^{-15}$
$1.830928978001428 \ 10^{-16}$	$1.563060792700191 \ 10^{-15}$
$4.736944660441535 \ 10^{-16}$	$4.892519930308358 \ 10^{-15}$
$3.262840326174143 \ 10^{-16}$	$2.283723508317851 \ 10^{-15}$
$4.674783789891258 \ 10^{-16}$	$3.174213198624942 \ 10^{-15}$
0	$6.678063847742964 \ 10^{-15}$
$1.243378783474516 \ 10^{-16}$	$2.326778245074579 \ 10^{-15}$

Table 10.6 Generalized Laguerre weight function with $\alpha = -0.75$ on $[0, \infty)$, $n = 10$, computation from the moments, comparison with Concus et al. [69]

Relative error on nodes	Relative error on weights
$3.090815280665558 \ 10^{-10}$	$6.855827303908786 \ 10^{-11}$
$3.062097785080427 \ 10^{-10}$	$6.615014267187470 \ 10^{-11}$
$2.978726514140397 \ 10^{-10}$	$3.513227631926626 \ 10^{-10}$
$2.817927312559359 \ 10^{-10}$	$7.611583847316584 \ 10^{-10}$
$2.583118544459888 \ 10^{-10}$	$1.243582733322137 \ 10^{-9}$
$2.302667879234079 \ 10^{-10}$	$1.742482685575520 \ 10^{-9}$
$2.008898160422158 \ 10^{-10}$	$2.220646922285363 \ 10^{-9}$
$1.722976880568588 \ 10^{-10}$	$2.662697763114570 \ 10^{-9}$
$1.452306468713077 \ 10^{-10}$	$3.068116897502631 \ 10^{-9}$
$1.189059294560865 \ 10^{-10}$	$3.447605939952962 \ 10^{-9}$

of the eigenvectors even though we just need the first ones. This is what is done is the OPQ package from Gautschi [132]. For the generalized Laguerre polynomials the results are given in table 10.7. They are only slightly better than the ones using the Golub and Welsch algorithm.

Table 10.7 Generalized Laguerre weight function with $\alpha = -0.75$ on $[0, \infty)$, $n = 10$, Matlab 6 solver, comparison with Concus et al. [69]

Relative error on nodes	Relative error on weights
0	$1.730151039708872 \ 10^{-16}$
$9.764828955780807 \ 10^{-16}$	0
$1.606195509109565 \ 10^{-16}$	$1.547711985935508 \ 10^{-15}$
$3.134032053543904 \ 10^{-16}$	$2.988535734651490 \ 10^{-16}$
$1.830928978001428 \ 10^{-16}$	$2.344591189050285 \ 10^{-15}$
$4.736944660441535 \ 10^{-16}$	$3.409938133245221 \ 10^{-15}$
0	$2.854654385397276 \ 10^{-15}$
0	0
$1.708921545886483 \ 10^{-16}$	$3.672935116258631 \ 10^{-15}$
$3.730136350423549 \ 10^{-16}$	$1.329587568614044 \ 10^{-15}$

10.4 Another Implementation of QR

In his paper [220], Laurie described another implementation of the QR algorithm to compute the nodes and weights. According to a suggestion from Parlett [266], this algorithm first computes the eigenvalues (nodes) and then computes the weights by using one QR sweep with the eigenvalues as "exact" shifts. When coding these suggestions, we obtain the results described in table 10.8. The relative errors are of the same magnitude as with the Golub and Welsch algorithm.

Table 10.8 Generalized Laguerre weight function with $\alpha = -0.75$ on $[0, \infty)$, $n = 10$, QRSWEEP from Laurie [220], comparison with Concus et al. [69]

Relative error on nodes	Relative error on weights
$9.028957432437417 \ 10^{-15}$	$2.076181247650646 \ 10^{-15}$
$1.952965791156162 \ 10^{-15}$	$4.163257543663525 \ 10^{-15}$
$1.606195509109565 \ 10^{-16}$	$2.738259667424361 \ 10^{-15}$
$4.701048080315856 \ 10^{-16}$	$5.379364322372682 \ 10^{-15}$
$3.661857956002857 \ 10^{-16}$	$4.220264140290514 \ 10^{-15}$
$2.368472330220767 \ 10^{-16}$	$5.189036289720990 \ 10^{-15}$
$3.262840326174143 \ 10^{-16}$	$1.341687561136720 \ 10^{-14}$
$2.337391894945629 \ 10^{-16}$	$3.508340903743356 \ 10^{-15}$
0	$7.011967040130113 \ 10^{-15}$
$1.243378783474516 \ 10^{-16}$	$4.985953382302666 \ 10^{-16}$

10.5 Using the QL Algorithm

Parlett [266] has advocated the use of the QL algorithm instead of QR. In this algorithm L is a lower triangular matrix and the Wilkinson shift is computed from the upper 2×2 block of the tridiagonal matrix which is deflated from the top. This algorithm has also been recommended by Laurie [220], since it allows an easy computation of the weights and moreover it can be "reversed" to compute the Jacobi matrix from the nodes and weights in the inverse problem.

Implementing QL and computing the weights at the same time as the eigenvalues, we obtain the results in table 10.9. Some errors on the weights are larger than, for instance, with the QR Matlab implementation.

Table 10.9 Generalized Laguerre weight function with $\alpha = -0.75$ on $[0, \infty)$, $n = 10$, QLSWEEP from Laurie [220], comparison with Concus et al. [69]

Relative error on nodes	Relative error on weights
$6.019304954958278 \; 10^{-15}$	$1.730151039708872 \; 10^{-16}$
$1.220603619472601 \; 10^{-15}$	$5.742424198156587 \; 10^{-16}$
$3.212391018219129 \; 10^{-16}$	$2.857314435573246 \; 10^{-15}$
$4.701048080315856 \; 10^{-16}$	$4.482803601977235 \; 10^{-15}$
$3.661857956002857 \; 10^{-16}$	$5.001794536640609 \; 10^{-15}$
$8.289653155772686 \; 10^{-16}$	$5.633810828839931 \; 10^{-15}$
$8.157100815435358 \; 10^{-16}$	$1.570059911968502 \; 10^{-14}$
$1.168695947472815 \; 10^{-16}$	$3.341277051184149 \; 10^{-15}$
$5.126764637659450 \; 10^{-16}$	$4.922233910575146 \; 10^{-12}$
$3.730136350423549 \; 10^{-16}$	$8.476120749914532 \; 10^{-15}$

As with QR, the weights can be computed at the end of the process by a QL sweep using the eigenvalues (nodes) as perfect shifts. The interest of the QL algorithm is that it turns out to be necessary only to have the value of the cosine of the last rotation to compute the first element of the eigenvector. It is not necessary to apply all the rotations to the vector of weights. The errors for the weights are in table 10.10. We see that they are better than when computing the weights simultaneously with the nodes. However, the main conclusion, at least for this example, is that there is not much difference between the different implementations.

10.6 Gauss–Radau Quadrature Rules

To obtain a Gauss–Radau rule with one prescribed node at either end of the integration interval we extend the Jacobi matrix J_k; see chapter 6. Let us compute first Gauss–Radau rules for the measure $d\lambda$ on $[-1, 1]$ corresponding to the Legendre polynomials. We compare the Golub and Welsch algorithm results [160] (table 10.11) to those obtained with the Matlab package OPQ from Gautschi [132] (table 10.12). We first fix a node at $z = -1$. The results are the same up to the last two decimal digits. Note that since we know that the node -1 is prescribed we could have replace the value computed by the Golub and Welsch algorithm by -1.

Table 10.10 Generalized Laguerre weight function with $\alpha = -0.75$ on $[0, \infty)$, $n = 10$,
QLSWEEP from Laurie [220] with perfect shifts, comparison with Concus et
al. [69]

Relative error on weights
$2.768241663534195 \ 10^{-15}$
$1.320757565576015 \ 10^{-14}$
$1.642955800454616 \ 10^{-14}$
$3.287389308116639 \ 10^{-15}$
$5.001794536640609 \ 10^{-15}$
$8.154199883847268 \ 10^{-15}$
$6.993903244223327 \ 10^{-15}$
$9.021448038197202 \ 10^{-15}$
$8.681483002065855 \ 10^{-15}$
$3.323968921535110 \ 10^{-15}$

However, this shows how accurate is the modification of the Jacobi matrix.

Table 10.11 Legendre weight function on $[-1, 1]$, $n = 10$, node fixed at -1, Golub and
Welsch

Nodes	Weights
$-9.999999999999996 \ 10^{-1}$	$1.999999999999979 \ 10^{-2}$
$-9.274843742335808 \ 10^{-1}$	$1.202966705574827 \ 10^{-1}$
$-7.638420424200024 \ 10^{-1}$	$2.042701318789991 \ 10^{-1}$
$-5.256460303700790 \ 10^{-1}$	$2.681948378411793 \ 10^{-1}$
$-2.362344693905885 \ 10^{-1}$	$3.058592877244225 \ 10^{-1}$
$7.605919783797777 \ 10^{-2}$	$3.135824572269377 \ 10^{-1}$
$3.806648401447248 \ 10^{-1}$	$2.906101648329181 \ 10^{-1}$
$6.477666876740096 \ 10^{-1}$	$2.391934317143801 \ 10^{-1}$
$8.512252205816072 \ 10^{-1}$	$1.643760127369219 \ 10^{-1}$
$9.711751807022468 \ 10^{-1}$	$7.361700548675848 \ 10^{-2}$

Table 10.12 Legendre weight function on $[-1, 1]$, $n = 10$, node fixed at -1, OPQ

Nodes	Weights
-1.000000000000000	$1.999999999999983 \ 10^{-2}$
$-9.274843742335811 \ 10^{-1}$	$1.202966705574817 \ 10^{-1}$
$-7.638420424200026 \ 10^{-1}$	$2.042701318790005 \ 10^{-1}$
$-5.256460303700794 \ 10^{-1}$	$2.681948378411785 \ 10^{-1}$
$-2.362344693905883 \ 10^{-1}$	$3.058592877244226 \ 10^{-1}$
$7.605919783797817 \ 10^{-2}$	$3.135824572269378 \ 10^{-1}$
$3.806648401447244 \ 10^{-1}$	$2.906101648329185 \ 10^{-1}$
$6.477666876740094 \ 10^{-1}$	$2.391934317143814 \ 10^{-1}$
$8.512252205816080 \ 10^{-1}$	$1.643760127369209 \ 10^{-1}$
$9.711751807022472 \ 10^{-1}$	$7.361700548675876 \ 10^{-2}$

Let us now consider the Chebyshev weight function of the first kind on $[-1, 1]$

with a node fixed at $z = 1$. Again, the results are the same up to the last two decimal digits (tables 10.13 and 10.14).

Table 10.13 Chebyshev weight function on $[-1, 1]$, $n = 10$, node fixed at 1, Golub and Welsch

Nodes	Weights
$-9.863613034027220 \ 10^{-1}$	$3.306939635357689 \ 10^{-1}$
$-8.794737512064897 \ 10^{-1}$	$3.306939635357680 \ 10^{-1}$
$-6.772815716257409 \ 10^{-1}$	$3.306939635357673 \ 10^{-1}$
$-4.016954246529697 \ 10^{-1}$	$3.306939635357679 \ 10^{-1}$
$-8.257934547233214 \ 10^{-2}$	$3.306939635357673 \ 10^{-1}$
$2.454854871407992 \ 10^{-1}$	$3.306939635357680 \ 10^{-1}$
$5.469481581224269 \ 10^{-1}$	$3.306939635357675 \ 10^{-1}$
$7.891405093963938 \ 10^{-1}$	$3.306939635357679 \ 10^{-1}$
$9.458172417006351 \ 10^{-1}$	$3.306939635357693 \ 10^{-1}$
$9.999999999999994 \ 10^{-1}$	$1.653469817678827 \ 10^{-1}$

Table 10.14 Chebyshev weight function on $[-1, 1]$, $n = 10$, node fixed at 1, OPQ

Nodes	Weights
$-9.863613034027223 \ 10^{-1}$	$3.306939635357677 \ 10^{-1}$
$-8.794737512064891 \ 10^{-1}$	$3.306939635357674 \ 10^{-1}$
$-6.772815716257411 \ 10^{-1}$	$3.306939635357679 \ 10^{-1}$
$-4.016954246529694 \ 10^{-1}$	$3.306939635357677 \ 10^{-1}$
$-8.257934547233234 \ 10^{-2}$	$3.306939635357675 \ 10^{-1}$
$2.454854871407993 \ 10^{-1}$	$3.306939635357681 \ 10^{-1}$
$5.469481581224266 \ 10^{-1}$	$3.306939635357668 \ 10^{-1}$
$7.891405093963936 \ 10^{-1}$	$3.306939635357658 \ 10^{-1}$
$9.458172417006349 \ 10^{-1}$	$3.306939635357681 \ 10^{-1}$
1.000000000000000	$1.653469817678845 \ 10^{-1}$

10.7 Gauss–Lobatto Quadrature Rules

To obtain the Gauss–Lobatto rule with two prescribed nodes at both ends of the integration interval $[a, b]$, we first need to solve two tridiagonal linear systems,

$$(J_k - aI)\omega = e^k, \quad (J_k - bI)\rho = e^k.$$

Then, the two values ω_{k+1} and β_k that extend the matrix J_k are given by solving

$$\begin{pmatrix} 1 & -\omega_k \\ 1 & -\rho_k \end{pmatrix} \begin{pmatrix} \omega_{k+1} \\ \beta_k^2 \end{pmatrix} = \begin{pmatrix} a \\ b \end{pmatrix}.$$

Let us consider the Legendre weight function on $[-1, 1]$ with two fixed nodes at both ends of the interval. Results in tables 10.15 and 10.16 show that we have differences in the last two decimal digits.

Table 10.15 Legendre weight function on $[-1, 1]$, $n = 10$, nodes fixed at -1 and 1, Golub and Welsch

Nodes	Weights
$-9.999999999999999 \ 10^{-1}$	$2.222222222222217 \ 10^{-2}$
$-9.195339081664596 \ 10^{-1}$	$1.333059908510702 \ 10^{-1}$
$-7.387738651055050 \ 10^{-1}$	$2.248893420631271 \ 10^{-1}$
$-4.779249498104440 \ 10^{-1}$	$2.920426836796831 \ 10^{-1}$
$-1.652789576663868 \ 10^{-1}$	$3.275397611838977 \ 10^{-1}$
$1.652789576663872 \ 10^{-1}$	$3.275397611838972 \ 10^{-1}$
$4.779249498104442 \ 10^{-1}$	$2.920426836796839 \ 10^{-1}$
$7.387738651055051 \ 10^{-1}$	$2.248893420631256 \ 10^{-1}$
$9.195339081664585 \ 10^{-1}$	$1.333059908510702 \ 10^{-1}$
1.000000000000000	$2.222222222222224 \ 10^{-2}$

Table 10.16 Legendre weight function on $[-1, 1]$, $n = 10$, nodes fixed at -1 and 1, OPQ

Nodes	Weights
$-9.999999999999996 \ 10^{-1}$	$2.222222222222240 \ 10^{-2}$
$-9.195339081664586 \ 10^{-1}$	$1.333059908510701 \ 10^{-1}$
$-7.387738651055048 \ 10^{-1}$	$2.248893420631266 \ 10^{-1}$
$-4.779249498104444 \ 10^{-1}$	$2.920426836796839 \ 10^{-1}$
$-1.652789576663870 \ 10^{-1}$	$3.275397611838975 \ 10^{-1}$
$1.652789576663869 \ 10^{-1}$	$3.275397611838974 \ 10^{-1}$
$4.779249498104443 \ 10^{-1}$	$2.920426836796844 \ 10^{-1}$
$7.387738651055049 \ 10^{-1}$	$2.248893420631251 \ 10^{-1}$
$9.195339081664588 \ 10^{-1}$	$1.333059908510702 \ 10^{-1}$
$9.999999999999998 \ 10^{-1}$	$2.222222222222210 \ 10^{-2}$

10.8 Anti-Gauss Quadrature Rule

We compute the results for the anti-Gauss rule by multiplying by a factor $\sqrt{2}$ two elements of the Jacobi matrix for the Legendre weight function and then by using the Golub and Welsch algorithm. We choose $n = 11$ since this rule must give errors which are of opposite sign to the error of the Gauss rule for $n = 10$ (see table 10.17).

Table 10.17 Anti-Gauss, Legendre weight function on $[-1, 1]$, $n = 11$

Nodes	Weights
$-9.959918853818236 \ 10^{-1}$	$2.257839165513059 \ 10^{-2}$
$-9.297956389113654 \ 10^{-1}$	$1.091543623802435 \ 10^{-1}$
$-7.809379654082114 \ 10^{-1}$	$1.863290923563876 \ 10^{-1}$
$-5.626785950628905 \ 10^{-1}$	$2.469272555985873 \ 10^{-1}$
$-2.944199592771482 \ 10^{-1}$	$2.855813256108908 \ 10^{-1}$
0	$2.988591447975199 \ 10^{-1}$
$2.944199592771473 \ 10^{-1}$	$2.855813256108902 \ 10^{-1}$
$5.626785950628914 \ 10^{-1}$	$2.469272555985887 \ 10^{-1}$
$7.809379654082104 \ 10^{-1}$	$1.863290923563861 \ 10^{-1}$
$9.297956389113663 \ 10^{-1}$	$1.091543623802462 \ 10^{-1}$
$9.959918853818245 \ 10^{-1}$	$2.257839165512853 \ 10^{-2}$

10.9 Gauss–Kronrod Quadrature Rule

Let us give an example of a Gauss–Kronrod quadrature rule for the Legendre weight function. We choose $N = 10$ and compute a rule with 21 nodes because this rule must have the 10-point Gauss rule nodes as a subset. When we want to compute a quadrature rule with a small number of nodes we can simplify a little bit the algorithm of Calvetti, Golub, Gragg and Reichel [59]. Assume N is even; we first compute the eigenvalues and eigenvectors of J_N and $J_{N+2:\frac{3N}{2}+1}$. Then we compute the first components of the eigenvectors of $\breve{J}_N$ by

$$\breve{w}_k = \sum_{j=1}^{N/2} l_k(\theta_j^*)w_j^*, \ k = 1, \ldots, N.$$

We compute the eigenvalues and eigenvector matrix U_{2N+1} of the matrix

$$\begin{pmatrix} \Theta_N & \gamma_N Z_N^T e^N & 0 \\ \gamma_N (e^N)^T Z_N & \omega_{N+1} & \gamma_{N+1}(e^1)^T \breve{Z}_N \\ 0 & \gamma_{N+1} \breve{Z}_N^T e^1 & \Theta_N \end{pmatrix},$$

which is similar to $\hat{J}_{2N+1}$. This gives the nodes of the Gauss–Kronrod rule. The first components of the eigenvectors are obtained by computing $(u^T \quad 0) U_{2N+1}$ where u^T is the first row of the matrix of the eigenvectors of J_N. The results are given in table 10.18. We can check that every other node is a node of the 10-point Gauss rule.

Table 10.18 Gauss–Kronrod, Legendre weight function on $[-1, 1]$, $N = 10, n = 21$

Nodes	Weights
$-9.956571630258079 \ 10^{-1}$	$1.169463886737180 \ 10^{-2}$
$-9.739065285171706 \ 10^{-1}$	$3.255816230796485 \ 10^{-2}$
$-9.301574913557080 \ 10^{-1}$	$5.475589657435226 \ 10^{-2}$
$-8.650633666889848 \ 10^{-1}$	$7.503967481091979 \ 10^{-2}$
$-7.808177265864176 \ 10^{-1}$	$9.312545458369767 \ 10^{-2}$
$-6.794095682990247 \ 10^{-1}$	$1.093871588022972 \ 10^{-1}$
$-5.627571346686043 \ 10^{-1}$	$1.234919762620656 \ 10^{-1}$
$-4.333953941292472 \ 10^{-1}$	$1.347092173114734 \ 10^{-1}$
$-2.943928627014605 \ 10^{-1}$	$1.427759385770600 \ 10^{-1}$
$-1.488743389816314 \ 10^{-1}$	$1.477391049013385 \ 10^{-1}$
$-5.985584791552117 \ 10^{-1}$	$1.494455540029168 \ 10^{-1}$
$1.488743389816317 \ 10^{-1}$	$1.477391049013382 \ 10^{-1}$
$2.943928627014603 \ 10^{-1}$	$1.427759385770603 \ 10^{-1}$
$4.333953941292473 \ 10^{-1}$	$1.347092173114733 \ 10^{-1}$
$5.627571346686046 \ 10^{-1}$	$1.234919762620654 \ 10^{-1}$
$6.794095682990238 \ 10^{-1}$	$1.093871588022977 \ 10^{-1}$
$7.808177265864170 \ 10^{-1}$	$9.312545458369791 \ 10^{-2}$
$8.650633666889850 \ 10^{-1}$	$7.503967481091990 \ 10^{-2}$
$9.301574913557086 \ 10^{-1}$	$5.475589657435200 \ 10^{-2}$
$9.739065285171715 \ 10^{-1}$	$3.255816230796519 \ 10^{-2}$
$9.956571630258081 \ 10^{-1}$	$1.169463886737200 \ 10^{-2}$

10.10 Computation of Integrals

Let us consider the computation of some integrals over the interval $[-1, 1]$. Some examples are from Trefethen [327]. The first function is a monomial λ^{20} (Example Q1). Its integral is of course equal to $2/21 = 9.523809523809523 \ 10^{-2}$. The approximate values computed with the Gauss–Legendre quadrature rule are given in table 10.19 as a function of the number n of integration points. We see that the error (the exact value minus the approximation) decreases significantly (i.e., is almost zero up to machine precision) for $n = 11$ for which the Gauss rule is exact for polynomials of degree less than $2n - 1 = 21$. For $n = 10$, the rule is only exact for polynomials of degree 19.

We now compute the integral of e^{λ} (Example Q2) which is an entire function. The rounded value of the integral is 2.350402387287603. The results are given in table 10.20. The error is zero up to machine precision for $n = 7$. The Gauss rule gives a lower bound of the integral (except when we are at the roundoff level) since the derivatives of the function are all positive in $[-1, 1]$.

Then we consider the function $1/(1 + 10\lambda^2)$ (Example Q3). The rounded value of the integral over $[-1, 1]$ is 0.7997520101115316. The $\log_{10}$ of the absolute value of the error is given in figure 10.1. We see that we need almost 55 points to obtain an error close to machine precision. The signs of the derivatives alternate when n increases. Therefore, we have a lower bound of the integral when the number of nodes n is even and an upper bound when n is odd.

Let us compute the integral of e^{-1/λ^2} over $[-1, 1]$ (Example Q4) whose rounded

Table 10.19 Integral of λ^{20} on $[-1, 1]$ as a function of n, Gauss rule, Golub and Welsch

n	Integral	Error
2	$3.387017561686067 \ 10^{-5}$	$9.520422506247837 \ 10^{-2}$
3	$6.718463999999998 \ 10^{-3}$	$8.851963123809524 \ 10^{-2}$
4	$3.498372981825397 \ 10^{-2}$	$6.025436541984126 \ 10^{-2}$
5	$6.606306950648783 \ 10^{-2}$	$2.917502573160741 \ 10^{-2}$
6	$8.481758799621919 \ 10^{-2}$	$1.042050724187604 \ 10^{-2}$
7	$9.252257541458878 \ 10^{-2}$	$2.715519823506457 \ 10^{-3}$
8	$9.474699295632059 \ 10^{-2}$	$4.911022817746386 \ 10^{-4}$
9	$9.518280330178025 \ 10^{-2}$	$0.529193631498286 \ 10^{-5}$
10	$9.523516964776493 \ 10^{-2}$	$2.925590330299377 \ 10^{-6}$
11	$9.523809523809357 \ 10^{-2}$	$1.665334536937735 \ 10^{-15}$
12	$9.523809523809479 \ 10^{-2}$	$4.16333634234433710^{-16}$
13	$9.523809523809480 \ 10^{-2}$	$4.44089209850062610^{-16}$
14	$9.523809523809566 \ 10^{-2}$	$-4.163336342344337 \ 10^{-16}$
15	$9.523809523809551 \ 10^{-2}$	$-2.914335439641036 \ 10^{-16}$

Table 10.20 Integral of e^{λ} on $[-1, 1]$ as a function of n, Gauss rule, Golub and Welsch

n	Integral	Error
1	2	$3.504023872876032 \ 10^{-1}$
2	2.342696087909731	$7.706299377872039 \ 10^{-3}$
3	2.350336928680010	$6.545860759343825 \ 10^{-5}$
4	2.350402092156377	$2.951312261245676 \ 10^{-7}$
5	2.350402386462827	$8.247758032098318 \ 10^{-10}$
6	2.350402387286034	$1.568967178400271 \ 10^{-12}$
7	2.350402387287602	$8.881784197001252 \ 10^{-16}$
8	2.350402387287603	0
9	2.350402387287603	0
10	2.350402387287603	$4.440892098500626 \ 10^{-16}$
11	2.350402387287602	$1.332267629550188 \ 10^{-15}$
12	2.350402387287604	$-4.440892098500626 \ 10^{-16}$

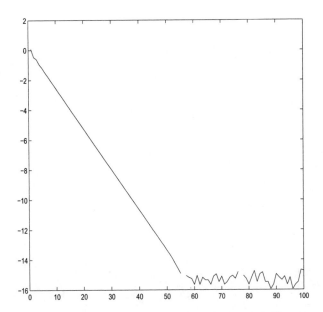

Figure 10.1 $\log_{10}$ of the error for the integral of $1/(1 + 10\lambda^2)$, Gauss rule

value is 0.1781477117815611. The $\log_{10}$ of the absolute value of the error is given in figure 10.2. For this example the error is oscillating, so we do not always obtain better results when the number of nodes is increased.

Another example is the function $(1 - \lambda^2)^{-1/2}$ over $[-1, 1]$ corresponding to Chebyshev polynomials of the first kind. We have at least two ways of computing this integral, whose value is π. We can compute the approximate value using the Legendre weight function (Example Q5). With 1500 nodes we obtain the value 3.140432120277716; this is not very satisfactory. Using the Chebyshev weight function and integrating the constant function equal to 1 (Example Q6), we obtain the (rounded) exact value with only one integration point. Of course, this is an extreme example because the function corresponds to a weight function for which we know the orthogonal polynomials, but this illustrates the interest of being able to compute orthogonal polynomials corresponding to other measures. This can sometimes be done by modification algorithms.

We now compute the integral of $(1 - \lambda^2)^{1/2}/(2 + \lambda)^{1/2}$ over $[-1, 1]$. Figure 10.3 shows the $\log_{10}$ of the absolute value of the error when integrating this function with the Gauss–Legendre rule (Example Q7) and when integrating $1/(2 + \lambda)^{1/2}$ using the measure corresponding to the Chebyshev polynomials of the second kind (that is $(1 - \lambda^2)^{1/2} d\lambda$, Example Q8). One can see the large difference in the errors.

Let us consider the Gauss–Radau and Gauss–Lobatto rules where some nodes are prescribed. Tables 10.21 and 10.22 give the errors as a function of the number of nodes for Example Q1. The results in table 10.21 might seem strange when compared with those of table 10.19 since one might think that the Gauss–Radau

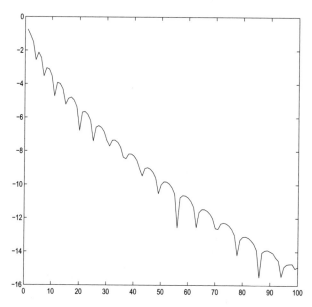

Figure 10.2 $\log_{10}$ of the error for the integral of e^{-1/λ^2}

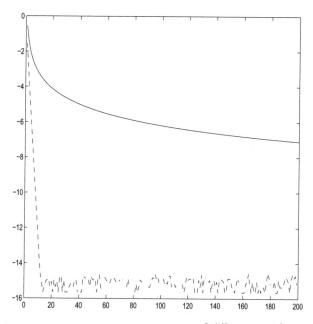

Figure 10.3 $\log_{10}$ of the error for the integral of $(1 - \lambda^2)^{1/2}/(2 + \lambda)^{1/2}$, Gauss–Legendre (solid line), Chebyshev second kind (dashed line)

rule is less precise than the Gauss rule. However, the column labeled n gives the total number of nodes. In the Gauss–Radau rule we have $n - 1$ free nodes and one prescribed node. Therefore, when $n = 10$, the rule is exact for polynomials of degree at most $2(n-1) = 18$. When $n = 11$, the rule must be exact for polynomials of degree 20, which is what we observe. Similarly, for the Gauss–Lobatto rule, for $n = 11$, it must be exact for polynomials of degree $2(n - 2) + 1 = 19$. For $n = 12$ the rule is exact for polynomials of degree at most 21, which is what we can see in the table.

Table 10.21 Integral of λ^{20} on $[-1, 1]$ as a function of n, Gauss–Radau rule

n	Error Gauss–Radau, node=1	Error Gauss–Radau, node=-1
2	$-4.047619051920983 \cdot 10^{-1}$	$-4.047619051920959 \cdot 10^{-1}$
3	$-1.274332654308090 \cdot 10^{-1}$	$-1.274332654308121 \cdot 10^{-1}$
4	$-3.869514350464778 \cdot 10^{-2}$	$-3.869514350464628 \cdot 10^{-2}$
5	$-1.081482914060589 \cdot 10^{-2}$	$-1.081482914060511 \cdot 10^{-2}$
6	$-2.653678531877202 \cdot 10^{-3}$	$-2.653678531878090 \cdot 10^{-3}$
7	$-5.196757636501403 \cdot 10^{-4}$	$-5.196757636500848 \cdot 10^{-4}$
8	$-7.449570767370517 \cdot 10^{-5}$	$-7.449570767464886 \cdot 10^{-5}$
9	$-6.894455820208312 \cdot 10^{-6}$	$-6.894455820111167 \cdot 10^{-6}$
10	$-3.079568777486497 \cdot 10^{-7}$	$-3.079568764441376 \cdot 10^{-7}$
11	$-1.942890293094024 \cdot 10^{-16}$	$9.714451465470120 \cdot 10^{-17}$
12	$-1.942890293094024 \cdot 10^{-16}$	$2.775557561562891 \cdot 10^{-17}$
13	$9.298117831235686 \cdot 10^{-16}$	$8.049116928532385 \cdot 10^{-16}$
14	$-5.551115123125783 \cdot 10^{-17}$	$-7.355227538141662 \cdot 10^{-16}$
15	$5.273559366969494 \cdot 10^{-16}$	$3.469446951953614 \cdot 10^{-16}$

Table 10.22 Integral of λ^{20} on $[-1, 1]$ as a function of n, Gauss–Lobatto rule

n	Error Gauss–Lobatto
2	-1.904761904761895
3	$-5.714285714285746 \cdot 10^{-1}$
4	$-2.380954087619067 \cdot 10^{-1}$
5	$-1.049895275367407 \cdot 10^{-1}$
6	$-4.166734114303916 \cdot 10^{-2}$
7	$-1.340136107659812 \cdot 10^{-2}$
8	$-3.276609531666960 \cdot 10^{-3}$
9	$-5.681860545760920 \cdot 10^{-4}$
10	$-6.212577114121654 \cdot 10^{-5}$
11	$-3.218149364950240 \cdot 10^{-6}$
12	$4.024558464266193 \cdot 10^{-16}$
13	$2.220446049250313 \cdot 10^{-16}$
14	$3.053113317719181 \cdot 10^{-16}$
15	$1.526556658859590 \cdot 10^{-16}$

The results for Example Q2 are in tables 10.23 and 10.24. In this example the derivatives of the function are positive. Therefore, when the prescribed node is located at the end of the interval we obtain an upper bound (the error, defined as the exact value minus the approximation, is negative, except when we are at the roundoff level). When the prescribed node is at the beginning of the interval we obtain a lower bound of the integral. The Gauss–Lobatto rule gives an upper bound.

Table 10.23 Integral of e^λ on $[-1, 1]$ as a function of n, Gauss–Radau rule

n	Error Gauss–Radau, node=1	Error Gauss–Radau, node=-1
2	$-8.353549280260264 \ 10^{-2}$	$7.304402907274943 \ 10^{-2}$
3	$-9.488216490005641 \ 10^{-4}$	$8.960763974479313 \ 10^{-4}$
4	$-5.416475832742407 \ 10^{-6}$	$5.247179327039220 \ 10^{-6}$
5	$-1.837145857663813 \ 10^{-8}$	$1.800400228901822 \ 10^{-8}$
6	$-4.113198670552265 \ 10^{-11}$	$4.056177616007517 \ 10^{-11}$
7	$-6.528111384795921 \ 10^{-14}$	$6.483702463810914 \ 10^{-14}$
8	$4.440892098500626 \ 10^{-16}$	$8.881784197001252 \ 10^{-16}$
9	$-8.881784197001252 \ 10^{-16}$	$8.881784197001252 \ 10^{-16}$
10	$-4.440892098500626 \ 10^{-16}$	0
11	$8.881784197001252 \ 10^{-16}$	$4.440892098500626 \ 10^{-16}$
12	$4.440892098500626 \ 10^{-16}$	$-1.332267629550188 \ 10^{-15}$

Table 10.24 Integral of e^λ on $[-1, 1]$ as a function of n, Gauss–Lobatto rule

n	Error Gauss–Lobatto
2	$-7.357588823428833 \ 10^{-1}$
3	$-1.165136925589216 \ 10^{-2}$
4	$-8.752023187019731 \ 10^{-5}$
5	$-3.693924663927817 \ 10^{-7}$
6	$-9.904339570709908 \ 10^{-10}$
7	$-1.831423901421658 \ 10^{-12}$
8	0
9	$-1.332267629550188 \ 10^{-15}$
10	$2.220446049250313 \ 10^{-15}$
11	$-2.220446049250313 \ 10^{-15}$
12	$-4.440892098500626 \ 10^{-16}$

We also consider Example Q1 with the anti-Gauss rule using 11 nodes to be able to compare with the Gauss rule using 10 nodes. The approximate value of the integral obtained by using the Gauss rule is $9.523516964776493 \ 10^{-2}$ and the error is $2.925590330299377 \ 10^{-6}$, while the anti-Gauss rule with 11 nodes gives a value $9.524102082842620 \ 10^{-2}$ and the error is $-2.925590330965511 \ 10^{-6}$. So, to 10 decimal digits, the error is the opposite of the error of the Gauss rule with one fewer node. This gives an estimation of the error for the Gauss rule of $2.925590330632444 \ 10^{-6}$ which is correct to 10 decimal digits. Taking the average of the Gauss and anti-Gauss values we obtain $9.523809523809557 \ 10^{-2}$ with an error of $-3.330669073875470 \ 10^{-16}$.

Using a Gauss–Kronrod rule with 21 nodes on Example Q1, we obtain an approximate value of the integral of $9.523809523809526 \ 10^{-2}$ when the exact value is $2/21 = 9.523809523809523 \ 10^{-2}$. Thus the error is at the roundoff level. This gives an estimation for the error of the Gauss rule of $2.925590330327132 \ 10^{-6}$.

For Example Q3 with 30 nodes, the Gauss rule gives 0.7997519988056411 with an error $1.130589055708953 \ 10^{-8}$. The anti-Gauss rule with 31 nodes gives 0.7997520214173953 with an error of $-1.130586368969233 \ 10^{-8}$, which is the negative of the Gauss rule error up to 6 decimal digits. The average of the two rules gives 0.7997520101115182 with an error of $1.343369859796439 \ 10^{-14}$.

The Gauss–Kronrod rule with 61 nodes gives 0.7997520101115313 with an error $3.330669073875470 \ 10^{-16}$. The approximation of the Gauss rule error is $1.130589022402262 \ 10^{-8}$ with 8 correct decimal digits.

10.11 Modification Algorithms

Let us illustrate the use of modification algorithms by computing Gauss rules for the Legendre measure multiplied by a polynomial. We first multiply by a linear factor $\lambda + 2$. This is particularly simple since we compute the Cholesky factorization of $J_N + 2I = LL^T$, and then we form $\tilde{J}_N = L^T L - 2I$. The Jacobi matrix we are seeking is the leading submatrix of order $N - 1$ of $\tilde{J}_N$. Then we use the Golub and Welsch algorithm to compute the nodes and weights. Note that this factorization can also be done using the QD algorithm. The results of the 9-point rule for $N = 10$ using the Cholesky factorization are given in table 10.25. Up to the last two decimal digits they are equal to the nodes and weights given by the modification algorithm of the OPQ package of Gautschi.

Table 10.25 Multiplication of the Legendre weight by $\lambda + 2$, Gauss rule, Golub and Welsch

Nodes	Weights
$-9.656446552418069 \ 10^{-1}$	$9.048144617079827 \ 10^{-2}$
$-8.247634420027529 \ 10^{-1}$	$2.243999204425018 \ 10^{-1}$
$-5.922714172905702 \ 10^{-1}$	$3.788698771441218 \ 10^{-1}$
$-2.971217556807707 \ 10^{-1}$	$5.373601521445760 \ 10^{-1}$
$2.735769852646871 \ 10^{-2}$	$6.641696281904350 \ 10^{-1}$
$3.466713414373445 \ 10^{-1}$	$7.169055718980852 \ 10^{-1}$
$6.279518601817928 \ 10^{-1}$	$6.629658604495480 \ 10^{-1}$
$8.427336504825537 \ 10^{-1}$	$4.936926435406916 \ 10^{-1}$
$9.695179097981458 \ 10^{-1}$	$2.311549000192406 \ 10^{-1}$

Then we modify the Legendre weight by a quadratic factor $r(\lambda) = (\lambda + 2)^2$. To obtain the nodes and weights corresponding to $r \, d\lambda$, we compute the spectral decomposition $J_N = Z\Theta Z^T$ of the Jacobi Legendre matrix and the diagonal matrix $D^2 = r(\Theta)$. A QR factorization $UR = DZ^T$ gives the solution, which is a submatrix of $\tilde{J} = U^T \Theta U$. The results of the 8-point rule are given in table 10.26.

To check our codes we computed a modification of the Laguerre weight by a quadratic polynomial $r(\lambda)w(\lambda) = \lambda^2 e^{-\lambda}$ on $[0, \infty)$. The result can also be directly computed because we know the recurrence coefficients for the generalized Laguerre weight $\lambda^\alpha e^{-\lambda}$. The results of a 10-point rule by modification are given in table 10.27. The relative error with the nodes and weights computed directly is at most 10^{-14}.

Table 10.26 Multiplication of the Legendre weight by $(\lambda + 2)^2$, Gauss rule, Golub and Welsch

Nodes	Weights
$-9.528330676642192\ 10^{-1}$	$1.308154992912135\ 10^{-1}$
$-7.652771465436233\ 10^{-1}$	$3.790280369125821\ 10^{-1}$
$-4.716614027220163\ 10^{-1}$	$7.715671502593013\ 10^{-1}$
$-1.210428334349832\ 10^{-1}$	1.281180888296799
$2.386080141552248\ 10^{-1}$	1.748935115599083
$5.630550164503659\ 10^{-1}$	1.928253609533150
$8.143685870664167\ 10^{-1}$	1.618035092187539
$9.639225083058827\ 10^{-1}$	$8.088512745870043\ 10^{-1}$

Table 10.27 Multiplication of the Laguerre weight by λ^2, Gauss rule, Golub and Welsch

Nodes	Weights
$5.763138581163630\ 10^{-1}$	$1.419969396448682\ 10^{-1}$
1.559343460467388	$6.182793763589497\ 10^{-1}$
3.003710358249312	$7.540018044944447\ 10^{-1}$
4.942019073857106	$3.836098427417975\ 10^{-1}$
7.422707108754277	$9.131042204305420\ 10^{-2}$
$1.051881997561453\ 10^1$	$1.027704776239767\ 10^{-2}$
$1.434451458602408\ 10^1$	$5.146265531306630\ 10^{-4}$
$1.909178003200538\ 10^1$	$9.887226319049070\ 10^{-6}$
$2.513064781033858\ 10^1$	$5.313992045295483\ 10^{-8}$
$3.341014373657300\ 10^1$	$3.511750313883932\ 10^{-11}$

Finally let us consider the inverse Cholesky algorithm and then the Golub and Welsch algorithm for computing the nodes and weights associated with $d\lambda/(\lambda+2)$. Results for the 11-point rule are given in table 10.28. As a check we then computed the coefficients for a multiplication by $\lambda + 2$. They are different from the exact coefficients for the Legendre polynomials by 1 to 4 times the roundoff unit.

10.12 Inverse Eigenvalue Problems

The first experiment uses the Laguerre nodes and weights as in Laurie [220]. We first computed the coefficients of the Laguerre polynomials. Then, using a variable precision version of the QR algorithm we computed the eigenvalues and first components of the eigenvectors with 32 decimal digits. They are converted to double precision. From this data we use the different methods we have discussed in chapter 6. They are:

- The Lanczos algorithm with full double reorthogonalization

- The MGS variant of the Lanczos algorithm advocated by Paige

Table 10.28 Modification of the Legendre weight by division with $\lambda + 2$, Gauss rule, Golub and Welsch

Nodes	Weights
$-9.795263491601517 \ 10^{-1}$	$5.136126150560710 \ 10^{-2}$
$-8.932832431004641 \ 10^{-1}$	$1.078344955970998 \ 10^{-1}$
$-7.431029041780251 \ 10^{-1}$	$1.428070637989318 \ 10^{-1}$
$-5.381390419625562 \ 10^{-1}$	$1.560509479502007 \ 10^{-1}$
$-2.921725981087949 \ 10^{-1}$	$1.527298117657011 \ 10^{-1}$
$-2.289051245982113 \ 10^{-2}$	$1.387539521769125 \ 10^{-1}$
$2.495089713520985 \ 10^{-1}$	$1.186674611104695 \ 10^{-1}$
$5.040698745878063 \ 10^{-1}$	$9.537918021179864 \ 10^{-2}$
$7.209133166287860 \ 10^{-1}$	$7.055691745705198 \ 10^{-2}$
$8.829588256941766 \ 10^{-1}$	$4.507803619921015 \ 10^{-2}$
$9.774119492229954 \ 10^{-1}$	$1.939316089512510 \ 10^{-2}$

- The standard Lanczos algorithm

- The rational Lanczos algorithm as described in Gragg and Harrod [164]

- convqr from Laurie [220]

- RKPW from Gragg and Harrod [164] as in OPQ

- pftoqd from Laurie [220]

- The Stieltjes algorithm

Table 10.29 gives the relative error for the diagonal coefficient α_k divided by the roundoff unit u, that is,

$$\frac{1}{u} \max_i \left| 1 - \frac{\alpha_i}{\alpha_i^{ex}} \right|,$$

where α_i^{ex} are the exact values. Table 10.30 displays similar quantities for the subdiagonal coefficient β_k. The results are somewhat different from those that have been published in the literature [164], [220]. The Lanczos algorithms perform well except for the rational Lanczos algorithm, for which the coefficient ρ_k^2 overflows after $k = 100$. The standard Lanczos algorithm is a little bit worse than the versions with reorthogonalization and MGS. Surprisingly, convqr does not perform too well, even though it is supposed to be equivalent to RKPW. pftoqd is not as good as the Lanczos algorithms and RKPW. Finally, the Stieltjes algorithm works but its results are worse than those of the Lanczos algorithm.

In fact this problem is too easy for the Lanczos algorithms because of the eigenvalue distribution. Let us consider an example for which it is known that there is a rapid increase of the roundoff errors in the Lanczos algorithm. This example is due to Z. Strakoš. The matrix Λ of dimension n is diagonal with eigenvalues

$$\lambda_i = \lambda_1 + \left(\frac{i-1}{n-1} \right) (\lambda_n - \lambda_1) \rho^{n-i}, \ i = 1, \ldots, n.$$

Table 10.29 Inverse problem for the Laguerre polynomials, relative error / u for α_k

k	Lanc. reorth	Lanc. MGS	Lanc.	Rat. Lanc.	convqr	RKPW	pftoqd	Stielt.
1	0	0	0	0	0	0	0	0
5	4	2	2	2	5	2	2	2
10	6	8	8	6	10	4	12	18
15	5	10	15	4	11	6	10	13
20	4	11	21	13	19	4	11	21
25	12	8	19	22	18	10	12	36
30	6	10	19	24	38	10	16	26
35	8	10	26	11	30	7	37	54
40	18	13	32	28	38	16	84	34
45	11	12	12	14	60	17	30	48
50	14	18	39	28	77	12	50	80
55	10	16	39	36	78	14	45	46
60	13	13	38	39	72	13	26	46
65	14	16	34	26	106	20	36	62
70	20	17	36	49	180	18	36	148
75	20	20	62	34	137	14	52	152
80	22	16	55	44	118	14	35	72
85	22	20	32	70	160	34	48	88
90	20	15	60	42	91	20	50	113
95	22	18	80	55	134	28	38	84
100	14	20	44	Inf	108	26	78	130
110	20	22	39	Inf	146	24	80	159
120	16	23	51	Inf	315	25	78	113
130	18	28	44	Inf	126	22	108	102
140	24	20	76	Inf	276	24	128	129
150	22	26	132	Inf	274	36	134	144

Table 10.30 Inverse problem for the Laguerre polynomials, relative error / u for β_k

k	Lanc. reorth	Lanc. MGS	Lanc.	Rat. Lanc.	convqr	RKPW	pftoqd	Stielt.
1	0	0	0	0	0	0	0	0
5	2	2	2	1	3	2	1	2
10	3	5	3	7	16	5	9	11
15	4	6	4	4	10	4	10	12
20	3	6	5	6	16	7	8	16
25	6	8	8	10	36	15	14	23
30	6	6	4	6	26	11	24	13
35	12	8	6	6	40	11	20	33
40	12	10	10	10	52	10	39	23
45	10	8	8	6	58	8	19	39
50	14	14	10	10	78	16	28	48
55	12	10	14	12	71	16	32	37
60	16	14	12	14	112	17	44	32
65	13	12	14	10	68	14	28	37
70	10	12	10	8	84	14	46	75
75	14	9	12	14	116	16	39	84
80	16	14	14	16	104	18	45	57
85	22	22	18	19	148	22	40	70
90	18	14	13	15	94	16	60	72
95	20	18	18	17	149	26	34	74
100	20	16	16	Inf	204	22	53	69
110	17	18	20	Inf	144	27	82	102
120	21	20	20	Inf	242	27	86	71
130	24	20	22	Inf	210	30	93	81
140	20	20	16	Inf	219	26	93	78
150	23	24	22	Inf	179	30	72	98

The parameters λ_1 and λ_n are respectively the smallest and largest eigenvalues. The parameter ρ controls the distribution of the eigenvalues. We use $\lambda_1 = 0.1, \lambda_n = 100$ and a value $\rho = 0.9$ which gives well-separated large eigenvalues. Then we choose $A = Q\Lambda Q^T$ where Q is an orthonormal matrix. The eigenvalues and the first components of the eigenvectors are explicitly known. The tridiagonal matrix which is the solution of the inverse problem is computed with the Lanczos algorithm with double reorthogonalization which (for this particular problem) gives results that are close to those of exact arithmetic; see [239]. Results are given in tables 10.31 and 10.32. The stars mean that the elements of the Jacobi matrix are completely wrong. We see that the Lanczos methods are not able to correctly reconstruct the Jacobi matrix for this problem, except at the very beginning when the orthogonality is still preserved. The method that seems the most reliable is pftoqd. Note that the results for β_k are worst than those for the diagonal coefficient α_k.

Table 10.31 Inverse problem for the Strakoš matrix, relative error $/\ u$ for α_k

k	Lanc. MGS	Lanc.	Rat. Lanc.	convqr	RKPW	pftoqd	Stielt.
1	0	0	0	0	0	0	0
5	3	14	8	4	4	5	16
10	8	22	24	4	4	2	83
15	8	68	71	16	18	7	347
20	*	*	*	12	10	6	*
25	*	*	*	17	12	14	*
30	*	*	*	28	31	16	*
35	*	*	*	14	27	12	*
40	*	*	*	36	60	22	*
45	*	*	*	26	62	34	*
50	*	*	*	23	72	28	*
55	*	*	*	33	94	38	*
60	*	*	*	36	58	38	*
65	*	*	*	44	102	38	*
70	*	*	*	54	150	36	*
75	*	*	*	30	102	22	*
80	*	*	*	42	55	32	*
85	*	*	*	34	156	16	*
90	*	*	*	28	73	17	*
95	*	*	*	32	114	21	*
100	*	*	*	29	118	21	*

Table 10.32 Inverse problem for the Strakoš matrix, relative error / u for β_k

k	Lanc. MGS	Lanc.	Rat. Lanc.	convqr	RKPW	pftoqd	Stielt.
1	0	0	0	0	0	0	0
5	1	3	2	4	3	2	6
10	4	4	4	6	8	2	40
15	6	8	6	12	6	6	492
20	*	*	*	9	12	6	*
25	*	*	*	16	34	8	*
30	*	*	*	16	42	8	*
35	*	*	*	38	32	14	*
40	*	*	*	22	52	14	*
45	*	*	*	25	134	31	*
50	*	*	*	23	58	27	*
55	*	*	*	34	119	50	*
60	*	*	*	169	195	116	*
65	*	*	*	179	324	106	*
70	*	*	*	324	388	168	*
75	*	*	*	137	202	89	*
80	*	*	*	1076	328	526	*
85	*	*	*	420	531	411	*
90	*	*	*	1129	701	223	*
95	*	*	*	3746	1060	926	*
100	*	*	*	3492	1513	646	*

Chapter Eleven

Bounds and Estimates for Elements of Functions of Matrices

11.1 Introduction

In this chapter we consider the computation of elements of a function $f(A)$ of a symmetric matrix A. If $A = Q\Lambda Q^T$ is the spectral decomposition of A with Q orthogonal and Λ diagonal, then $f(A) = Qf(\Lambda)Q^T$. Diagonal elements can be estimated by considering

$$[f(A)]_{i,i} = (e^i)^T f(A) e^i,$$

where e^i is the ith column of the identity matrix. Following chapter 7, we can apply the Lanczos algorithm to the matrix A with a starting vector e^i. This generates tridiagonal Jacobi matrices J_k. The estimate of $[f(A)]_{i,i}$ given by the Gauss quadrature rule at iteration k is $(e^1)^T f(J_k) e^1$, that is, the $(1,1)$ entry of the matrix $f(J_k)$. This estimate is a lower or upper bound if the derivative of order $2k$ of the function f has a constant sign on the interval $[\lambda_1, \lambda_n]$. Other bounds can be obtained with Gauss–Radau and Gauss–Lobatto quadrature rules by suitably modifying some elements of J_k as we have seen in chapter 6.

Off-diagonal elements of $f(A)$ correspond to

$$[f(A)]_{i,j} = (e^i)^T f(A) e^j, \quad i \neq j.$$

For this case, we cannot directly use the Lanczos algorithm to generate the orthogonal polynomials. Nevertheless, as we have already seen in chapter 7, there are several ways to deal with this problem. First of all, one can use the identity

$$(e^i)^T f(A) e^j = \frac{1}{4}[(e^i + e^j)^T f(A)(e^i + e^j) - (e^i - e^j)^T f(A)(e^i - e^j)].$$

Then by using the Lanczos algorithm twice with starting vectors $(1/2)(e^i + e^j)$ and $(1/2)(e^i - e^j)$, one can obtain estimates of $[f(A)]_{i,j}$. Signs of the derivatives of f permitting, bounds can be obtained by combining upper and lower bounds from the Gauss and Gauss–Radau rules.

Another possibility is to use the symmetric variant of the nonsymmetric Lanczos algorithm, see chapter 4. However, we cannot directly use $v^1 = e^i$ and $\hat{v}^1 = e^j$ as starting vectors because these vectors are orthogonal and therefore, the algorithm will break down immediately. A way to avoid this difficulty is to choose $v^1 = e^i/\delta$ and $\hat{v}^1 = \delta e^i + e^j$ where δ is a positive parameter. Then, $(v^1)^T \hat{v}^1 = 1$. Introducing this parameter δ has the added advantage of being able to choose it to have the product of the off-diagonal elements of J_k to be positive. In this case the two sets of polynomials are the same, simplifying the algorithm.

The third possibility is to use the block Lanczos algorithm with a starting matrix $X_0 = (e^i, e^j)$. The difficulty with this approach is that we only obtain estimates and not bounds.

Bounds for the entries of matrix functions have been used by Benzi and Golub [27] to construct preconditioners for iterative methods. Benzi and Razouk [28] obtained decay bounds for elements of functions of sparse matrices.

11.2 Analytic Bounds for the Elements of the Inverse

We consider obtaining analytical bounds for the entries of the inverse of a given matrix by doing one or two iterations of the Lanczos algorithm. This is obtained by considering the function

$$f(\lambda) = \frac{1}{\lambda}, \quad 0 < a \le \lambda \le b,$$

for which the derivatives are

$$f^{(2k+1)}(\lambda) = -(2k+1)! \, \lambda^{-(2k+2)}, \quad f^{(2k)}(\lambda) = (2k)! \, \lambda^{-(2k+1)}.$$

Therefore, the even derivatives are positive on $[a, b]$ when $a > 0$ and the odd derivatives are negative, which implies that we can use the results of chapter 6, mainly theorems 6.3, 6.4 and 6.5, which show that the Gauss rule gives a lower bound, the Gauss–Radau rule gives lower and upper bounds and the Gauss–Lobatto rule gives an upper bound.

Performing analytically two Lanczos iterations, we are able to obtain bounds for the entries of the inverse.

THEOREM 11.1 *Let A be a symmetric positive definite matrix. Let*

$$s_i^2 = \sum_{j \ne i} a_{ji}^2, \quad i = 1, \dots, n;$$

we have the following bounds for the diagonal entries of the inverse given respectively by the Gauss, Gauss–Radau and Gauss–Lobatto rules

$$\frac{\sum_{k \ne i} \sum_{l \ne i} a_{k,i} a_{k,l} a_{l,i}}{a_{i,i} \sum_{k \ne i} \sum_{l \ne i} a_{k,i} a_{k,l} a_{l,i} - \left(\sum_{k \ne i} a_{k,i}^2 \right)^2} \le (A^{-1})_{i,i},$$

$$\frac{a_{i,i} - b + \frac{s_i^2}{b}}{a_{i,i}^2 - a_{i,i} b + s_i^2} \le (A^{-1})_{i,i} \le \frac{a_{i,i} - a + \frac{s_i^2}{a}}{a_{i,i}^2 - a_{i,i} a + s_i^2},$$

$$(A^{-1})_{i,i} \le \frac{a + b - a_{ii}}{ab}.$$

Proof. We denote the elements of the matrix A as $a_{i,j}$. We choose the initial vector $v^1 = e^i$ and we apply the Lanczos algorithm. The first step of the Lanczos algorithm gives

$$\alpha_1 = (e^i)^T A e^i = a_{ii},$$

$$\eta_1 v^2 = (A - \alpha_1 I) e^i.$$

Let s_i be defined by

$$s_i^2 = \sum_{j \neq i} a_{ji}^2,$$

and

$$d^i = (a_{1,i}, \ldots, a_{i-1,i}, 0, a_{i+1,i}, \ldots, a_{n,i})^T.$$

Then

$$\eta_1 = s_i, \quad v^2 = \frac{1}{s_i} d^i.$$

From this, we have

$$\alpha_2 = (Av^2, v^2) = \frac{1}{s_i^2} \sum_{k \neq i} \sum_{l \neq i} a_{k,i} a_{k,l} a_{l,i}.$$

We can now compute the Gauss rule and obtain a lower bound on the diagonal element by considering the matrix

$$J_2 = \begin{pmatrix} \alpha_1 & \eta_1 \\ \eta_1 & \alpha_2 \end{pmatrix},$$

and its inverse

$$J_2^{-1} = \frac{1}{\alpha_1 \alpha_2 - \eta_1^2} \begin{pmatrix} \alpha_2 & -\eta_1 \\ -\eta_1 & \alpha_1 \end{pmatrix}.$$

The lower bound is given by $(e^1)^T J_2^{-1} e^1$, the $(1,1)$ entry of the inverse,

$$(e^1)^T J_2^{-1} e^1 = \frac{\alpha_2}{\alpha_1 \alpha_2 - \eta_1^2} = \frac{\sum_{k \neq i} \sum_{l \neq i} a_{k,i} a_{k,l} a_{l,i}}{a_{i,i} \sum_{k \neq i} \sum_{l \neq i} a_{k,i} a_{k,l} a_{l,i} - \left(\sum_{k \neq i} a_{k,i}^2 \right)^2}.$$

Note that this bound does not depend on the extreme eigenvalues of A. To obtain an upper bound we consider the Gauss–Radau rule. Then, we have to modify the $(2,2)$ element of the Lanczos matrix

$$\tilde{J}_2 = \begin{pmatrix} \alpha_1 & \eta_1 \\ \eta_1 & \xi \end{pmatrix},$$

and the eigenvalues λ of $\tilde{J}_2$ are the roots of $(\alpha_1 - \lambda)(\xi - \lambda) - \eta_1^2 = 0$, which gives the relation

$$\xi = \lambda + \frac{\eta_1^2}{\alpha_1 - \lambda}.$$

To obtain an upper bound we impose the requirement of an eigenvalue equal to the lower bound of the eigenvalues of A, $\lambda = a$. The solution is

$$\xi = \xi_a = a + \frac{\eta_1^2}{\alpha_1 - a},$$

from which we can compute the $(1,1)$ element of the inverse of $\tilde{J}_2$,

$$(e^1)^T \tilde{J}_2^{-1} e^1 = \frac{\xi}{\alpha_1 \xi - \eta_1^2}.$$

For the Gauss–Lobatto rule we want $\tilde{J}_2$ to have a and b as eigenvalues. This leads to solving the linear system

$$\begin{pmatrix} \alpha_1 - a & -1 \\ \alpha_1 - b & -1 \end{pmatrix} \begin{pmatrix} \xi \\ \eta_1^2 \end{pmatrix} = \begin{pmatrix} a\alpha_1 - a^2 \\ b\alpha_1 - b^2 \end{pmatrix}.$$

Solving this system gives $\xi = a + b - \alpha_1$ and, since the determinant should be ab, computing the $(1,1)$ element of the inverse gives the upper bound

$$\frac{a + b - \alpha_1}{ab}.$$

Note that in this case all the eigenvalues are prescribed. $\qquad\square$

Of course, these bounds are not sharp since they can be improved by doing more Lanczos iterations, except if the Lanczos algorithm converges in one iteration. More iterations can eventually be done analytically by using a symbolic calculation software. For off-diagonal entries it is not too easy to derive analytical bounds from the block Lanczos algorithm since we have to compute repeated inverses of 2×2 matrices. It is much easier to use the nonsymmetric Lanczos method with the Gauss–Radau rule. We are looking at the sum of the (i, i) and (i, j) elements of the inverse. The computations are essentially the same as for the diagonal case.

THEOREM 11.2 *Let A be a symmetric positive definite matrix and*

$$t_i = \sum_{k \neq i} a_{k,i}(a_{k,i} + a_{k,j}) - a_{i,j}(a_{i,j} + a_{i,i}).$$

For $(A^{-1})_{i,j} + (A^{-1})_{i,i}$ we have the two following estimates

$$\frac{a_{i,i} + a_{i,j} - a + \frac{t_i}{a}}{(a_{i,i} + a_{i,j})^2 - a(a_{i,i} + a_{i,j}) + t_i}, \qquad \frac{a_{i,i} + a_{i,j} - b + \frac{t_i}{b}}{(a_{i,i} + a_{i,j})^2 - b(a_{i,i} + a_{i,j}) + t_i}.$$

If $t_i \geq 0$, the first expression with a gives an upper bound and the second one with b a lower bound. Then, we have to subtract the bounds for the diagonal term to obtain bounds on $(A^{-1})_{i,j}$.

The previous results can be compared with those obtained using variational methods by Robinson and Wathen [285]. More precise results can also be obtained for sparse matrices taking into account the sparsity structure since in this case some terms in the sums arising in theorems 11.1 and 11.2 are zero.

11.3 Analytic Bounds for Elements of Other Functions

If we would like to obtain analytical bounds of diagonal elements for other functions, we see from the derivation for the inverse that all we have to do is to compute $f(J)$ for a symmetric matrix J of order 2 whose coefficients are known. Let

$$J = \begin{pmatrix} \alpha & \eta \\ \eta & \xi \end{pmatrix}.$$

If we are interested in the exponential we have to compute $\exp(J)$ (in fact only the $(1, 1)$ element). We use, for instance, a symbolic mathematics package. In the Matlab symbolic toolbox there is a function giving the exponential of a symbolic matrix. The result is the following.

We have $\alpha = a_{i,i}$, $\eta = s_i$, using the notations of the previous section. The element ξ is either α_2 or ξ_a.

PROPOSITION 11.3 *Let*

$$\delta = (\alpha - \xi)^2 + 4\eta^2,$$

$$\gamma = \exp\left(\frac{1}{2}(\alpha + \xi - \sqrt{\delta})\right),$$

$$\omega = \exp\left(\frac{1}{2}(\alpha + \xi + \sqrt{\delta})\right).$$

Then, the $(1, 1)$ element of the exponential of J is

$$\frac{1}{2}\left[\gamma + \omega + \frac{\omega - \gamma}{\sqrt{\delta}}(\alpha - \xi)\right].$$

Although these expressions are quite complicated, if we substitute the values of the parameters we obtain a lower bound from the Gauss rule and an upper bound from the Gauss–Radau rule.

For other functions which are not available in the symbolic packages we can compute analytically the eigenvalues and eigenvectors of J. In fact we just need the first components of the eigenvectors. The eigenvalues are

$$\lambda_+ = \frac{1}{2}(\alpha + \xi + \sqrt{\delta}), \quad \lambda_- = \frac{1}{2}(\alpha + \xi - \sqrt{\delta}).$$

The matrix of the unnormalized eigenvectors is

$$Q = \begin{pmatrix} \theta & \mu \\ 1 & 1 \end{pmatrix},$$

where

$$\theta = -\frac{1}{2\eta}(\alpha - \xi + \sqrt{\delta}), \quad \mu = -\frac{1}{2\eta}(\alpha - \xi - \sqrt{\delta}).$$

The first components of the normalized eigenvectors are $\theta/\sqrt{1 + \theta^2}$ and $\mu/\sqrt{1 + \mu^2}$. Then we have to compute the $(1, 1)$ element of $\tilde{Q}f(\Lambda)\tilde{Q}^T$ where Λ is the diagonal matrix of the eigenvalues λ_+ and λ_- and $\tilde{Q}$ is the matrix of the normalized eigenvectors. We need the values $\theta^2/(1 + \theta^2)$ and $\mu^2/(1 + \mu^2)$.

LEMMA 11.4 *We have*

$$\frac{\theta^2}{1+\theta^2} = \frac{\alpha - \xi + \sqrt{\delta}}{2\sqrt{\delta}}, \quad \frac{\mu^2}{1+\mu^2} = -\frac{\alpha - \xi - \sqrt{\delta}}{2\sqrt{\delta}}.$$

From this lemma we obtain the $(1,1)$ element of $f(J)$.

THEOREM 11.5 *The* $(1,1)$ *element of* $f(J)$ *is*

$$\frac{1}{2\sqrt{\delta}}\left[(\alpha - \xi)(f(\lambda_+) - f(\lambda_-)) + \sqrt{\delta}(f(\lambda_+) + f(\lambda_-))\right].$$

Proof. Clearly the $(1,1)$ element is

$$\frac{\theta^2}{1+\theta^2}f(\lambda_+) + \frac{\mu^2}{1+\mu^2}f(\lambda_-).$$

Using the expressions of lemma 11.4 and simplifying we obtain the result. □

We see that if f is the exponential function we recover the results of proposition 11.3. From the last theorem we can obtain analytic bounds for the (i, i) element of $f(A)$ for any function for which we can compute $f(\lambda_+)$ and $f(\lambda_-)$.

11.4 Computing Bounds for Elements of $f(A)$

A way to compute bounds or estimates of elements of the matrix $f(A)$ for a symmetric matrix A is to apply the general framework described in chapter 7. For diagonal entries we are interested in $(e^i)^T f(A)e^i$ and we apply the Lanczos algorithm to A with a starting vector $v^1 = e^i$. At each iteration k we augment the Jacobi matrix J_{k-1} to obtain J_k. Then we can use quadrature rules to obtain the bounds if the signs of the derivatives of the function f are constant on the integration interval. If, for instance, we are interested in the diagonal elements of the inverse of A, then we can compute incrementally the quantities $(e^1)^T J_k^{-1} e^1$ that give the bounds; see chapter 3. We have a lower bound from the Gauss rule. If we know lower and upper bounds of the smallest and largest eigenvalues respectively, then the Gauss–Radau rule gives lower and upper bounds for the diagonal entries of the inverse of A.

If we are interested in the off-diagonal elements, we may use the nonsymmetric Lanczos algorithm. In some cases we can obtain bounds. Another possibility is to use the block Lanczos algorithm, but in this case we are not able to tell if the computed quantities are lower or upper bounds.

11.5 Solving $Ax = c$ and Looking at $d^T x$

In some applications it may be interesting to look at quantities like $d^T x$, where d is a vector and x is the solution of a linear system $Ax = c$, without explicitly solving for x. For instance, one may be interested only in the lth component of the solution, in which case $d = e^l$. For a symmetric matrix, this can be done by using the framework of chapter 7. We use the Lanczos or block Lanczos algorithms with only a very small storage and we can obtain bounds or estimates of $d^T x$. Of course, this can also be done by first solving the linear system $Ax = c$ and then computing $d^T x$. But, in some problems this process is less stable and accurate than computing the estimates using the Lanczos algorithms.

11.6 Estimates of tr(A^{-1}) and det(A)

In [15] Bai and Golub studied how to estimate the trace of the inverse $\text{tr}(A^{-1})$ and the determinant $\det(A)$ of symmetric positive definite matrices; see also [16]. There are applications arising in the study of fractals [291], [351], lattice quantum chromodynamics (QCD) [296], [88] and crystals [253], [254]; see also [227]. Estimates of determinants have also been considered in Ipsen and Lee [197]. The estimates are based on the following result.

LEMMA 11.6 *Let A be a symmetric positive definite matrix. Then,*

$$\ln(\det(A)) = \text{tr}(\ln(A)). \tag{11.1}$$

Proof. The matrix A being symmetric positive definite, we have $A = Q\Lambda Q^T$ and $\ln(A) = Q\ln(\Lambda)Q^T$. Therefore, if we denote by q^i the columns of Q that are vectors of norm 1,

$$\text{tr}(\ln(A)) = \sum_{i=1}^{n}[(q_i^1)^2 \ln(\lambda_1) + \cdots + (q_i^n)^2 \ln(\lambda_n)] = \ln(\lambda_1) + \cdots + \ln(\lambda_n).$$

On the other hand, it is obvious that

$$\det(A) = \det(\Lambda) = \prod_{i=1}^{n} \lambda_i.$$

Therefore

$$\ln(\det(A)) = \sum_{i=1}^{n} \ln(\lambda_i),$$

which proves the result. $\qquad\square$

Lemma 11.6 shows that to bound $\det(A)$ we can bound $\ln(\det(A))$, which is the same as bounding $\text{tr}(\ln(A))$. Therefore, in this section, we consider the problem of estimating $\text{tr}(f(A))$ with $f(\lambda) = 1/\lambda$ and $f(\lambda) = \ln(\lambda)$.

11.6.1 tr(A^{-1})

In [15] Bai and Golub obtained bounds using Gauss quadrature analytically. Let

$$\mu_r = \text{tr}(A^r) = \sum_{i=1}^{n} \lambda_i^r = \int_a^b \lambda^r \, d\alpha$$

be the moments related to α, the measure (which we do not know explicitly) with steps at the eigenvalues of A. The first three moments are easily computed

$$\mu_0 = n, \quad \mu_1 = \text{tr}(A) = \sum_{i=1}^{n} a_{i,i}, \quad \mu_2 = \text{tr}(A^2) = \sum_{i,j=1}^{n} a_{i,j}^2 = \|A\|_F^2.$$

A Gauss–Radau quadrature rule is written as

$$\mu_r = \int_a^b \lambda^r \, d\alpha = \bar{\mu}_r + R_r.$$

The approximation of the integral is

$$\bar{\mu}_r = w_0 t_0^r + w_1 t_1^r, \tag{11.2}$$

where the weights w_0, w_1 and the node t_1 are to be determined. The node t_0 is prescribed to be a or b, the ends of the integration interval. From chapter 6 we know that t_0 and t_1 are the eigenvalues of a 2×2 matrix. Hence, they are solutions of a quadratic equation that we write as $c\xi^2 + d\xi - 1 = 0$. Because of equation (11.2), this implies that

$$c\bar{\mu}_r + d\bar{\mu}_{r-1} - \bar{\mu}_{r-2} = 0. \tag{11.3}$$

For $r = 0, 1, 2$ the quadrature rule is exact, $\bar{\mu}_r = \mu_r$ and t_0 is a root of the quadratic equation. This gives two equations for c and d,

$$c\mu_2 + d\mu_1 - \mu_0 = 0,$$
$$ct_0^2 + dt_0 - 1 = 0.$$

By solving this linear system we obtain the values of c and d,

$$\begin{pmatrix} c \\ d \end{pmatrix} = \begin{pmatrix} \mu_2 & \mu_1 \\ t_0^2 & t_0 \end{pmatrix}^{-1} \begin{pmatrix} \mu_0 \\ 1 \end{pmatrix}.$$

The unknown root t_1 of the quadratic equation is obtained by using the product of the roots, $t_1 = -1/(t_0 c)$. The weights are found by solving

$$w_0 t_0 + w_1 t_1 = \mu_1,$$
$$w_0 t_0^2 + w_1 t_1^2 = \mu_2.$$

This gives

$$\begin{pmatrix} w_0 \\ w_1 \end{pmatrix} = \begin{pmatrix} t_0 & t_1 \\ t_0^2 & t_1^2 \end{pmatrix}^{-1} \begin{pmatrix} \mu_1 \\ \mu_2 \end{pmatrix}.$$

To bound $\mathrm{tr}(A^{-1})$ we use equation (11.3) with $r = 1$,

$$c\bar{\mu}_1 + d\bar{\mu}_0 - \bar{\mu}_{-1} = 0.$$

But $\bar{\mu}_0 = \mu_0$ and $\bar{\mu}_1 = \mu_1$. Hence,

$$\bar{\mu}_{-1} = \begin{pmatrix} \mu_1 & \mu_0 \end{pmatrix} \begin{pmatrix} c \\ d \end{pmatrix},$$

which gives

$$\bar{\mu}_{-1} = \begin{pmatrix} \mu_1 & \mu_0 \end{pmatrix} \begin{pmatrix} \mu_2 & \mu_1 \\ t_0^2 & t_0 \end{pmatrix}^{-1} \begin{pmatrix} \mu_0 \\ 1 \end{pmatrix}.$$

Then,

$$\mu_{-1} = \bar{\mu}_{-1} + R_{-1}(\lambda),$$

and the remainder is

$$R_{-1}(\lambda) = -\frac{1}{\eta^4} \int_a^b (\lambda - t_0)(\lambda - t_1)^2 \, d\alpha,$$

for some $a < \eta < b$. If the prescribed node is $t_0 = a$ the remainder is negative and $\bar{\mu}_{-1}$ is an upper bound of μ_{-1}. It is a lower bound if $t_0 = b$. This leads to the following result.

THEOREM 11.7 *Let A be a symmetric positive definite matrix, $\mu_1 = \text{tr}(A)$, $\mu_2 = \|A\|_F^2$, the spectrum of A being contained in $[a, b]$, then*

$$
\begin{pmatrix} \mu_1 & n \end{pmatrix} \begin{pmatrix} \mu_2 & \mu_1 \\ b^2 & b \end{pmatrix}^{-1} \begin{pmatrix} n \\ 1 \end{pmatrix} \le \text{tr}(A^{-1}) \le \begin{pmatrix} \mu_1 & n \end{pmatrix} \begin{pmatrix} \mu_2 & \mu_1 \\ a^2 & a \end{pmatrix}^{-1} \begin{pmatrix} n \\ 1 \end{pmatrix}.
$$
(11.4)

11.6.2 det(A)

For bounding $\text{tr}(\ln(A))$ we use the same method as for the trace of the inverse. We have

$$
\text{tr}(\ln(A)) = \sum_{i=1}^{n} \ln \lambda_i = \int_a^b \ln \lambda \, d\alpha.
$$

Slight modifications of what we did before lead to the following theorem.

THEOREM 11.8 *Let A be a symmetric positive definite matrix, $\mu_1 = \text{tr}(A)$, $\mu_2 = \|A\|_F^2$, the spectrum of A being contained in $[a, b]$, then*

$$
\begin{pmatrix} \ln a & \ln \underline{t} \end{pmatrix} \begin{pmatrix} a & \underline{t} \\ a^2 & \underline{t}^2 \end{pmatrix}^{-1} \begin{pmatrix} \mu_1 \\ \mu_2 \end{pmatrix} \le \text{tr}(\ln(A)) \le \begin{pmatrix} \ln b & \ln \bar{t} \end{pmatrix} \begin{pmatrix} b & \bar{t} \\ b^2 & \bar{t}^2 \end{pmatrix}^{-1} \begin{pmatrix} \mu_1 \\ \mu_2 \end{pmatrix},
$$
(11.5)

where

$$
\underline{t} = \frac{a\mu_1 - \mu_2}{an - \mu_1}, \quad \bar{t} = \frac{b\mu_1 - \mu_2}{bn - \mu_1}.
$$

Another possibility to compute the trace is to consider the diagonal elements of A^{-1}. From chapters 6 and 11 we know how to estimate $(e^i)^T A^{-1} e^i$. However, this approach requires computing n such estimates. This might be too costly if n is large. In Bai, Fahey, Golub, Menon and Richter [18] it was proposed to use a Monte Carlo technique based on the following proposition; see Hutchinson [195] and also Bai, Fahey and Golub [17].

PROPOSITION 11.9 *Let B be a symmetric matrix of order n with $\text{tr}(B) \neq 0$. Let $\mathcal{Z}$ be a discrete random variable with values 1 and -1 with equal probability 0.5 and let z be a vector of n independent samples from $\mathcal{Z}$. Then $z^T B z$ is an unbiased estimator of $\text{tr}(B)$,*

$$
E(z^T B z) = \text{tr}(B),
$$

$$
\text{var}(z^T B z) = 2 \sum_{i \neq j} b_{i,j}^2,
$$

where $E(\cdot)$ denotes the expected value and var *denotes the variance.*

The method proposed in [18] is to first generate p sample vectors z^k, $k = 1, \ldots, p \ll n$ and then to estimate $(z^k)^T f(A) z^k$. This gives p estimates σ_k from which an unbiased estimate of the trace is derived as

$$
\text{tr}(f(A)) \approx \frac{1}{p} \sum_{k=1}^{p} \sigma_k.
$$

If we have p lower bounds σ_k^L and p upper bounds σ_k^U, we obtain lower and upper bounds by computing the means

$$\frac{1}{p}\sum_{k=1}^{p}\sigma_k^L \leq \frac{1}{p}\sum_{k=1}^{p}(z^k)^T f(A)z^k \leq \frac{1}{p}\sum_{k=1}^{p}\sigma_k^U.$$

The quality of such an estimation was assessed in Bai, Fahey and Golub [17] by using the following Hoeffding exponential inequality.

PROPOSITION 11.10 *Let $w_1, w_2, \ldots, w_p$ be p independent random variables with zero means and ranges $a_i \leq w_i \leq b_i$. Then for each $\eta > 0$ we have the following inequalities for the probabilities*

$$P(w_1 + \cdots + w_p \geq \eta) \leq \exp\left(\frac{-2\eta^2}{\sum_{i=1}^{p}(b_i - a_i)^2}\right)$$

and

$$P(|w_1 + \cdots + w_p| \geq \eta) \leq 2\exp\left(\frac{-2\eta^2}{\sum_{i=1}^{p}(b_i - a_i)^2}\right).$$

We apply this result with $w_i = (z^i)^T(\ln(A))z^i - \mathrm{tr}(\ln(A))$. Let $\sigma^L = \min \sigma_k^L$ and $\sigma^U = \max \sigma_k^U$. Then we have the bounds

$$\sigma^L - \mathrm{tr}(\ln(A)) \leq w_i \leq \sigma^U - \mathrm{tr}(\ln(A)).$$

Hence, by proposition 11.10 we have

$$P\left(\left|\frac{1}{p}\sum_{k=1}^{p}(z^k)^T \ln(A)z^k - \mathrm{tr}(\ln(A))\right| \geq \frac{\eta}{p}\right) \leq 2\exp\left(\frac{-2\eta^2}{d}\right),$$

where $d = p(\sigma^U - \sigma^L)^2$ and η is a given positive tolerance value. This means that

$$P\left(\frac{1}{p}\sum_{k=1}^{p}\sigma_k^L - \frac{\eta}{p} < \mathrm{tr}(\ln(A)) < \frac{1}{p}\sum_{k=1}^{p}\sigma_k^U + \frac{\eta}{p}\right) > 1 - 2\exp\left(\frac{-2\eta^2}{d}\right).$$

Therefore the trace of $\ln(A)$ is in the interval

$$\left[\frac{1}{p}\sum_{k=1}^{p}\sigma_k^L - \frac{\eta}{p}, \frac{1}{p}\sum_{k=1}^{p}\sigma_k^U + \frac{\eta}{p}\right],$$

with probability $q = 1 - 2\exp(-2\eta^2/d)$. If the probability q is specified we have

$$\frac{\eta}{p} = \sqrt{-\frac{1}{2p}(\sigma^U - \sigma^L)^2 \ln\left(\frac{1-q}{2}\right)}.$$

For a fixed value of q, $\eta/p \to 0$ when $p \to \infty$. Therefore, for a large value of p, the confidence interval is determined by the means of the lower and upper bounds. Of course, it does not make sense to have $p \geq n$.

For the case of a nonsingular and nonsymmetric matrix A which arises in some lattice QCD problems [296], one has to estimate

$$\det(A^T A) = (\det(A))^2 = \exp(\mathrm{tr}(\ln(A^T A))).$$

Therefore, we can estimate $\mathrm{tr}(f(A^T A))$ with $f(\lambda) = \ln(\lambda)$ using the same techniques as before. Finally, we have

$$\det(A) = \pm\sqrt{\exp(\mathrm{tr}(\ln(A^T A)))}.$$

The drawback of this technique is that we cannot obtain the sign of the determinant. However, there are many problems for which this is known in advance.

11.6.3 Partial Eigensums

Some applications in solid state physics for computing the total energy of an electronic structure require solving the following problem; see Bai, Fahey, Golub, Menon and Richter [18]. Let H and S be two symmetric matrices of order n, S being positive definite, involved in the generalized eigenvalue problem

$$H\psi = \lambda S\psi,$$

with eigenvalues λ_i. Let μ be a real number such that

$$\lambda_1 \leq \lambda_2 \cdots \leq \lambda_m < \mu < \lambda_{m+1} \leq \cdots \leq \lambda_n.$$

Then, one wants to compute the partial eigenvalue sum

$$\tau_\mu = \lambda_1 + \cdots + \lambda_m.$$

To solve this problem, let us construct a function f such that

$$f(\lambda_i) = \begin{cases} \lambda_i, & \text{if } \lambda_i < \mu, \\ 0, & \text{if } \lambda_i > \mu. \end{cases}$$

A simple choice is $f(\lambda) = \lambda h(\lambda)$, h being a step function:

$$h(\lambda) = \begin{cases} 1, & \text{if } \lambda < \mu. \\ 0, & \text{if } \lambda > \mu. \end{cases}$$

However, for our purposes, it is better to use a continuously differentiable function. Let $f(\lambda) = \lambda g(\lambda)$ where

$$g(\lambda) = \frac{1}{1 + \exp\left(\frac{\lambda - \mu}{\kappa}\right)},$$

κ being a given parameter. The smaller is κ, the closer g is to the step function h. Hence, as an approximation of the partial eigensum of A, we use

$$\sum_{i=1}^{n} (e^i)^T f(A) e^i.$$

The problem has been transformed to computing $(e^i)^T f(A) e^i$, $i = 1, \ldots, n$. In [18] this algorithm is applied to the matrix $A = L^{-1} H L^{-T}$ where L arises from a Cholesky factorization of $S = LL^T$.

This can be expensive if n is large. Another possibility that we will explore in the numerical experiments is to use the modified Chebyshev algorithm to compute the trace of $f(A)$.

11.7 Krylov Subspace Spectral Methods

Another application of Gauss quadrature rules for estimating bilinear forms was developed by J. Lambers [211], [212], [213], [215] and [215]. The goal is to solve time-dependent linear partial differential equations with spectral methods. Consider, for instance, the following problem in one spatial dimension

$$\frac{\partial u}{\partial t} + L(x, D)u = 0, \quad u(x, 0) = u_0$$

in $[0, 2\pi]$ with periodic boundary conditions

$$u(0, t) = u(2\pi, t),$$

where D is a differential operator of order m. To solve such a PDE, the important operator is $S = \exp(-L)$. The Krylov subspace spectral (KSS) method computes an approximation of the solution at discrete times $t_n = n\Delta t$. It uses Gauss quadrature to compute the Fourier components of the approximation $\tilde{u}$ of the solution

$$\hat{u}(\omega, t_{n+1}) = \left\langle \frac{1}{\sqrt{2\pi}} e^{i\omega x}, S(x, D)\tilde{u}(x, t_n) \right\rangle.$$

Space is discretized with a uniform grid of step size $h = 2\pi/N$. Then,

$$\hat{u}_{n+1} \approx (\hat{e}_\omega)^H S_N(\Delta t) u(t_n),$$

where the components of the vectors are

$$[\hat{e}_\omega]_j = \frac{1}{\sqrt{2\pi}} e^{i\omega jh}, \quad [u(t_n)]_j = u(jh, t_n),$$

and the matrix is $S_N = \exp(-L_N)$ where L_N is the operator restricted to the spatial grid.

The bilinear form is estimated by using our techniques with Gauss or Gauss–Radau quadrature rules. In practical applications the number of quadrature nodes is small, say 2 or 3. The main reason for the success of this type of method is that a different Krylov subspace is used for each Fourier component. Note that is is different from the methods using exponential integrators, see for instance Hochbruck and Lubich [192].

The practical implementation of KSS algorithms for different types of PDEs is studied in [212]. Numerous numerical examples of the efficiency of these methods are given in Lambers' articles. More recently, block quadrature rules have been investigated in [215].

11.8 Numerical Experiments

In this section we provide numerical results concerning the problems we have described in the previous sections. First we consider computing bounds for elements of a function of a symmetric matrix A. We consider $f(\lambda) = 1/\lambda$, which corresponds to the inverse of A, $f(\lambda) = e^\lambda$ corresponding to the exponential of the matrix and $f(\lambda) = \sqrt{\lambda}$ for the square root of the matrix. For applications in physics, see Haydock [183], [182], Haydock and Te [184] and Nex [247], [248]. We use some examples from [149].

11.8.1 Description of the examples

We first consider examples of small dimension for which the inverses are explicitly known. Then, we will turn to larger examples arising from the discretization of partial differential equations. The numerical computations were done with Matlab 6 on different personal computers using IEEE floating point arithmetic.

Example F1

This is an example of dimension 10,

$$
A = \frac{1}{11}
\begin{pmatrix}
10 & 9 & 8 & 7 & 6 & 5 & 4 & 3 & 2 & 1 \\
9 & 18 & 16 & 14 & 12 & 10 & 8 & 6 & 4 & 2 \\
8 & 16 & 24 & 21 & 18 & 15 & 12 & 9 & 6 & 3 \\
7 & 14 & 21 & 28 & 24 & 20 & 16 & 12 & 8 & 4 \\
6 & 12 & 18 & 24 & 30 & 25 & 20 & 15 & 10 & 5 \\
5 & 10 & 15 & 20 & 25 & 30 & 24 & 18 & 12 & 6 \\
4 & 8 & 12 & 16 & 20 & 24 & 28 & 21 & 14 & 7 \\
3 & 6 & 9 & 12 & 15 & 18 & 21 & 24 & 16 & 8 \\
2 & 4 & 6 & 8 & 10 & 12 & 14 & 16 & 18 & 9 \\
1 & 2 & 3 & 4 & 5 & 6 & 7 & 8 & 9 & 10
\end{pmatrix}.
$$

It is known (see [234]) that the inverse of A is a tridiagonal matrix

$$
A^{-1} =
\begin{pmatrix}
2 & -1 & & & \\
-1 & 2 & -1 & & \\
& \ddots & \ddots & \ddots & \\
& & -1 & 2 & -1 \\
& & & -1 & 2
\end{pmatrix}.
$$

The eigenvalues of A are therefore distinct, explicitly known as the inverses of those of A^{-1}. To four decimal digits the minimum and maximum eigenvalues are 0.2552 and 12.3435.

Example F2

We use a tridiagonal matrix whose nonzero elements in a row are -1, 2, -1 except for the first row where the diagonal element is 3 and the last row for which the diagonal element is 1. The inverse is one half a matrix whose elements of the first row and column are equal to 1. In row and column i, the elements (i, j) for $j \geq i$ in the row and (j, i) for $j \geq i$ in the column are equal to $2(i - 1) + 1$.

An example of order 5 is

$$
A =
\begin{pmatrix}
3 & -1 & & & \\
-1 & 2 & -1 & & \\
& -1 & 2 & -1 & \\
& & -1 & 2 & -1 \\
& & & -1 & 1
\end{pmatrix},
$$

whose inverse is

$$
A^{-1} = \frac{1}{2}
\begin{pmatrix}
1 & 1 & 1 & 1 & 1 \\
1 & 3 & 3 & 3 & 3 \\
1 & 3 & 5 & 5 & 5 \\
1 & 3 & 5 & 7 & 7 \\
1 & 3 & 5 & 7 & 9
\end{pmatrix}.
$$

The minimum and maximum eigenvalues of the matrix A are 0.0979 and 3.9021.

Example F3

We consider an example proposed by Strakoš [315]. This is a diagonal matrix Λ whose diagonal elements are

$$\lambda_i = \lambda_1 + \left(\frac{i-1}{n-1}\right)(\lambda_n - \lambda_1)\rho^{n-i}, \; i = 1, \ldots, n.$$

The parameters λ_1 and λ_n are respectively the smallest and largest eigenvalues. The parameter ρ controls the distribution of the eigenvalues. We will use $\lambda_1 = 0.1, \lambda_n = 100$ and a value $\rho = 0.9$ which gives well-separated large eigenvalues. Let Q be the orthogonal matrix of the eigenvectors of the tridiagonal matrix $(-1, 2, -1)$. Then the matrix is $A = Q^T \Lambda Q$. It has the same eigenvalues as Λ.

Example F4

This example is the matrix arising from the five-point finite difference approximation of the Poisson equation in a unit square with an $m \times m$ mesh. This gives a linear system $Ax = c$ of order m^2, where

$$A = \begin{pmatrix} T & -I & & & \\ -I & T & -I & & \\ & \ddots & \ddots & \ddots & \\ & & -I & T & -I \\ & & & -I & T \end{pmatrix},$$

each block being of order m, and

$$T = \begin{pmatrix} 4 & -1 & & & \\ -1 & 4 & -1 & & \\ & \ddots & \ddots & \ddots & \\ & & -1 & 4 & -1 \\ & & & -1 & 4 \end{pmatrix}.$$

For $m = 6$, the minimum and maximum eigenvalues are 0.3961 and 7.6039.

Example F5

This example arises from the five-point finite difference approximation of the following diffusion equation in a unit square,

$$-\text{div}(a\nabla u)) = f,$$

with homogeneous Dirichlet boundary conditions. The diffusion coefficient $a(x, y)$ is constant and equal to 1000 in the square $]1/4, 3/4[\times]1/4, 3/4[$ and equal to 1 otherwise. For $m = 6$, the minimum and maximum eigenvalues are 0.4354 and 6828.7.

In the captions of the tables we will denote the use of Gauss quadrature with the Lanczos algorithm by GL, the nonsymmetric Gauss quadrature with the nonsymmetric Lanczos algorithm by GNS and the block Gauss quadrature with the block Lanczos algorithm by GB.

11.8.2 Bounds for the Elements of the Inverse

11.8.2.1 Diagonal Elements

In the following results, Nit denotes the number of iterations of the Lanczos algorithm. This corresponds to N for the Gauss and Gauss–Radau rules and $N - 1$ for the Gauss–Lobatto rule.

Example F1

We are looking for bounds for $(A^{-1})_{5,5}$ whose exact value is 2. Results (rounded to four decimal digits) are given in table 11.1. In this example five or six iterations should be sufficient, so we are a little off the theory. For Gauss–Radau and Gauss–Lobatto we use the "exact" smallest and largest eigenvalues.

Table 11.1 Example F1, GL, $A_{5,5}^{-1} = 2$

Rule	Nit=1	2	3	4	5	6	7
G	0.3667	1.3896	1.7875	1.9404	1.9929	1.9993	2
G-R b_L	1.3430	1.7627	1.9376	1.9926	1.9993	2.0000	2
G-R b_U	3.0330	2.2931	2.1264	2.0171	2.0020	2.0001	2
G-L	3.1341	2.3211	2.1356	2.0178	2.0021	2.0001	2

Example F2

Let us first consider a small example of order $n = 5$. We look at bounds for $(A^{-1})_{5,5}$, whose exact value is 4.5. Results are given in table 11.2.

Table 11.2 Example F2, GL, $n = 5$, $A_{5,5}^{-1} = 4.5$

Rule	Nit=1	2	3	4	5
G	1	2	3	4	4.5
G-R b_L	1.3910	2.4425	3.4743	4.5	4.5
G-R b_U	5.8450	4.7936	4.5257	4.5	4.5
G-L	7.8541	5.2361	4.6180	4.5	4.5

Now, we consider the same example with $n = 100$. We are interested in the $(50, 50)$ element of the inverse whose value is 49.5. Results are given in table 11.3. These results do not seem to be very encouraging since we obtain good bounds only after a large number of iterations, around 80 or 90 iterations. However, the eigenvalue distribution for the matrix A is such that the convergence of the Ritz values towards the eigenvalues is very slow and this is what we see also for the lower and upper bounds for the quadratic form. With this kind of matrix, since there is no convergence in the early iterations, orthogonality of the Lanczos vectors is usually quite good. We have seen in chapter 7 that the convergence of the bounds for the quadratic form is closely linked to the convergence of the Ritz values towards the eigenvalues of A. For this example, it could be interesting to use a preconditioner.

Table 11.3 Example F2, GL, $n = 100$, $A_{50,50}^{-1} = 49.5$

Nit	G	G-R b_L	G-R b_U	G-L
10	5.0000	5.2503	196.4856	205.9564
20	10.0000	10.2507	105.6575	107.9577
30	15.0000	15.2510	76.5260	77.4667
40	20.0000	20.2515	63.3448	63.8080
50	24.8333	24.9187	55.6146	56.3129
60	29.9783	30.2094	52.2493	52.3811
70	34.9884	35.2308	50.5347	50.5955
80	39.9921	40.2407	49.7693	49.7925
90	44.9940	45.2526	49.5253	49.5300
100	49.5000	49.5000	49.5000	49.5000

Example F3

We choose $n = 100$ and we consider the $(50, 50)$ element whose value is 4.2717. The results are given in table 11.4. The convergence is much faster than with the previous example. Due to the rapid convergence of some eigenvalues at the beginning of the iterations, there is a large growth of the rounding errors and appearance of multiple copies of the already converged eigenvalues; see [239]. But we obtain good bounds after only 30 iterations.

Table 11.4 Example F3, GL, $n = 100$, $A_{50,50}^{-1} = 4.2717$

Nit	G	G-R b_L	G-R b_U	G-L
10	2.7850	3.0008	5.1427	5.1664
20	4.0464	4.0505	4.4262	4.4643
30	4.2545	4.2553	4.2883	4.2897
40	4.2704	4.2704	4.2728	4.2733
50	4.2716	4.2716	4.2718	4.2718
60	4.2717	4.2717	4.2717	4.2717

Example F4

We first consider $m = 6$. Then we have a small matrix of order $n = 36$ and we look for bounds on $(A^{-1})_{18,18}$ whose value is 0.3515. This matrix has 19 distinct eigenvalues, therefore we should get the exact answer in about 10 iterations for Gauss and Gauss–Radau and 9 iterations for Gauss–Lobatto. Results are given in table 11.5.

Now, we consider $m = 30$ which gives a matrix of order 900. We want to compute bounds for the $(150, 150)$ element whose value is 0.3602. We show the

Table 11.5 Example F4, GL, $n = 36$, $A^{-1}_{18,18} = 0.3515$

Rule	Nit=1	2	3	4	8	9
G	0.25	0.3077	0.3304	0.3411	0.3512	0.3515
G-R b_L	0.2811	0.3203	0.3366	0.3443	0.3514	0.3515
G-R b_U	0.6418	0.4178	0.3703	0.3572	0.3515	0.3515
G-L	1.3280	0.4990	0.3874	0.3619	0.3515	-

results in table 11.6. We have very good estimates much sooner than expected. This is because there are distinct eigenvalues which are very close to each other.

Table 11.6 Example F4, GL, $n = 900$, $A^{-1}_{150,150} = 0.3602$

Nit	G	G-R b_L	G-R b_U	G-L
10	0.3578	0.3581	0.3777	0.3822
20	0.3599	0.3599	0.3608	0.3609
30	0.3601	0.3601	0.3602	0.3602
40	0.3602	0.3602	0.3602	0.3602

Example F5

We take $m = 6$ as in the previous example. So we have a matrix of order 36. The $(2, 2)$ element of the inverse has an "exact" value of 0.3088 and there are 23 distinct eigenvalues so that the exact answer should be obtained after 12 iterations but the matrix is ill conditioned. We get the results in table 11.7. Then, we use $m = 30$ which gives a matrix of order $n = 900$ and look for the element $(200, 200)$ whose value is 0.4347. Results are given in table 11.8. Convergence is slower than for Example F4.

Table 11.7 Example F5, GL, $n = 36$, $A^{-1}_{2,2} = 0.3088$

Rule	Nit=1	2	4	6	8	10	12	15
G	0.25	0.2503	0.2525	0.2609	0.2837	0.2889	0.3036	0.3088
G-R b_L	0.2504	0.2516	0.2583	0.2821	0.2879	0.2968	0.3044	0.3088
G-R b_U	0.5375	0.5202	0.5080	0.5039	0.5013	0.3237	0.3098	0.3088
G-L	2.2955	0.5765	0.5156	0.5065	0.5020	0.3237	0.3098	0.3088

In conclusion, we have seen that the convergence of the bounds or estimates obtained with the Lanczos algorithm is closely linked to the convergence of the Ritz values toward the eigenvalues of A.

Table 11.8 Example F5, GL, $n = 900$, $A_{200,200}^{-1} = 0.4347$

Nit	G	G-R b_L	G-R b_U	G-L
10	0.2992	0.3005	6.4655	6.5569
60	0.3393	0.3400	0.9783	0.9800
110	0.3763	0.3763	0.4791	0.4800
160	0.4085	0.4085	0.4371	0.4374
210	0.4292	0.4292	0.4351	0.4351
260	0.4343	0.4343	0.4348	0.4348
310	0.4347	0.4347	0.4347	0.4347

11.8.2.2 Off-Diagonal Elements with the Nonsymmetric Lanczos Algorithm

Here, we use the nonsymmetric Lanczos algorithm to obtain estimates or bounds of off-diagonal elements.

Example F1

We are looking for estimates for the sum of the $(2, 2)$ and $(2, 1)$ elements whose exact value is 1. First, we use $\delta = 1$ for which the measure is positive but not increasing. Results are in table 11.9.

Table 11.9 Example F1, GNS, $A_{2,2}^{-1} + A_{2,1}^{-1} = 1$

Rule	Nit=1	2	3	4	5	6	7	8
G	0.4074	0.6494	0.8341	0.9512	0.9998	1.0004	1	-
G-R b_L	0.6181	0.8268	0.9488	0.9998	1.0004	1.0001	1	-
G-R b_U	2.6483	1.4324	1.0488	1.0035	1.0012	0.9994	1	-
G-L	3.2207	1.4932	1.0529	1.0036	1.0012	0.9993	0.9994	1

We have a small problem at the end near convergence where we obtain estimates and not bounds, but the estimates are quite good. Note that for $\delta = 4$ the measure is positive and increasing.

Example F2

This example illustrates some of the problems that can happen with the nonsymmetric Lanczos algorithm. We would like to compute the sum of the $(2, 2)$ and $(2, 1)$ elements which is equal to 2. After two iterations we have a breakdown of the Lanczos algorithm since $\gamma\beta = 0$. The same happens at the first iteration for the Gauss–Radau rule and at the second one for the Gauss–Lobatto rule. Choosing a value of δ different from 1 cures the breakdown problem. We can obtain bounds with a value $\delta = 10$ (with a positive and increasing measure). Then the value we are looking for is 1.55 and the results of table 11.10 follow. Table 11.11 shows the results for $n = 100$. We are computing approximations of the sum of elements $(50, 50)$ and $(50, 1)/10$ whose value is 49.55. As with the symmetric Lanczos algorithm the convergence is slow. At the beginning the results are very close to those for the symmetric case because $A_{50,1}^{-1} = 0.5$, thus the correction to $A_{50,50}^{-1} = 49.5$ is small.

Table 11.10 Example F2, GNS, $n = 5$, $A_{2,2}^{-1} + A_{2,1}^{-1}/10 = 1.55$

Rule	Nit=1	2	3	4	5
G	0.5263	0.8585	1.0333	1.4533	1.55
G-R b_L	-	1.0011	1.2771	1.55	-
G-R b_U	-	1.9949	1.5539	1.55	-
G-L	-	2.2432	1.5696	1.55	-

Table 11.11 Example F2, GNS, $n = 100$, $A_{50,50}^{-1} + A_{50,1}^{-1}/10 = 49.55$

Nit	G	G-R b_L	G-R b_U	G-L
10	5.0000	5.2503	196.4856	205.9564
20	10.0000	10.2507	105.6575	107.9577
30	15.0263	15.2771	76.6303	77.5740
40	20.0263	20.2775	63.4181	63.8826
50	24.8585	24.9436	55.6715	56.3736
60	30.0044	30.2352	52.2992	52.4315
70	35.0146	35.2564	50.5807	50.6418
80	40.1023	40.5801	49.8167	49.8283
90	45.1135	45.3149	49.5733	49.5791
100	49.5500	49.5500	49.5500	49.5500

Example F3

We use $n = 100$ and we compute $A_{50,50}^{-1} + A_{50,49}^{-1} = 1.4394$. The results are given in table 11.12. We see that we obtain good bounds with a few iterations.

Table 11.12 Example F3, GNS, $n = 100$, $A_{50,50}^{-1} + A_{50,49}^{-1} = 1.4394$

Nit	G	G-R b_L	G-R b_U	G-L
10	0.8795	0.9429	2.2057	2.2327
20	1.3344	1.3362	1.5535	1.5839
30	1.4301	1.4308	1.4510	1.4516
40	1.4386	1.4387	1.4404	1.4404
50	1.4394	1.4394	1.4395	1.4395
60	1.4394	1.4394	1.4394	1.4394

Example F4

We consider $m = 6$; then, we have a system of order 36 and we look for estimates of the sum of the $(2, 2)$ and $(2, 1)$ elements which is 0.4471. Remember there are 19 distinct eigenvalues. Results are given in table 11.13. Then, in table 11.14, we use $n = 900$ and we compute the sum $A_{150,150}^{-1} + A_{150,50}^{-1} = 0.3665$.

Table 11.13 Example F4, GNS, $n = 36$, $A_{2,2}^{-1} + A_{2,1}^{-1} = 0.4471$

Rule	Nit=1	2	4	6	7	8	9	10
G	0.3333	0.4000	0.4369	0.4446	0.4461	0.4468	0.4471	-
G-R b_L	0.3675	0.4156	0.4390	0.4456	0.4466	0.4470	0.4471	-
G-R b_U	0.7800	0.5319	0.4537	0.4476	0.4472	0.4472	0.4471	-
G-L	1.6660	0.6238	0.4596	0.4480	0.4473	0.4472	0.4472	0.4471

Table 11.14 Example F4, GNS, $n = 900$, $A_{150,150}^{-1} + A_{150,50}^{-1} = 0.3665$

Nit	G	G-R b_L	G-R b_U	G-L
10	0.3611	0.3615	0.3917	0.3979
20	0.3656	0.3657	0.3678	0.3680
30	0.3663	0.3664	0.3666	0.3666
40	0.3665	0.3665	0.3665	0.3665

Example F5

We first take $m = 6$. The sum of the $(2, 2)$ and $(2, 1)$ elements of the inverse is 0.3962 and there are 23 distinct eigenvalues. We obtain the results in table 11.15. Then we use $n = 900$ and we compute the sum $A_{200,200}^{-1} + A_{200,172}^{-1} = 0.5625$. The results are given in table 11.16. We see that the convergence is slow.

Table 11.15 Example F5, GNS, $n = 36$, $A_{2,2}^{-1} + A_{2,1}^{-1} = 0.3962$

Rule	Nit=1	2	4	6	8	10	12	15
G	0.3333	0.3336	0.3348	0.3396	0.3607	0.3689	0.3899	0.3962
G-R b_L	-	0.3337	0.3355	0.3460	0.3672	0.3803	0.3912	0.3962
G-R b_U	-	0.6230	0.5793	0.5698	0.5660	0.4078	0.3970	0.3962
G-L	2.2959	0.6898	0.5850	0.5703	0.5664	0.4078	0.3970	0.3962

11.8.2.3 Off-Diagonal Elements with the Block Lanczos Algorithm

When using block quadrature rules for computing elements of the inverse we are faced with the problem of computing $e^T J_k^{-1} e$ where e is a $2k \times 2$ matrix which is

Table 11.16 Example F5, GNS, $n = 900$, $A_{200,200}^{-1} + A_{200,172}^{-1} = 0.5625$

Nit	G	G-R b_L	G-R b_U	G-L
10	0.3163	0.3186	6.6552	6.7077
60	0.3859	0.3867	1.3049	1.3077
110	0.4592	0.4593	0.6562	0.6570
160	0.5153	0.5154	0.5693	0.5694
210	0.5336	0.5346	0.5635	0.5635
260	0.5607	0.5607	0.5626	0.5626
310	0.5623	0.5623	0.5625	0.5625
360	0.5625	0.5625	0.5625	0.5625

made of the two first columns of the identity and J_k is block tridiagonal. This can be done incrementally as for the point tridiagonal case; see chapter 3. Let f be a matrix of dimension $2k \times 2$ whose columns are the last two columns of the identity and denote by $(J_k^{-1})_{1,1}$ the 2×2 $(1,1)$ block of the inverse of J_k with

$$
J_k = \begin{pmatrix}
\Omega_1 & \Gamma_1^T & & & \\
\Gamma_1 & \Omega_2 & \Gamma_2^T & & \\
& \ddots & \ddots & \ddots & \\
& & \Gamma_{k-2} & \Omega_{k-1} & \Gamma_{k-1}^T \\
& & & \Gamma_{k-1} & \Omega_k
\end{pmatrix}.
$$

The matrix J_{k+1} is written as

$$
J_{k+1} = \begin{pmatrix} J_k & f\Gamma_k^T \\ \Gamma_k f^T & \Omega_{k+1} \end{pmatrix}.
$$

To obtain the block $(1,1)$ element we are interested in the inverse of the Schur complement $S_k = J_k - f\Gamma_k^T \Omega_{k+1}^{-1} \Gamma_k f^T$. We use the Sherman–Morrison–Woodbury formula (see [154]) which gives

$$
S_k^{-1} = J_k^{-1} + J_k^{-1} f\Gamma_k^T (\Omega_{k+1} - \Gamma_k f^T J_k^{-1} f\Gamma_k^T)^{-1} \Gamma_k f^T J_k^{-1}.
$$

For the 2×2 matrix we are interested in we obtain

$$
(J_{k+1}^{-1})_{1,1} = (J_k^{-1})_{1,1} + (e^T J_k^{-1} f)\Gamma_k^T (\Omega_{k+1} - \Gamma_k f^T J_k^{-1} f\Gamma_k^T)^{-1} \Gamma_k (f^T J_k^{-1} e).
$$

Hence, we have to compute $e^T J_k^{-1} f$ which is the $(1, k)$ block of the inverse of J_k and $(\Omega_{k+1} - \Gamma_k f^T J_k^{-1} f\Gamma_k^T)^{-1}$. This is done using the block LU factorization of J_k whose block diagonal elements are given by the recurrence

$$
\Delta_1 = \Omega_1, \quad \Delta_i = \Omega_i - \Gamma_{i-1}\Omega_{i-1}^{-1}\Gamma_{i-1}^T, \quad i = 2, \dots, k.
$$

With these notations, we have

$$
e^T J_k^{-1} f = (-1)^{k-1} \Delta_1^{-1} \Gamma_1^T \Delta_2^{-1} \Gamma_2^T \cdots \Delta_{k-1}^{-1} \Gamma_{k-1}^T \Delta_k^{-1},
$$

$f^T J_k^{-1} f = \Delta_k^{-1}$ and $\Omega_{k+1} - \Gamma_k \Delta_k^{-1} \Gamma_k^T = \Delta_{k+1}$. Let

$$
C_k = \Delta_1^{-1} \Gamma_1^T \Delta_2^{-1} \Gamma_2^T \cdots \Delta_{k-1}^{-1} \Gamma_{k-1}^T \Delta_k^{-1} \Gamma_k^T.
$$

Putting these formulas together we obtain

$$(J_{k+1}^{-1})_{1,1} = (J_k^{-1})_{1,1} + C_k \Delta_{k+1}^{-1} C_k^T.$$

Going from step k to step $k+1$ we compute C_{k+1} incrementally. Note that we can reuse $C_k \Delta_{k+1}^{-1}$ to compute C_{k+1}.

Since it is difficult to compute small examples using 2×2 blocks, we start with Example F3.

Example F3

This example uses $n = 100$. The $(2, 1)$ element of the inverse is -3.2002. We obtain the figures in table 11.17. We see that we obtain good approximations but not always bounds. As an added bonus we also obtain estimates of $A_{1,1}^{-1}$ and $A_{2,2}^{-1}$.

Table 11.17 Example F3, GB, $n = 100$, $A_{2,1}^{-1} = -3.2002$

Nit	G	G-R b_L	G-R b_U	G-L
2	-3.0808	-3.0948	-3.9996	-4.1691
3	-3.1274	-3.1431	-3.5655	-3.6910
4	-3.2204	-3.2187	-3.2637	-3.5216
5	-3.2015	-3.2001	-3.1974	-3.2473
6	-3.1969	-3.1966	-3.1964	-3.1969
7	-3.1970	-3.1972	-3.1995	-3.1994
8	-3.1993	-3.1995	-3.2008	-3.1999
9	-3.2001	-3.2001	-3.2005	-3.2008
10	-3.2002	-3.2002	-3.2002	-3.2004

Example F4

We consider a problem of order $n = 900$ and look for the $(400, 100)$ element of the inverse which is equal to 0.0597. Results are given in table 11.18. Note that for this problem everything works well. The Gauss rule gives a lower bound, Gauss–Radau a lower and an upper bound.

Table 11.18 Example F4, GB, $n = 900$, $A_{400,100}^{-1} = 0.0597$

Nit	G	G-R b_L	G-R b_U	G-L
10	0.0172	0.0207	0.0632	0.0588
20	0.0527	0.0532	0.0616	0.0621
30	0.0590	0.0591	0.0597	0.0597
40	0.0597	0.0597	0.0597	0.0597

Example F5

As before we use $n = 900$. We would like to obtain estimates of the $(2, 1)$ element of the inverse whose value is 0.1046. We get the results in table 11.19.

Table 11.19 Example F5, GB, $n = 900$, $A_{2,1}^{-1} = 0.1046$

Nit	G	G-R b_L	G-R b_U	G-L
2	0.0894	0.0894	0.7349	1.8810
4	0.1008	0.1008	0.2032	0.3383
6	0.1033	0.1033	0.1280	0.1507
8	0.1040	0.1040	0.1119	0.1173
10	0.1044	0.1044	0.1074	0.1090
12	0.1045	0.1045	0.1058	0.1064
14	0.1046	0.1046	0.1054	0.1054
16	0.1046	0.1046	0.1054	0.1054

Note that in this example we obtain bounds. Now, we illustrate the fact that some estimates can be 0 for some iterations. This is one of the reasons for which we cannot always obtain bounds with the block Lanczos algorithm. We take $n = 36$ and we would like to estimate the $(36, 1)$ element of the inverse, which is 0.005. For the first iterations, the computed approximations are 0 which means that the 2×2 matrices which provide them are diagonal; see table 11.20.

Table 11.20 Example F5, GB, $n = 36$, $A_{36,1}^{-1} = 0.005$

Rule	Nit=2	4	6	8	10	11
G	0.	0.	0.0002	0.0037	0.0049	0.0050
G-R b_L	0.	0.	0.0023	0.0037	0.0049	0.0050
G-R b_U	0.	0.	0.0024	0.0050	0.0050	0.0050
G-L	0.	0.	0.0022	0.0044	0.0050	0.0050

11.8.2.4 Dependence on the Eigenvalue Estimates

In this part, we investigate numerically how the bounds and estimates of the Gauss–Radau rules depend on the accuracy of the estimates of the extreme eigenvalues of A. We take Example F4 with $m = 6$ and look at the results given by the Gauss–Radau rule as a function of a and b. Remember that in the previous experiments we took for a and b the values returned by the EIG function of Matlab, that is $a = 0.3961, b = 7.6039$. Using these values, we need nine Lanczos iterations to obtain the correct result up to four decimal digits.

It turns out that (for this example) the estimates are only weakly dependent on the values of a and b. We look at the number of Lanczos iterations needed to obtain an upper bound for the element $(18, 18)$ with four exact digits. The results are given in table 11.21. Note that it works even when $a > \lambda_{\min}$. We have the same properties when b is varied.

Therefore, we see that the estimation of the extreme eigenvalues does not seem to matter too much. In any case, better estimates of the smallest and largest eigenvalues can be obtained after a few iterations of the Lanczos algorithm or with the

Table 11.21 Example F4, GL, $n = 36$

$a = 10^{-4}$	10^{-2}	0.1	0.3	0.4	1	6
15	13	11	11	8	8	9

Gerschgorin disks.

11.8.3 Bounds for the Elements of the Exponential

In this section we are looking for bounds of elements of the exponential of the matrices for some of the examples. We will see that the convergence of the bounds to the exact values is quite fast.

11.8.3.1 Diagonal Elements

We first compute diagonal elements of the exponential for some of our examples.

Example F1
   The $(5, 5)$ entry is $4.0879 \; 10^4$. The Gauss rule obtains the "exact" value in four iterations, the Gauss–Radau and Gauss–Lobatto rules in three iterations.

Example F3
   We use $n = 100$ and compute the $(50, 50)$ element, whose value is $5.3217 \; 10^{41}$. Results are given in table 11.22. We obtain good bounds very rapidly.

Table 11.22 Example F3, GL, $n = 100$, $\exp(A)_{50,50} = 5.3217 \; 10^{41}$. Results $\times 10^{-41}$

Nit	G	G-R b_L	G-R b_U	G-L
2	0.0000	0.0000	7.0288	8.8014
3	0.0075	0.2008	5.6649	6.0776
4	1.0322	2.5894	5.3731	5.4565
5	3.9335	4.7779	5.3270	5.3385
6	5.1340	5.2680	5.3235	5.3232
7	5.3070	5.3178	5.3218	5.3219
8	5.3203	5.3209	5.3218	5.3218
9	5.3212	5.3213	5.3217	5.3217
10	5.3215	5.3217	5.3217	5.3217
11	5.3217	5.3217	5.3217	5.3217

Example F4
   We use $n = 36$ and we consider the $(18, 18)$ element, whose value is 197.8311. We obtain the results in table 11.23. We remark that to compute diagonal elements of the exponential the convergence rate is quite fast. Then we take $n = 900$ and

compute the $(50, 50)$ element whose value is 277.4061. Results are given in table 11.24.

Table 11.23 Example F4, GL, $n = 36$, $\exp(A)_{18,18} = 197.8311$

Rule	Nit=2	3	4	5	6	7
G	159.1305	193.4021	197.5633	197.8208	197.8308	197.8311
G-R b_L	182.2094	196.6343	197.7779	197.8296	197.8311	197.8311
G-R b_U	217.4084	199.0836	197.8821	197.8325	197.8311	197.8311
G-L	273.8301	203.4148	198.0978	197.8392	197.8313	197.8311

Table 11.24 Example F4, GL, $n = 900$, $\exp(A)_{50,50} = 277.4061$

Rule	Nit=2	3	4	5	6	7	8
G	205.4089	270.6459	276.9261	277.3863	277.4055	277.4060	277.4061
G-R b_L	248.6974	275.1781	277.2898	277.4021	277.4060	277.4060	277.4061
G-R b_U	319.2222	280.3322	277.5413	277.4105	277.4062	277.4061	277.4061
G-L	409.7618	292.5355	278.1514	277.4350	277.4068	277.4061	277.4061

11.8.3.2 Off-Diagonal Elements

We consider only Example F4 with $n = 36$ and we would like to compute the element $(2, 1)$, whose value is -119.6646. First, we use the block Lanczos algorithm, which gives the results in table 11.25.

Table 11.25 Example F4, GB, $n = 36$, $\exp(A)_{2,1} = -119.6646$

Rule	Nit=2	3	4	5	6
G	-111.2179	-119.0085	-119.6333	-119.6336	-119.6646
G-R b_L	-115.9316	-119.4565	-119.6571	-119.6644	-119.6646
G-R b_U	-122.2213	-119.7928	-119.6687	-119.6647	-119.6646
G-L	-137.7050	-120.6801	-119.7008	-119.6655	-119.6646

Then, we use the nonsymmetric Lanczos algorithm. The sum of the $(2, 2)$ and $(2, 1)$ elements of the exponential is 73.9023. Results are in table 11.26.

Finally, we take $n = 900$ and consider $\exp(A)_{50,50} + \exp(A)_{50,49} = 83.8391$. Results are in table 11.27. Again, convergence is quite fast. We obtain good results after four iterations.

Table 11.26 Example F4, GNS, $n = 36$, $\exp(A)_{2,2} + \exp(A)_{2,1} = 73.9023$

Rule	Nit=2	3	4	5	6	7
G	54.3971	71.6576	73.7637	73.8962	73.9021	73.9023
G-R b_L	65.1847	73.2896	73.8718	73.9014	73.9023	-
G-R b_U	84.0323	74.6772	73.9323	73.9014	73.9023	-
G-L	113.5085	77.2717	74.0711	73.9070	73.9024	73.9023

Table 11.27 Example F4, GNS, $n = 900$, $\exp(A)_{50,50} + \exp(A)_{50,49} = 83.8391$

Rule	Nit=2	3	4	5	6	7
G	63.4045	81.4124	83.6607	83.8318	83.8389	83.8391
G-R b_L	76.1266	83.7668	83.7781	83.8383	83.8391	83.8391
G-R b_U	108.0918	86.3239	83.8796	83.8420	83.8392	83.8391
G-L	163.8043	90.9304	84.1878	83.8530	83.8395	83.8391

11.8.4 Bounds for the Elements of the Square Root

The last function we consider as an example is the square root. We use the same numerical examples as for the exponential to be able to compare the speed of convergence.

11.8.4.1 Diagonal Elements

Results for Example F1 are given in table 11.28. We compute bounds for the $(5, 5)$ entry of the square root. For this small example, we need one or two more iterations than for the exponential function to obtain the same precision.

Table 11.28 Example F1, GL, $n = 10$, $(\sqrt{A})_{5,5} = 1.2415$

Nit	G	G-R b_L	G-R b_U	G-L
2	1.2705	1.2328	1.2471	1.2311
3	1.2462	1.2392	1.2423	1.2390
4	1.2422	1.2413	1.2415	1.2413
5	1.2415	1.2415	1.2415	1.2415

The second example is F3 with $n = 100$. We compute bounds of the $(50, 50)$ element whose value is 1.8973. Results are in table 11.29. We need more iterations than for the exponential function. Note that the lower bound from the Gauss–Radau rule and the lower bound from the Gauss–Lobatto rule are slow to converge. However, convergence is faster than when computing elements of the inverse of A.

Results for Example F4 are displayed in table 11.30 for $n = 36$ and in table 11.31 for $n = 900$. For the case $n = 36$ the number of iterations is about the same as for

Table 11.29 Example F3, GL, $n = 100$, $(\sqrt{A})_{50,50} = 1.8973$

Nit	G	G-R b_L	G-R b_U	G-L
2	2.3440	1.7211	2.1962	1.6628
3	2.1328	1.7938	2.0637	1.7729
4	2.0385	1.8310	1.9992	1.8217
5	1.9875	1.8529	1.9628	1.8483
6	1.9569	1.8669	1.9407	1.8643
7	1.9373	1.8761	1.9279	1.8744
8	1.9247	1.8819	1.9214	1.8795
9	1.9182	1.8847	1.9143	1.8837
10	1.9133	1.8881	1.9092	1.8876
11	1.9085	1.8905	1.9069	1.8899
12	1.9060	1.8917	1.9042	1.8914
13	1.9039	1.8933	1.9023	1.8931
14	1.9020	1.8942	1.9013	1.8939
15	1.9011	1.8949	1.9001	1.8948
16	1.9000	1.8956	1.8995	1.8955
17	1.8994	1.8959	1.8990	1.8958
18	1.8988	1.8963	1.8988	1.8960
19	1.8984	1.8965	1.8984	1.8964
20	1.8982	1.8968	1.8981	1.8966
21	1.8979	1.8969	1.8979	1.8968
22	1.8978	1.8970	1.8978	1.8969
23	1.8976	1.8971	1.8976	1.8970
24	1.8976	1.8972	1.8976	1.8971
25	1.8975	1.8972	1.8975	1.8972
26	1.8974	1.8973	1.8974	1.8972

the exponential, but the convergence is slower when we increase the dimension of the matrix.

Table 11.30 Example F4, GL, $n = 36$, $(\sqrt{A})_{18,18} = 1.9438$

Nit	G	G-R b_L	G-R b_U	G-L
2	1.9501	1.9391	1.9468	1.9292
3	1.9452	1.9429	1.9445	1.9418
4	1.9442	1.9436	1.9440	1.9434
5	1.9439	1.9438	1.9439	1.9437
6	1.9438	1.9438	1.9438	1.9438

Table 11.31 Example F4, GL, $n = 900$, $(\sqrt{A})_{50,50} = 1.9189$

Nit	G	G-R b_L	G-R b_U	G-L
2	1.9319	1.8945	1.9255	1.8697
3	1.9220	1.9112	1.9209	1.9038
4	1.9201	1.9160	1.9197	1.9140
5	1.9195	1.9176	1.9193	1.9169
6	1.9192	1.9183	1.9191	1.9180
7	1.9191	1.9186	1.9190	1.9185
8	1.9190	1.9187	1.9190	1.9187
9	1.9190	1.9188	1.9190	1.9188
10	1.9190	1.9189	1.9190	1.9189
11	1.9190	1.9189	1.9190	1.9189
12	1.9190	1.9189	1.9189	1.9189
13	1.9189	1.9189	1.9189	1.9189

11.8.4.2 Off-Diagonal Elements

We first use the block Gauss quadrature rule for Example F4 with $n = 36$. The results are given in table 11.32. We obtain a nice result quite rapidly.

Table 11.32 Example F4, GB, $n = 36$, $(\sqrt{A})_{2,1} = -0.2627$.

Nit	G	G-R b_L	G-R b_U	G-L
2	-0.2612	-0.2618	-0.2638	-0.2669
3	-0.2623	-0.2625	-0.2629	-0.2633
4	-0.2626	-0.2626	-0.2628	-0.2628
5	-0.2627	-0.2627	-0.2627	-0.2627

Then, we use the nonsymmetric Lanczos algorithm for computing the sum of the elements $(2, 1)$ and $(2, 2)$ of the square root of the matrix of Example F4. Table 11.33 gives the results for $n = 36$ and table 11.34 for $n = 900$. As for the symmetric Lanczos algorithm, convergence is slower than for the exponential function. Nevertheless we obtain good bounds in a few iterations.

Table 11.33 Example F4, GNS, $n = 36$, $(\sqrt{A})_{2,1} + (\sqrt{A})_{2,2} = 1.6819$

Nit	G	G-R b_L	G-R b_U	G-L
2	1.6882	1.6778	1.6854	1.6664
3	1.6832	1.6808	1.6825	1.6796
4	1.6823	1.6817	1.6821	1.6815
5	1.6821	1.6819	1.6820	1.6818
6	1.6820	1.6819	1.6820	1.6819
7	1.6819	1.6819	1.6819	1.6819

Table 11.34 Example F4, GNS, $n = 900$, $(\sqrt{A})_{50,49} + (\sqrt{A})_{50,50} = 1.6411$

Nit	G	G-R b_L	G-R b_U	G-L
2	1.6559	1.6061	1.6484	1.5751
3	1.6451	1.6287	1.6433	1.6209
4	1.6428	1.6374	1.6424	1.6344
5	1.6419	1.6391	1.6416	1.6382
6	1.6415	1.6402	1.6414	1.6397
7	1.6413	1.6406	1.6413	1.6404
8	1.6412	1.6408	1.6412	1.6407
9	1.6412	1.6410	1.6412	1.6409
10	1.6412	1.6410	1.6412	1.6410
11	1.6412	1.6411	1.6411	1.6410
12	1.6411	1.6411	1.6411	1.6411

11.8.5 Linear Combination of the Solution of a Linear System

We are interested in $d^T x$ where x is the solution of the linear system $Ax = c$. Therefore, the value we would like to compute is $d^T A^{-1} c$. This can be done using the same techniques and codes as in the previous sections using the nonsymmetric Lanczos algorithm. Let us consider Example F4. We use $n = 900$, c is a random vector and d is a vector whose all components are equal to 1. This means that we are interested in the sum of all the components of x whose value is 1.6428 10^4. The results are in table 11.35. For solving this problem we can also use the block Lanczos algorithm. Results are given in table 11.36. They are almost equivalent to those of the nonsymmetric Lanczos algorithm in terms of the number of iterations.

Another interesting application is to compute only one component (say the ith one) of the solution of the linear system $Ax = c$. We have to estimate $(e^i)^T A^{-1} c$. Let us use the same example with the nonsymmetric Lanczos algorithm and look for the 10th component of the solution, whose value is 4.6884. Results are in table 11.37. Note that, in this case, since the matrix A is positive definite the solution could have been computed with the conjugate gradient algorithm. Without

Table 11.35 Example F4, GNS, $n = 900$, $d^T A^{-1} c = 1.6428 \ 10^4$. Results $\times 10^{-4}$

Nit	G	G-R b_L	G-R b_U	G-L
2	0.5998	0.6653	1.9107	1.9783
4	0.9339	0.9937	1.7798	1.8094
6	1.1766	1.2240	1.7093	1.7237
8	1.3512	1.3839	1.6725	1.6799
10	1.4766	1.4955	1.6550	1.6589
12	1.5537	1.5666	1.6471	1.6486
14	1.6012	1.6081	1.6441	1.6447
16	1.6265	1.6302	1.6431	1.6432
18	1.6383	1.6399	1.6429	1.6430
20	1.6419	1.6422	1.6428	1.6429
22	1.6425	1.6425	1.6428	1.6428
24	1.6428	1.6428	1.6428	1.6428

Table 11.36 Example F4, GB, $n = 900$, $d^T A^{-1} c = 1.6428 \ 10^4$. Results $\times 10^{-4}$

Nit	G	G-R b_L	G-R b_U	G-L
2	0.6207	0.6611	1.8844	1.9925
4	0.9719	1.0247	1.7595	1.7883
6	1.2127	1.2547	1.6980	1.7114
8	1.3822	1.4114	1.6671	1.6736
10	1.4955	1.5151	1.6524	1.6553
12	1.5670	1.5787	1.6460	1.6472
14	1.6086	1.6152	1.6437	1.6441
16	1.6304	1.6335	1.6430	1.6431
18	1.6397	1.6407	1.6429	1.6429
20	1.6422	1.6423	1.6428	1.6429
22	1.6425	1.6426	1.6428	1.6428
24	1.6427	1.6428	1.6428	1.6428
26	1.6428	1.6428	1.6428	1.6428

preconditioning, it takes 65 CG iterations to obtain $x(10)$ with the same accuracy. However, remember that there is only one matrix-vector multiplication per iteration in CG, whereas, we have two multiplications in the nonsymmetric Lanczos algorithm.

Table 11.37 Example F4, GNS, $n = 900$, $x(10) = 4.6884$

Nit	G	G-R b_L	G-R b_U	G-L
5	1.9024	1.8461	9.3913	7.4680
10	3.1996	3.2251	4.9937	5.3832
15	4.1500	4.1770	4.7265	4.7584
20	4.5654	4.6255	4.6951	4.6955
25	4.6812	4.6820	4.6915	4.6919
30	4.6856	4.6817	4.6881	4.6881
35	4.6883	4.6893	4.6882	4.6882
40	4.6900	4.6907	4.6885	4.6885
45	4.6883	4.6893	4.6884	4.6884
50	4.6885	4.6884	4.6884	4.6884

We can also improve one component of an approximate solution $\tilde{x}$ of $Ax = c$. Let $r = c - A\tilde{x}$ be the residual. The error $\epsilon = x - \tilde{x}$ satisfies the equation $A\epsilon = r$. Assume we want to improve the ith component of the solution. Then we estimate the ith component of ϵ by considering $(e^i)^T A^{-1} r$. Finally, we add the estimate of the bilinear form to the ith component of $\tilde{x}$. Let us use the same example as before. After five CG iterations we have $x_{CG}(10) = 4.0977$. We compute the residual vector and then use the block Lanczos algorithm. We obtain the results in table 11.38, which displays the sum of $x_{CG}(10)$ and the estimates of the error. Note that the first iterations do not always improve the solution.

Table 11.38 Example F4, GB, $n = 900$, $x(10) = 4.6884$

Nit	G	G-R b_L	G-R b_U	G-L
6	3.1890	3.2047	3.1493	2.9772
8	3.2996	3.3639	3.8110	3.6118
10	3.5631	3.6605	4.1239	4.0717
12	3.9683	4.0541	4.4261	4.3705
14	4.2816	4.3474	4.5523	4.5330
16	4.4759	4.5050	4.6330	4.6172
18	4.5761	4.5969	4.6723	4.6682
20	4.6504	4.6616	4.6848	4.6837
22	4.6816	4.6843	4.6884	4.6879
24	4.6874	4.6877	4.6891	4.6891
26	4.6878	4.6881	4.6887	4.6887
28	4.6883	4.6883	4.6886	4.6886
30	4.6884	4.6883	4.6885	4.6885

11.8.6 Estimates of Traces and Determinants

Let us first consider the analytic bounds of Bai and Golub [15], that we have re-called in section 11.6, for Example F4 with $n = 36$. The trace of the inverse is 13.7571. The lower and upper bounds obtained using the first three moments are 10.2830 and 24.3776. However, if we consider a larger problem with $n = 900$ for which the trace of the inverse is 512.6442, the bounds computed from three moments are 261.0030 and 8751.76; the upper bound is a large overestimate.

One can also compute more moments, which are the traces $\mathrm{tr}(A^i)$, $i > 2$, and from the moments recover (with the Chebyshev algorithm) the Jacobi matrix whose eigenvalues and eigenvectors allows us to compute an approximation of the trace of the inverse, which is the moment of order -1, using the Gauss quadrature rule. Results are given for $n = 36$ in table 11.39. They seem fine after $k = 4$ which corresponds to the computation of eight moments. However, the moment matrices are ill-conditioned and if we continue the computations after $k = 10$ they are no longer positive definite. Table 11.40 gives the results for $n = 900$. The ill-conditioning of the moment matrices does not allow us to go further. Hence, this method is not feasible for large matrices.

Table 11.39 Example F4, $n = 36$, Chebyshev, $\mathrm{tr}(A^{-1}) = 13.7571$

k	Estimate
1	9.0000
2	11.3684
3	12.5714
4	13.1581
5	13.4773
6	13.6363
7	13.7139
8	13.7452
9	13.7550
10	13.7568

Table 11.40 Example F4, $n = 900$, Chebyshev, $\mathrm{tr}(A^{-1}) = 512.6442$

k	Estimate
1	225.0000
2	296.7033
3	344.6869
4	375.8398
5	400.0648
6	418.2138
7	433.1216
8	444.9913
9	455.0122
10	463.2337

Since the moment matrices are too ill-conditioned, it is tempting to see if we can use modified moments to solve this problem; see [241]. We use the shifted Chebyshev polynomials of the first kind as the auxiliary orthogonal polynomials. The drawback is that we need to have estimates of the smallest and largest eigenvalues of A. On the interval $[\lambda_{\min}, \lambda_{\max}]$ these polynomials satisfy the following three-term recurrence

$$C_0(\lambda) \equiv 1, \quad \left(\frac{\lambda_{\max} - \lambda_{\min}}{2}\right) C_1(\lambda) = \lambda - \left(\frac{\lambda_{\max} + \lambda_{\min}}{2}\right),$$

$$\left(\frac{\lambda_{\max} - \lambda_{\min}}{4}\right) C_{k+1}(\lambda) = \left(\lambda - \frac{\lambda_{\max} + \lambda_{\min}}{2}\right) C_k(\lambda)$$
$$- \left(\frac{\lambda_{\max} - \lambda_{\min}}{4}\right) C_{k-1}(\lambda).$$

From these relations we can compute the trace of the matrices $C_i(A)$, $i = 0, \ldots, k$, which are the modified moments. The modified Chebyshev algorithm (see chapter 5) generates the coefficients of monic polynomials corresponding to the measure related to the problem. We symmetrize this Jacobi matrix and obtain the nodes and weights of the Gauss quadrature rule from the Golub and Welsch algorithm. The function to consider is $f(x) = 1/x$. Results are displayed in tables 11.41 for $n = 36$ and 11.42 for $n = 900$. Note that upper bounds can be obtained by using the Gauss–Radau rule. Using the modified moments there are no breakdowns in the computations and we obtain quite good results for the trace of the inverse. This example illustrates the benefits of using modified moments.

Table 11.41 Example F4, $n = 36$, modified moments, $\mathrm{tr}(A^{-1}) = 13.7571$

k	Estimate
1	9.0000
2	11.3684
3	12.5714
4	13.1581
5	13.4773
6	13.6363
7	13.7139
8	13.7452
9	13.7550
10	13.7568
11	13.7571

Estimates of the trace of the inverse can also be computed using Monte Carlo techniques; see Hutchinson [195]. We use p random vectors z^i with components 1 and -1 and we compute lower and upper bounds of $(z^i)^T A^{-1} z^i$ with the quadrature rules. The estimates of the trace are the averages of the bounds over the p computations. Table 11.43 gives the results for Example F4 with $n = 36$ and five iterations of the Lanczos algorithm. Results for $= 900$ with 30 iterations are given

Table 11.42 Example F4, $n = 900$, modified moments, $\mathrm{tr}(A^{-1}) = 512.6442$

k	Estimate
5	400.0648
10	463.2560
15	489.5383
20	502.0008
25	508.0799
30	510.9301
35	512.1385
40	512.5469

Table 11.43 Example F4, $n = 36$, Monte Carlo, 5 it., $\mathrm{tr}(A^{-1}) = 13.7571$

p	G	G-R b_L	G-R b_U	G-L
1	12.8274	12.8749	12.9087	13.1169
2	14.7464	14.8440	14.9300	15.1671
3	14.8973	14.9681	15.0277	15.2448
4	13.6203	13.6777	13.7226	13.8941
5	13.9216	13.9918	14.0495	14.1970

Table 11.44 Example F4, $n = 900$, Monte Carlo, 30 it., $\mathrm{tr}(A^{-1}) = 512.6442$

p	G	G-R b_L	G-R b_U	G-L
1	478.1734	478.3272	478.4955	479.6967
2	466.4618	466.5600	466.6667	467.6658
3	458.1058	458.1850	458.2703	459.0539
4	466.5929	466.6975	466.8028	467.7714
5	511.1780	511.2772	511.3732	512.2220

in table 11.44. The results are good even though we do not always obtain lower and upper bounds, but not as good as with the modified Chebyshev algorithm.

Since the matrices that have to be computed when using the modified moments become denser and denser as k increases, it can be costly to compute and to store them. Therefore it is tempting to combine the modified moments algorithm and the Monte Carlo estimates of the trace of a matrix to compute approximate modified moments. Instead of computing the matrices $C_i(A)$ and their traces, we can choose p random vectors z^j, $j = 1, \ldots, p$, compute $C_i(A)z^j$ by three-term vector recurrences and obtain an estimate of the trace of $C_i(A)$ by averaging the values $(z^j)^T C_i(A)z^j$. The results for $n = 36$ are given in table 11.45. Of course, the results are not as good as when using the exact traces of the matrices $C_i(A)$. They are of the same order of accuracy as those obtained with the Monte Carlo method on A^{-1}. The best result is given by $p = 5$.

Table 11.45 Example F4, $n = 36$, modified moments + Monte Carlo, $\operatorname{tr}(A^{-1}) = 13.7571$

p	G
1	12.8274
2	14.7289
3	14.8535
4	13.5780
5	13.8215
6	14.1153
7	14.1134
8	14.5652
9	14.9474
10	14.7008

We now turn to numerical experiments for the computation of the determinant of a matrix A. The analytic bounds of Bai and Golub [15] for Example F4 with $n = 36$ using the first three moments are $6.2482 \ 10^9$ and $3.2863 \ 10^{20}$ when the exact value is $1.9872 \ 10^{19}$. The results using more moments and the Chebyshev algorithm are given in table 11.46. Again we have a breakdown after 10 iterations. The results using the modified Chebyshev algorithm are essentially the same except that we can go beyond $k = 10$. Monte Carlo results are in table 11.47. There are large variations in the estimates and, in fact, the best ones are given for $p = 1$.

For this example we cannot use $n = 900$ since the determinant overflows. There are 827 eigenvalues larger than 1, so the product of the eigenvalues is very large. One way to get around this problem is to normalize the matrix. It turns out that by dividing the matrix by $\lambda_{\max}/2.45 \simeq 3.2569$ we obtain a determinant of order 1, precisely 9.9174. For this matrix $\tilde{A}$ the analytic bounds are useless since they are $1.6042 \ 10^{-256}$ and $7.5059 \ 10^{45}$. The Chebyshev algorithm works until $k = 11$ for which we obtain a value of 32.0947. The modified Chebyshev algorithm works much better, as we can see with the results in table 11.48.

Let us consider a smaller problem with $n = 400$ for which the determinant is $7.7187 \ 10^{206}$. The bounds computed from three moments are $2.1014 \ 10^{118}$ and

Table 11.46 Example F4, $n = 36$, Chebyshev, $\det(A) = 1.9872\ 10^{19}$, results $\times 10^{-19}$

k	Estimate
1	472.2366
2	7.0457
3	2.9167
4	2.2840
5	2.0900
6	2.0233
7	1.9982
8	1.9899
9	1.9877
10	1.9873

Table 11.47 Example F4, $n = 36$, Monte Carlo, 10 it., $\det(A) = 1.9872\ 10^{19}$, results $\times 10^{-19}$

p	G	G-R b_L	G-R b_U	G-L
1	1.9202	1.9202	1.9202	1.9140
2	0.1562	0.1562	0.1562	0.1556
3	0.1025	0.1025	0.1025	0.1022
4	0.7860	0.7860	0.7860	0.7843
5	1.1395	1.1395	1.1395	1.1369

Table 11.48 Example F4, $n = 900$, modified moments, $\det(\tilde{A}) = 9.9174$

k	Estimate
15	14.4863
16	13.3824
17	12.5776
18	11.9865
19	11.5371
20	11.1951
21	10.9282
22	10.7204
23	10.5556
24	10.4254

$5.9484 \ 10^{225}$. The results using the Chebyshev algorithm are in table 11.49. We cannot go beyond $k = 11$. As for the trace of the inverse, the results are much better using modified moments with the Chebyshev polynomials; see table 11.50.

When using the Monte Carlo estimates the results are not so good because the statistical variations are amplified by the exponential. The estimates of $\text{tr}(\ln(A))$ are reasonably good. We have $\text{tr}(\ln(A)) = 476.3762$. The statistical estimates of the trace are in table 11.51. They are not far from the exact answer. However, when we take the exponential, we do not even obtain the correct order of magnitude. Doing more iterations or using more samples does not improve the results significantly.

Table 11.49 Example F4, $n = 400$, Chebyshev, $\det(A) = 7.7187 \ 10^{206}$

k	Estimate
1	$6.6680 \ 10^{240}$
2	$1.8705 \ 10^{217}$
3	$1.7314 \ 10^{211}$
4	$1.4990 \ 10^{209}$
5	$1.5589 \ 10^{208}$
6	$4.9892 \ 10^{207}$
7	$2.5627 \ 10^{207}$
8	$1.7268 \ 10^{207}$
9	$1.3375 \ 10^{207}$
10	$1.1338 \ 10^{207}$
11	$1.0147 \ 10^{207}$

Table 11.50 Example F4, $n = 400$, modified moments, $\det(A) = 7.7187 \ 10^{206}$

k	Estimate
2	$1.8705 \ 10^{217}$
4	$1.4990 \ 10^{209}$
6	$4.9892 \ 10^{207}$
8	$1.7268 \ 10^{207}$
10	$1.1338 \ 10^{207}$
12	$9.3701 \ 10^{206}$
14	$8.5330 \ 10^{206}$
16	$8.1315 \ 10^{206}$
18	$7.9273 \ 10^{206}$
20	$7.8210 \ 10^{206}$

Let us finally consider computing partial eigensums. We first use Example F4 with $n = 36$. We would like to compute the sum of the eigenvalues smaller than $\mu = 2.8$. This corresponds to the sum of the 10 smallest eigenvalues, which is equal to 17.2125. We use the modified Chebyshev algorithm. The approximate

Table 11.51 Example F4, $n = 400$, Monte Carlo, 10 it., $\text{tr}(\ln(A)) = 476.3762$

p	G	G-R b_L	G-R b_U	G-L
1	466.6638	466.6206	466.6391	466.1251
2	459.7104	459.6521	459.6778	459.0748
3	457.4522	457.3549	457.3957	456.6670
4	466.2908	466.2037	466.2403	465.5342
5	469.5209	469.4212	469.4645	468.7859

step function separating the first 10 eigenvalues from the other ones is

$$f(\lambda) = \frac{\lambda}{1 + \exp\left(\frac{\lambda - \mu}{\kappa}\right)}.$$

The parameter κ is set equal to 0.01. Results are given in table 11.52. Even though the convergence is not monotone we obtain a good estimate with $k = 16$. Results for $n = 900$ and the sum of the first 50 eigenvalues are given in table 11.53.

Table 11.52 Example F4, $n = 36$, modified moments, $\sum_{i=1}^{10} \lambda_i = 17.2125$

k	Estimate
2	39.1366
4	3.7983
6	9.8230
8	14.9895
10	20.0729
12	12.4985
14	15.7451
16	17.9512
18	17.2278
20	17.2125

Table 11.53 Example F4, $n = 900$, modified moments, $\sum_{i=1}^{50} \lambda_i = 19.4656$

k	Estimate
10	29.8038
20	26.2463
30	22.6129
40	17.7532
50	17.3318
60	18.1218

In conclusion, we have seen that use of the modified Chebyshev algorithm with modified moments improves the results that were previously obtained for the trace of the inverse, the determinant and the computation of partial eigensums.

Chapter Twelve

Estimates of Norms of Errors in the Conjugate Gradient Algorithm

In this chapter we study how the techniques for computing bounds of quadratic forms can be applied to the computation of bounds for norms of the error in iterative methods for solving linear systems. We are particularly interested in the conjugate gradient algorithm since we will see that it is closely related to Gauss quadrature.

12.1 Estimates of Norms of Errors in Solving Linear Systems

Let A be a symmetric positive definite matrix of order n and suppose that an approximate solution $\tilde{x}$ of the linear system

$$Ax = c,$$

where c is a given vector, has been computed by either a direct or an iterative method. The residual r is defined as

$$r = c - A\tilde{x}.$$

The error ϵ being defined as $\epsilon = x - \tilde{x}$, we obviously have

$$\epsilon = A^{-1}r.$$

Therefore, if we consider the A-norm of the error, we see that it corresponds to a quadratic form involving the inverse of A,

$$\|\epsilon\|_A^2 = \epsilon^T A\epsilon = r^T A^{-1} A A^{-1} r = r^T A^{-1} r.$$

One can also consider the l_2 norm, for which $\|\epsilon\|^2 = r^T A^{-2} r$. Here, the matrix to consider is the square of the inverse of A. Note that when A is nonsymmetric we still have $(\epsilon, \epsilon) = (A^{-1}r)^T A^{-1} r = r^T (AA^T)^{-1} r$. Therefore we have a quadratic form with a symmetric matrix AA^T.

In order to bound or estimate $\|\epsilon\|_A$ or $\|\epsilon\|$, we must obtain bounds or estimates for the quadratic forms $r^T A^{-1} r$ or $r^T A^{-2} r$. This problem was considered in Dahlquist, Eisenstat and Golub [75], Dahlquist, Golub and Nash [76] and more recently in Golub and Meurant [150]. Note that r is easy to compute but, of course, we do not want to compute A^{-1} or even to solve a linear system $Ay = r$.

As we have seen several times in this book, the first step toward a solution is to express the quadratic form as a Riemann–Stieltjes integral and to apply the general framework of chapter 7. Let $A = Q\Lambda Q^T$ be the spectral decomposition of A, with Q orthonormal and Λ diagonal. For $i = 1, 2$ we have

$$r^T A^{-i} r = r^T Q\Lambda^{-i} Q^T r$$

$$= \sum_{j=1}^{n} \lambda_j^{-i} [\hat{r}_j]^2,$$

where $\hat{r} = Q^T r$. This last sum can be considered as a Riemann–Stieltjes integral

$$I[A, r] = r^T A^{-i} r = \int_a^b \lambda^{-i} \, d\alpha(\lambda), \qquad (12.1)$$

where the measure α is piecewise constant and defined (when the eigenvalues of A are distinct) by

$$\alpha(\lambda) = \begin{cases} 0, & \text{if } \lambda < a = \lambda_1, \\ \sum_{j=1}^{i} [\hat{r}_j]^2, & \text{if } \lambda_i \leq \lambda < \lambda_{i+1}, \\ \sum_{j=1}^{n} [\hat{r}_j]^2, & \text{if } b = \lambda_n \leq \lambda. \end{cases}$$

Then, quadrature rules can be used to approximate the integral in equation (12.1). For using the Gauss rule, the first step is to generate the orthogonal polynomials associated with the measure α. We saw in chapters 4 and 7 that this can be done by doing N iterations of the Lanczos algorithm with $v^1 = r/\|r\|$. This builds up the Jacobi matrix J_N, and the Gauss estimate of the integral is

$$\|r\|^2 (e^1)^T (J_N)^{-i} e^1.$$

This can be computed by solving $J_N y = e^1$, that is, computing (elements of) the first column of the inverse of J_N. Then, for $i = 1$ we obtain $\|r\|^2 y_1$. This means we need only the first component of the solution y. This is easily obtained by using a UL Cholesky factorization of the tridiagonal matrix; see Meurant [234] and chapter 3. Another possibility is to compute the $(1, 1)$ entry of the inverse of J_N incrementally as in chapter 3. For $i = 2$ the approximation is $\|r\|^2 y^T y$, but this can also be computed using the QR factorization of J_N; see Meurant [239]. When $i = 1$, the function to be considered is $f(\lambda) = 1/\lambda$. Hence, all the derivatives of f of even order are positive and the Gauss rule gives a lower bound. The same is true for $i = 2$. Upper bounds can be obtained with the Gauss–Radau or Gauss–Lobatto rules by suitably modifying the tridiagonal matrix J_N. But we need lower and upper bounds of the smallest and largest eigenvalues of A. As we will see, the anti-Gauss rule can also be used to obtain an estimate of the integral.

However, if the approximation $\tilde{x}$ arises from an iterative method like Jacobi or Gauss–Seidel (see, for instance, Golub and Van Loan [154] or Meurant [237]) it does not make too much sense to have to do some iterations of the Lanczos algorithm to obtain bounds for the norm of the error. It is, of course, much better to directly solve the linear system by using the CG algorithm. This is considered later on. Use of this estimation technique makes more sense if $\tilde{x}$ comes from a direct solver.

Stochastic estimates of norms of the error in iterative methods can also be considered; see Golub and Melbø[147], [148].

12.2 Formulas for the A-Norm of the Error

When we have an approximation x^k of the solution of $Ax = c$ and the corresponding residual r^k (wherever they come from), we have seen that we can obtain bounds of the A-norm of the error by running some iterations of the Lanczos algorithm. Of course, this does not make too much sense when x^k is obtained by the Lanczos algorithm itself or CG. This would correspond to a restarting of the algorithm and a kind of iterative refinement with, maybe, no improvement of the solution. Therefore, we use another strategy to compute bounds or approximations of the norms of the error during the CG iterates.

In Meurant [236] the following theorem was proved concerning the A-norm of the error $\epsilon^k = x - x^k$ in CG.

THEOREM 12.1 *The square of the A-norm of the error at CG iteration k is given by*

$$\|\epsilon^k\|_A^2 = \|r^0\|^2[(J_n^{-1}e^1, e^1) - (J_k^{-1}e^1, e^1)]. \tag{12.2}$$

where n is the order of the matrix A and J_k is the Jacobi matrix of the Lanczos algorithm whose coefficients can be computed from those of CG using relations (4.9) and (4.10). Moreover,

$$\|\epsilon^k\|_A^2 = \|r^0\|^2 \left[\sum_{j=1}^n \frac{[(z_{(n)}^j)_1]^2}{\lambda_j} - \sum_{j=1}^k \frac{[(z_{(k)}^j)_1]^2}{\theta_j^{(k)}} \right],$$

where $z_{(k)}^j$ is the jth normalized eigenvector of J_k corresponding to the eigenvalue $\theta_j^{(k)}$.

Proof. The first relation has been well known for quite a long time; see the papers of Golub and his coauthors [75], [76]. It is also mentioned in a slightly different form in a paper by Paige, Parlett and van der Vorst [263] and apparently Stiefel was aware of it. By using the definition of the A-norm and the relation between the Lanczos and CG algorithms we have $A\epsilon^k = r^k = r^0 - AV_k y^k$ where V_k is the matrix of the Lanczos vectors and y^k is the solution of $J_k y^k = \|r^0\|e^1$, see [239]. Then,

$$\|\epsilon^k\|_A^2 = (A\epsilon^k, \epsilon^k) = (A^{-1}r^0, r^0) - 2(r^0, V_k y^k) + (AV_k y^k, V_k y^k).$$

The first term of the right-hand side is easy to evaluate since $AV_n = V_n J_n$ assuming that the eigenvalues are distinct. The square matrix V_n of order n is orthogonal; hence this gives $A^{-1}V_n = V_n J_n^{-1}$. Now,

$$r^0 = \|r^0\|v^1 = \|r^0\|V_n e^1.$$

Therefore,

$$A^{-1}r^0 = \|r^0\|A^{-1}V_n e^1 = \|r^0\|V_n J_n^{-1}e^1$$

and

$$(A^{-1}r^0, r^0) = \|r^0\|^2(V_n J_n^{-1}e^1, V_n e^1) = \|r^0\|^2(J_n^{-1}e^1, e^1).$$

For the second term we have to compute $(r^0, V_k y^k)$. But since $r^0 = \|r^0\| v^1 = \|r^0\| V_k e^1$, this term is equal to $\|r^0\|^2 (e^1, J_k^{-1} e^1)$ by using the orthogonality of the Lanczos vectors. The third term is $(AV_k y^k, V_k y^k)$. Using $V_k^T A V_k = J_k$ we have

$$(AV_k y^k, V_k y^k) = (V_k^T A V_k y^k, y^k) = (J_k y^k, y^k).$$

Hence $(AV_k y^k, V_k y^k) = \|r^0\|^2 (J_k^{-1} e^1, e^1)$. This proves the formula in the theorem. The second relation is obtained by using the spectral decompositions of J_n and J_k.   □

The formula (12.2) is the link between CG and Gauss quadrature. It shows that the square of the A-norm of the error is the remainder of a Gauss quadrature rule. The inner product $\|r^0\|^2 (J_n^{-1} e^1, e^1) = (A^{-1} r^0, r^0)$ can be written as a Riemann–Stieltjes integral and $\|r^0\|^2 (J_k^{-1} e^1, e^1)$ is nothing other than the Gauss quadrature approximation to this integral, see [234]. It is interesting to consider this point of view because it allows the computation of lower and upper bounds (if we have estimates of λ_1 and λ_n) for the A-norm of the error. Therefore, estimating the A-norm of the error in CG is completely equivalent to computing an estimate of the remainder of a Gauss quadrature rule assuming only knowledge of the Jacobi matrix. We will elaborate on this point in the next sections.

We can also use the fact that the norm of the error is related to Gauss quadrature to obtain other expressions for $\|\epsilon^k\|_A$ as in the next theorem.

THEOREM 12.2 *For all k there exists $\xi_k, \lambda_1 \leq \xi_k \leq \lambda_n$ such that the A-norm of the error is given by*

$$\|\epsilon^k\|_A^2 = \frac{1}{\xi_k^{2k+1}} \sum_{i=1}^{n} \left[\prod_{j=1}^{k} (\lambda_i - \theta_j^{(k)})^2 \right] (r^0, q^i)^2,$$

where q^i is the ith eigenvector of A corresponding to the eigenvalue λ_i.

Proof. This is obtained by using the expression for the remainder of the Gauss quadrature.   □

An important consequence of the previous theorems is that in exact arithmetic, when an eigenvalue of J_k (a Ritz value) has converged to an eigenvalue of A, the corresponding component of the initial residual on the eigenvector of A has been eliminated from the norm of the error.

12.3 Estimates of the A-Norm of the Error

How can we approximately compute $\|\epsilon^k\|_A^2 = (r^k)^T A^{-1} r^k$? We can use the formula (12.2) that relates the A-norm of the error at step k and the inverse of matrix J_k. This formula has been used in Fischer and Golub [111] but the computations of $\|\epsilon^k\|_A$ were not performed below 10^{-5}. A partial analysis in finite precision arithmetic was done in Golub and Strakoš [151]. A more complete analysis was given by Strakoš and Tichý [317]. We will show below how reliable estimates of $\|\epsilon^k\|_A$ can be computed. In finite precision arithmetic we can still use the same formulas up to $O(u)$ perturbation terms (where u is the unit roundoff); see [317]. For variants of these estimates, see Calvetti, Morigi, Reichel and Sgallari [60]; for another point of view, see Knizhnerman [206].

For the sake of simplicity, let us first consider the lower bound computed by the Gauss rule. The formula (12.2) cannot be used directly since at CG iteration k we do not know $(J_n^{-1})_{1,1}$. But it is known (see [239]) that the absolute values of $(J_k^{-1})_{1,1}$ are an increasing sequence bounded by $|(J_n^{-1})_{1,1}|$. So we use the current value of $(J_k^{-1})_{1,1}$ to approximate the final value. Let d be a given delay integer; the approximation of the A-norm of the error at iteration $k - d$ is given by

$$\|\epsilon^{k-d}\|_A^2 \approx \|r^0\|^2((J_k^{-1})_{(1,1)} - (J_{k-d}^{-1})_{(1,1)}),$$

This can also be understood as writing

$$\|\epsilon^{k-d}\|_A^2 - \|\epsilon^k\|_A^2 = \|r^0\|^2((J_k^{-1})_{(1,1)} - (J_{k-d}^{-1})_{(1,1)}),$$

and supposing that $\|\epsilon^k\|_A$ is negligible against $\|\epsilon^{k-d}\|_A$. Another interpretation is to consider that, having a Gauss rule with $k - d$ nodes at iteration $k - d$, we use another more precise Gauss quadrature with k nodes to estimate the error of the quadrature rule. Usually, the larger is d, the better is the estimate.

Using the results of chapter 3, let us summarize for the convenience of the reader how to compute the difference $(J_k^{-1})_{(1,1)} - (J_{k-d}^{-1})_{(1,1)}$. Let α_i and η_i be the diagonal and off-diagonal nonzero entries of J_k and b_k be the computed value of $(J_k^{-1})_{1,1}$, which can be obtained in an additive way by using the Sherman–Morrison formula; see Golub and Van Loan [154] and chapter 3. Let $j_k = J_k^{-1}e^k$ be the last column of the inverse of J_k; then,

$$(J_{k+1}^{-1})_{1,1} = (J_k^{-1})_{1,1} + \frac{\eta_k^2(j_k j_k^T)_{1,1}}{\alpha_{k+1} - \eta_k^2(j_k)_k}.$$

The first and last elements of the last column of the inverse of J_k that we need can be computed using the Cholesky factorization of J_k whose diagonal elements are $\delta_1 = \alpha_1$ and

$$\delta_i = \alpha_i - \frac{\eta_{i-1}^2}{\delta_{i-1}}, \quad i = 2, \ldots, k.$$

Then,

$$(j_k)_1 = (-1)^{k-1} \frac{\eta_1 \cdots \eta_{k-1}}{\delta_1 \cdots \delta_k}, \quad (j_k)_k = \frac{1}{\delta_k}.$$

Using these results, we have

$$b_k = b_{k-1} + f_k, \quad f_k = \frac{\eta_{k-1}^2 c_{k-1}^2}{\delta_{k-1}(\alpha_k \delta_{k-1} - \eta_{k-1}^2)} = \frac{c_k^2}{\delta_k},$$

where

$$c_k = \frac{\eta_1 \cdots \eta_{k-2}}{\delta_1 \cdots \delta_{k-2}} \frac{\eta_{k-1}}{\delta_{k-1}} = c_{k-1} \frac{\eta_{k-1}}{\delta_{k-1}}.$$

Since J_k is positive definite, we have $\delta_k > 0$ and this shows that $f_k > 0$. Let s_k be the estimate of $\|\epsilon^k\|_A^2$ we are seeking and d be the given integer delay, at CG iteration number k, we set

$$s_{k-d} = \|r^0\|^2(b_k - b_{k-d}).$$

This gives an estimate of the error d iterations before the current one. It was shown in [235] that if we compute b_k in floating point arithmetic and use the formula for $\|\epsilon^{k-d}\|_A^2$ straightforwardly, there exists a $k_{\max}$ such that if $k > k_{\max}$, then $s_k = 0$. This happens when k is large enough because $\eta_k/\delta_k < 1$ and $c_k \to 0$; consequently $f_k \to 0$. Therefore, when $k > k_{\max}$, $b_k = b_{k_{max}}$. But fortunately, as noted in [235], we can compute s_{k-d} in another way since we just need to sum up the last d values of f_j.

Moreover, from [239] we have

$$c_k = \frac{\eta_1 \cdots \eta_{k-1}}{\delta_1 \cdots \delta_{k-1}} = \frac{\|r^{k-1}\|}{\|r^0\|},$$

and $\gamma_{k-1} = 1/\delta_k$. Therefore, $f_k = \gamma_{k-1}\|r^{k-1}\|^2/\|r^0\|^2$. The corresponding formula for the A-norm of the error was already given by Hestenes and Stiefel in [187]. This gives a simpler way of computing the Gauss lower bound. Of course, the Gauss quadrature framework is more general since we can also use the Gauss–Radau and Gauss–Lobatto rules to obtain other bounds by suitably modifying J_k and J_{k-d}.

If we let λ_m and λ_M to be approximations of the smallest and largest eigenvalues of A and d be a positive integer (whose choice is discussed later), the algorithm computing the iterates of CG and estimates from the Gauss (s_{k-d}), Gauss–Radau ($\underline{s}_{k-d}$ and $\bar{s}_{k-d}$) and Gauss–Lobatto ($\check{s}_{k-d}$) rules is given by the following algorithm whose name stands for Conjugate Gradient with Quadrature and Lanczos (with slight simplifications from [235]):

CGQL algorithm
Let x^0 be given, $r^0 = b - Ax^0$, $p^0 = r^0$, $\beta_0 = 0$, $\gamma_{-1} = 1$, $c_1 = 1$.
For $k = 1, \ldots$ until convergence

$$\gamma_{k-1} = \frac{(r^{k-1}, r^{k-1})}{(p^{k-1}, Ap^{k-1})},$$

$$\alpha_k = \frac{1}{\gamma_{k-1}} + \frac{\beta_{k-1}}{\gamma_{k-2}},$$

if $k = 1$ ─────────────────────────────────

$$f_1 = \frac{1}{\alpha_1},$$

$$\delta_1 = \alpha_1,$$

$$\bar{\delta}_1 = \alpha_1 - \lambda_m,$$

$$\underline{\delta}_1 = \alpha_1 - \lambda_M,$$

else ─────────────────────────────────

$$c_k = c_{k-1}\frac{\eta_{k-1}}{\delta_{k-1}} = \frac{\|r^{k-1}\|}{\|r^0\|},$$

$$\delta_k = \alpha_k - \frac{\eta_{k-1}^2}{\delta_{k-1}} = \frac{1}{\gamma_{k-1}},$$

$$f_k = \frac{\eta_{k-1}^2 c_{k-1}^2}{\delta_{k-1}(\alpha_k \delta_{k-1} - \eta_{k-1}^2)} = \gamma_{k-1} c_k^2,$$

$$\bar{\delta}_k = \alpha_k - \lambda_m - \frac{\eta_{k-1}^2}{\bar{\delta}_{k-1}} = \alpha_k - \bar{\alpha}_{k-1},$$

$$\underline{\delta}_k = \alpha_k - \lambda_M - \frac{\eta_{k-1}^2}{\underline{\delta}_{k-1}} = \alpha_k - \underline{\alpha}_{k-1}$$

end —————————————————————————————

$$x^k = x^{k-1} + \gamma_{k-1} p^{k-1},$$

$$r^k = r^{k-1} - \gamma_{k-1} A p^{k-1},$$

$$\beta_k = \frac{(r^k, r^k)}{(r^{k-1}, r^{k-1})},$$

$$\eta_k = \frac{\sqrt{\beta_k}}{\gamma_{k-1}},$$

$$p^k = r^k + \beta_k p^{k-1},$$

$$\bar{\alpha}_k = \lambda_m + \frac{\eta_k^2}{\bar{\delta}_k},$$

$$\underline{\alpha}_k = \lambda_M + \frac{\eta_k^2}{\underline{\delta}_k},$$

$$\breve{\alpha}_k = \frac{\bar{\delta}_k \underline{\delta}_k}{\underline{\delta}_k - \bar{\delta}_k} \left(\frac{\lambda_M}{\bar{\delta}_k} - \frac{\lambda_m}{\underline{\delta}_k} \right),$$

$$\breve{\eta}_k^2 = \frac{\bar{\delta}_k \underline{\delta}_k}{\underline{\delta}_k - \bar{\delta}_k} (\lambda_M - \lambda_m),$$

$$\bar{f}_k = \frac{\eta_k^2 c_k^2}{\delta_k (\bar{\alpha}_k \delta_k - \eta_k^2)},$$

$$\underline{f}_k = \frac{\eta_k^2 c_k^2}{\delta_k (\underline{\alpha}_k \delta_k - \eta_k^2)},$$

$$\breve{f}_k = \frac{\breve{\eta}_k^2 c_k^2}{\delta_k (\breve{\alpha}_k \delta_k - \breve{\eta}_k^2)},$$

if $k > d$ ————————————————————

$$g_k = \sum_{j=k-d+1}^{k} f_j,$$

$$s_{k-d} = \|r^0\|^2 g_k,$$

$$\bar{s}_{k-d} = \|r^0\|^2 (g_k + \bar{f}_k),$$

$$\underline{s}_{k-d} = \|r^0\|^2 (g_k + \underline{f}_k),$$

$$\check{s}_{k-d} = \|r^0\|^2 (g_k + \check{f}_k)$$

end ————————————————————

The algorithm CGQL computes lower bounds s_{k-d}, $\underline{s}_{k-d}$ and upper bounds $\bar{s}_{k-d}$, $\check{s}_{k-d}$ of $\|\epsilon^{k-d}\|_A^2$. The following result was proved in [235].

PROPOSITION 12.3 *Let J_k, $\underline{J}_k$, $\bar{J}_k$ and $\check{J}_k$ be the tridiagonal matrices of the Gauss, Gauss–Radau (with b and a as prescribed nodes) and Gauss–Lobatto rules. Then, if $0 < a = \lambda_m \leq \lambda_{\min}(A)$ and $b = \lambda_M \geq \lambda_{\max}(A)$, $\|r^0\|(J_k^{-1})_{1,1}$, $\|r^0\|(\underline{J}_k^{-1})_{1,1}$ are lower bounds of $\|e^0\|_A^2 = r^0 A^{-1} r^0$, $\|r^0\|(\bar{J}_k^{-1})_{1,1}$ and $\|r^0\|(\check{J}_k^{-1})_{1,1}$ are upper bounds of $r^0 A^{-1} r^0$.*

Proof. The proof is obtained since we know the sign of the remainder in the quadrature rules. Note that $\underline{J}_k$ and $\bar{J}_k$ are of order $k + 1$ as well as $\check{J}_k$. We have that $\bar{f}_k > \underline{f}_k$ and therefore, $\bar{\alpha}_k < \underline{\alpha}_k$. ☐

THEOREM 12.4 *At iteration number k of CGQL, s_{k-d} and $\underline{s}_{k-d}$ are lower bounds of $\|\epsilon^{k-d}\|_A^2$, $\bar{s}_{k-d}$ and $\check{s}_{k-d}$ are upper bounds of $\|\epsilon^{k-d}\|_A^2$.*

Proof. We have

$$\|\epsilon^{k-d}\|_A^2 = \|r^0\|^2 ((J_n^{-1})_{1,1} - (J_{k-d}^{-1})_{1,1})$$

and

$$s_{k-d} = \|r^0\|^2 ((J_k^{-1})_{1,1} - (J_{k-d}^{-1})_{1,1}).$$

Therefore,

$$\|\epsilon^{k-d}\|_A^2 - s_{k-d} = \|r^0\|^2 ((J_n^{-1})_{1,1} - (J_k^{-1})_{1,1}) > 0,$$

showing that s_{k-d} is a lower bound of $\|\epsilon^{k-d}\|_A^2$. A similar proof applies for the other cases since, for instance,

$$\bar{s}_{k-d} = \|r^0\|^2 ((\bar{J}_k^{-1})_{1,1} - (J_{k-d}^{-1})_{1,1}).$$

☐

Note that in the practical implementation we do not need to store all the f_k's but only the last d values. We can also compute only some of the estimates. The additional number of operations for CG is approximately $50 + d$ if we compute the four estimates, which is not significant compared to the $10\,n$ operations plus the matrix-vector product of CG.

An interesting question is to know how large d has to be to get a reliable estimate of the error. Unfortunately, in our experiments, the choice of d depends on the example. The faster is CG convergence, the smaller d has to be. In fact, this is closely linked to the convergence of the Ritz values toward the eigenvalues of A. When the smallest eigenvalues have converged d can be small but we do not change it during the CG iterations, although this can eventually be done adaptively. It is shown experimentally in [239] that the choice of d is related to the smoothness of $\|r^k\|$ as a function of k. Even though the A-norm of the error is monotonically decreasing, if the norm of the residual oscillates, then it is the same for the A-norm estimate. In this case a larger value of d allows the smoothing of these oscillations. When the residual norm curve is smooth, a small value of d gives good estimates. In the quadrature community, it is considered as a bad practice to estimate the error of a Gauss quadrature rule with $k - 1$ nodes by using a Gauss quadrature with k nodes. However, in many cases, $d = 1$ already gives good estimates of the norm of the error.

Nevertheless, if we accept storing a few more vectors whose lengths are the number of CG iterations, we can improve the bounds. For instance, for the Gauss lower bound at iteration k, we can compute f_k and sum it to the bounds we have computed at all the previous iterations. This will improve our previous bounds and as a result we shall have a vector containing bounds using $d = 1$ for iteration $k - 1$, $d = 2$ for iteration $k - 2$ and so on. This is interesting if we want to have an a posteriori look at the rate of convergence. Of course, it is not useful if we just want to use the bound as a stopping criterion. A similar idea was proposed by Strakoš and Tichý [318].

In the CGQL algorithm λ_m and λ_M are lower and upper bounds of the smallest and largest eigenvalues of A. Note that the value of s_k is independent of λ_m and λ_M, $\bar{s}_k$ depends only on λ_m and $\underline{s}_k$ only on λ_M. Experimentally the best bounds are generally the ones computed by the Gauss–Radau rule when using the exact extreme eigenvalues. It is unfortunate that estimates of the smallest eigenvalue are required to obtain upper bounds of the A-norm of the error. We have seen that the extreme eigenvalues of J_k are approximations of the extreme eigenvalues of A that are usually improving as k increases. Therefore, we propose the following adaptive algorithm. We begin the CGQL iterations with $\lambda_m = a_0$ an underestimate of $\lambda_{\min}(A)$. An estimate of the smallest eigenvalue can be easily obtained by inverse iteration using J_k (see Golub and Van Loan [154]) since, for computing the bounds of the norm, we already compute incrementally the Cholesky factorization of J_k. The smallest eigenvalue of J_k is obtained by repeatedly solving tridiagonal systems. We use a fixed number n_a of (inner) iterations of inverse iteration at every CG iteration, giving a value λ_m^k. When λ_m^k is such that

$$\frac{|\lambda_m^k - \lambda_m^{k-1}|}{\lambda_m^k} \le \varepsilon_a,$$

with a prescribed threshold ε_a, we set $\lambda_m = \lambda_m^k$, and stop computing the eigenvalue estimate. Then we continue with CGQL. Of course, this is cheating a little bit since the smallest Ritz value approximates the smallest eigenvalue from above and not from below as is required by the theorem.

For the preconditioned CG algorithm, the formula to consider is

$$\|\epsilon^k\|_A^2 = (z^0, r^0)((J_n^{-1})_{1,1} - (J_k^{-1})_{1,1}),$$

where $Mz^0 = r^0$, M being the preconditioner, a symmetric positive definite matrix that is chosen to speed up the convergence. This is easily obtained by a change of variables in CG applied to the linear system $M^{-1/2}AM^{-1/2}x = M^{-1/2}b$. So the Gauss rule estimate is

$$\|\epsilon^{k-d}\|_A^2 \approx \sum_{j=k-d}^{k-1} \gamma_j(z^j, r^j).$$

The case of preconditioned CG has been considered in Strakoš and Tichý [318] and Meurant [236] as well as in [239].

12.4 Other Approaches

Another possibility to obtain an upper bound for the error in CG is to use the anti-Gauss rule since the error at iteration $k - d + 1$ is of the opposite sign to that for the Gauss rule at iteration $k - d$. Something that has not been exploited so far is the use a Gauss–Kronrod rule.

Other formulas have been proposed by several authors. Let $r^k = b - Ax^k$ be the residual vector and ϵ^k the error. Brezinski [37] considered the first moments of $r = r^k$ (or $\varepsilon = \epsilon^k$),

$$m_0 = (r, r) = \|r\|^2 = (A\varepsilon, A\varepsilon),$$

$$m_1 = (r, Ar) = (A\varepsilon, A^2\varepsilon),$$

$$m_2 = (Ar, Ar) = (A^2\varepsilon, A^2\varepsilon),$$

$$m_{-1} = (r, A^{-1}r) = (A\varepsilon, \varepsilon) = \|\varepsilon\|_A^2.$$

The moments m_0, m_1 and m_2 are computable (but note that m_2 is not computed in CG). We would like to have estimates of m_{-1} and/or m_{-2}.

Using the first terms in singular value decomposition (SVD; see [154]) expansions, Brezinski obtained the following estimates e_i^2 for m_{-2}:

$$e_1^2 = m_1^4/m_2^3,$$
$$e_2^2 = m_0 m_1^2/m_2^2,$$
$$e_3^2 = m_0^2/m_2,$$
$$e_4^2 = m_0^3/m_1^2,$$
$$e_5^2 = m_0^4 m_2/m_1^4.$$

He proved that

$$e_1 \leq e_2 \leq e_3 \leq e_4 \leq e_5.$$

The value e_3 is usually the most appropriate estimate of $\|\varepsilon\|$. Moreover, it can be derived by other techniques (see Auchmuty [14]) and is valid for any consistent norm. Therefore, we consider

$$\|\varepsilon\|^2 \approx \frac{(r,r)^2}{(Ar,Ar)},$$

$$\|\varepsilon\|_A^2 \approx \frac{(r,Ar)^2}{(A^2r,Ar)}.$$

Unfortunately, we do not know if these estimates are lower or upper bounds. But they have the advantage that they can be used also for nonsymmetric problems. For more results on these error estimates, see Brezinski, Rodriguez and Seatzu [39], where the previous estimates were gathered in one single formula,

$$e_\nu^2 = m_0^{\nu-1}(m_1^2)^{3-\nu} m_2^{\nu-4},$$

where ν is a real parameter. Moreover e_ν is an increasing function of ν. See also [40] and [280].

12.5 Formulas for the l_2 Norm of the Error

Hestenes and Stiefel [187] proved the following result relating the l_2 norm and the A-norm of the error.

THEOREM 12.5

$$\|\epsilon^k\|^2 - \|\epsilon^{k+1}\|^2 = \frac{\|\epsilon^k\|_A^2 + \|\epsilon^{k+1}\|_A^2}{\mu(p^k)},$$

with

$$\mu(p^k) = \frac{(p^k, Ap^k)}{\|p^k\|^2}.$$

Proof. See [187] and another proof in Meurant [239]. $\qquad\qquad\square$

Expressions for the l_2 norm of the error can be obtained using the same techniques as for the A-norm. This leads to the following result [238].

THEOREM 12.6

$$\|\epsilon^k\|^2 = \|r^0\|^2[(e^1, J_n^{-2}e^1) - (e^1, J_k^{-2}e^1)] + (-1)^k 2\eta_k \frac{\|r^0\|}{\|r^k\|}(e^k, J_k^{-2}e^1)\|\epsilon^k\|_A^2.$$

Proof. See [238] and [239]. $\qquad\qquad\square$

COROLLARY 12.7

$$\|\epsilon^k\|^2 = \|r^0\|^2[(e^1, J_n^{-2}e^1) - (e^1, J_k^{-2}e^1)] - 2\frac{(e^k, J_k^{-2}e^1)}{(e^k, J_k^{-1}e^1)}\|\epsilon^k\|_A^2.$$

We see that the above formulas are not as nice as equation (12.2) for the A-norm. Besides the term with $(e^1, J_n^{-2}e^1) - (e^1, J_k^{-2}e^1)$, which could have been expected, there is another term involving $\|\epsilon^k\|_A^2$.

12.6 Estimates of the l_2 Norm of the Error

To obtain an estimate of the l_2 norm of the error using theorem 12.6 we can solve $J_k y = e^1$ with the Cholesky factorization but it is sometimes better to use a QR factorization of the tridiagonal Jacobi matrix J_k,

$$Q_k J_k = R_k,$$

where Q_k is an orthogonal matrix and R_k an upper triangular matrix. This gives the Cholesky factorization of $J_k^2 = R_k^T R_k$; hence

$$(e^1, J_k^{-2} e^1) = (R_k^{-T} e^1, R_k^{-T} e^1).$$

To compute this inner product we have to solve a linear system with matrix R_k^T and right-hand side e^1. For the QR factorization of J_k we use the results of Fischer [109]. We remark that the matrix R_k has only three nonzero diagonals whose entries are denoted as $r_{1,i}, r_{2,i}, r_{3,i}$. The general formulas are

$$\hat{r}_{1,1} = \alpha_1, \ \hat{r}_{1,2} = c_1 \alpha_2 - s_1 \eta_1, \ \hat{r}_{1,i} = c_{i-1} \alpha_i - s_{i-1} c_{i-2} \eta_{i-1}, \ i \geq 3,$$

$$r_{1,i} = \sqrt{\hat{r}_{1,i}^2 + \eta_i^2},$$

$$r_{3,i} = s_{i-2} \eta_{i-1}, \ i \geq 3,$$

$$r_{2,2} = c_1 \eta_1, \ r_{2,i} = c_{i-2} c_{i-1} \eta_{i-1} + s_{i-1} \alpha_i, \ i \geq 3,$$

$$c_i = \frac{\hat{r}_{1,i}}{r_{1,i}}, \ s_i = \frac{\eta_i}{r_{1,i}}.$$

To incrementally compute the solution of the linear systems $R_k^T w^k = e^1$ for $k = 1, 2, \ldots$ we have to be careful that, even though the other elements of R_k stay the same, the (k, k) element changes when we go from k to $k + 1$. Hence changing notations $w = w^k$ and with $\hat{w}$ an auxiliary vector, we define

$$\hat{w}_1 = \frac{1}{\hat{r}_{1,1}}, w_1 = \frac{1}{r_{1,1}},$$

$$\hat{w}_2 = -\frac{r_{2,2} w_1^2}{\hat{r}_{1,2}}, \quad w_2 = -\frac{r_{2,2} w_1^2}{r_{1,2}},$$

and more generally for $i \geq 3$

$$\hat{w}_i = -\frac{(r_{3,i} w_{i-2} + r_{2,i} w_{i-1})}{\hat{r}_{1,i}}, \quad w_i = -\frac{(r_{3,i} w_{i-2} + r_{2,i} w_{i-1})}{r_{1,i}}.$$

Therefore, $\hat{w}_k$ is the last component of the solution at iteration k and w_k is used in the subsequent steps. Then,

$$\|R_k^{-T} e^1\|^2 = \sum_{j=1}^{k-1} w_j^2 + \hat{w}_k^2.$$

As for the A-norm, we introduce an integer delay d and we approximate $(r^0, A^{-2}r^0) - \|r^0\|^2(e^1, J_{k-d}^{-2}e^1)$ at iteration k by the difference of the k and $k - d$ terms computed from the solutions, that is,

$$\hat{w}_k^2 - \hat{w}_{k-d}^2 + \sum_{j=k-d}^{k-1} w_j^2, \ k > d.$$

To approximate the last term

$$(-1)^{k-d}2\eta_{k-d}\frac{\|r^0\|}{\|r^{k-d}\|}(e^{k-d}, J_{k-d}^{-2}e^1)\|\epsilon^{k-d}\|_A^2$$

we use the lower bound of $\|\epsilon^{k-d}\|_A$ from Gauss quadrature and the value $(e^{k-d}, J_{k-d}^{-2}e^1)$ which is $\hat{w}_{k-d}/\hat{r}_{1,k-d}$. For more comments on this approximation and the computations using finite precision arithmetic, see Meurant [239].

12.7 Relation to Finite Element Problems

When the linear system to solve arises from the discretization of partial differential equations (PDEs), there are several sources of errors. Suppose we want to solve a PDE

$$\mathcal{L}u = f \quad \text{in } \Omega,$$

Ω being a two- or three-dimensional bounded domain, with appropriate boundary conditions on Γ the boundary of Ω. As a simple example, consider the PDE

$$-\Delta u = f, \quad u|_\Gamma = 0.$$

This problem is naturally formulated in the Hilbert space $H_0^1(\Omega)$ which is the space of square integrable functions (denoted as $L_2(\Omega)$) with square integrable derivatives (in the sense of distributions) and zero boundary traces. It is written in variational form for the solution u as

$$a(u, v) = (f, v), \ \forall v \in V = H_0^1(\Omega),$$

where $a(u, v)$ is a self-adjoint bilinear form

$$a(u, v) = \int_\Omega \nabla u \cdot \nabla v \, dx, \tag{12.3}$$

and

$$(f, v) = \int_\Omega f v \, dx.$$

The bilinear form is continuous and coercive, that is,

$$|a(u, v)| \leq C\|u\|_1 \|v\|_1, \quad a(u, u) \geq \gamma\|u\|_1^2,$$

where the H^1 norm is defined as

$$\|v\|_1^2 = \int_\Omega [u^2 + \nabla u^2] \, dx.$$

Equation (12.3) has a unique solution $u \in H_0^1(\Omega)$. Note that $a(v, v) = \|v\|_a^2$ defines a norm which is equivalent to the H^1 norm. An approximate solution u_h can be computed using the finite element method. The approximate solution is sought in a finite dimensional subspace $V_h \subset V$ as

$$a(u_h, v_h) = (f, v_h), \quad \forall v_h \in V_h.$$

The subspace V_h can be constructed in many different ways. The simplest one is to triangulate the domain Ω (with triangles or tetrahedrons of maximal diameter h) and to use functions which are linear on each element. Hence, a function $v_h \in V_h$ is piecewise linear and the unknowns are the values of u_h at the vertices of the triangulation. Using basis functions ϕ_i which are piecewise linear and have a value 1 at vertex i and 0 at all the other vertices,

$$v_h(x) = \sum_{j=1}^{n} v_j \phi_j(x).$$

The approximated problem is equivalent to a linear system $A\tilde{u} = c$, where

$$[A]_{i,j} = a(\phi_i, \phi_j), \quad c_i = (f, \phi_i).$$

The matrix A is symmetric and positive definite. The solution of the finite dimensional problem is

$$u_h(x) = \sum_{j=1}^{n} \tilde{u}_j \phi_j(x).$$

If the order of the linear system is large, it can be solved by an iterative method; the algorithm of choice is CG. Stopping at iteration k will give an approximate solution $u_h^{(k)}$.

Therefore, we have two sources of errors, the difference between the exact and approximate solutions $u - u_h$, and $u_h - u_h^{(k)}$, the difference between the approximate solution and its CG computed value (not speaking of rounding errors). Of course, we desire the norm of $u - u_h^{(k)}$ to be small. This depends on h and on the CG stopping criterion. The problem of finding an appropriate stopping criterion has been studied by Arioli and his coauthors [9], [10].

The rationale in [9] is based on the following inequality. Let $u_h^* \in V_h$ be such that

$$\|u_h - u_h^*\|_a \leq h^t \|u_h\|_a,$$

where the value of t is related to the stopping criterion and depends on the regularity of the solution. Then,

$$\|u - u_h^*\|_a \leq \|u - u_h\|_a + \|u_h - u_h^*\|_a$$
$$\leq h^t \|u\|_a + (1 + h^t)\|u - u_h\|_a.$$

If $t > 0$ and $h < 1$, we have

$$\|u - u_h^*\|_a \leq h^t \|u\|_a + 2\|u - u_h\|_a.$$

Therefore, if $u_h^* = u_h^{(k)}$ and we choose t and therefore $\|u_h - u_h^*\|_a$ such that $h^t \|u\|_a$ is of the same order as $\|u - u_h\|_a$ we have

$$\|u - u_h^*\|_a \approx \|u - u_h\|_a,$$

which is the best we can hope for; see examples in [9]. Now, it turns out that

$$\|v_h^{(k)}\|_a = \|v^k\|_A,$$

where v^k is the vector of the vertex unknowns. During CG iterations we know how to estimate the A-norm of the error. Let ζ_k be such an estimate of $\|\epsilon^k\|_A^2$ at iteration k (obtained with iterations up to $k + d$). Then, Arioli [9] proposed to compare ζ_k to $\|c\|_{A^{-1}}^2$. The proposed stopping criterion is

$$\|c - Au^{(k)}\|_{A^{-1}} \leq \eta\|c\|_{A^{-1}}.$$

But $\|c - Au^{(k)}\|_{A^{-1}} = \|u_h - u_h^{(k)}\|_a$ which is the A-norm of the error. We have

$$\|c\|_{A^{-1}}^2 = \tilde{u}^T A\tilde{u} \geq (u^{(k)})^T r^0 + c^T u^{(0)}.$$

Using this lower bound, the stopping test is

If $\zeta_k \leq \eta^2((u^k)^T r^0 + c^T u^0)$ then stop.

The parameter η is chosen as h or η^2 as the maximum area of the triangles in two dimensions. Numerical examples in [9] show that this stopping criterion is capable of stopping the CG iterations when $u_h^{(k)}$ is a reasonable approximation to $u(x)$, that is, when $\|u - u_h^{(k)}\|_a$ is of the same order as $\|u - u_h\|_a$.

12.8 Numerical Experiments

12.8.1 Examples

We will use Examples F3 and F4 from chapter 11. We also introduce two other examples. Example CG1 is the matrix Bcsstk01 from the Matrix Market (at the address *http://math.nist.gov/MatrixMarket/*). Example CG2 arises from the discretization of a diffusion problem similar to F5 except that the diffusion coefficient is given by

$$\lambda(x, y) = \frac{1}{(2 + p\sin\frac{x}{\eta})(2 + p\sin\frac{y}{\eta})}. \tag{12.4}$$

The function λ may have peaks. The parameter η allows us to choose the number of peaks and the value of the parameter p determines the heights of the peaks. We are interested in the values $p = 1.8$ and a value of $\eta = 0.1$ for which the diffusion coefficient has a single peak. A value of $\eta = 0.08$ would give four peaks.

12.8.2 Numerical Results

Example F3

We solve a linear system of order $n = 100$ with a random right-hand side. Figure 12.1 shows the $\log_{10}$ of the A-norm of the error (solid), the lower bound obtained with the Gauss quadrature rule (dashed) and the upper bound given by the Gauss–Radau rule with $a = \lambda_{\min}$ (dot-dashed). The lower bound is oscillating but the upper bound is very close to the exact error. This problem is difficult to solve with CG. We see that we need many more than 100 iterations to go down to the stagnation level of the norm of the error. This happens because of the large rounding errors, which delay convergence. Figure 12.1 is for a delay $d = 1$. When we increase the delay to $d = 5$ (figure 12.2) the oscillations of the lower bound are smoothed.

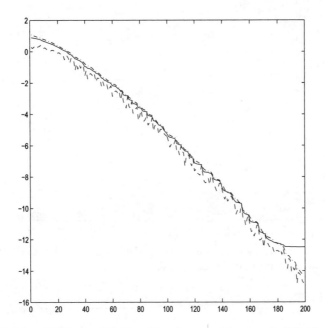

Figure 12.1 Example F3, $d = 1$, $\log_{10}$ of the A-norm of the error (solid), Gauss (dashed) and Gauss–Radau (dot-dashed)

Example F4

The order of the linear system is $n = 900$. The notations are the same as for Example F3. For the Gauss–Radau upper bound we use a value of $a = 0.02$, whence the smallest eigenvalue is $\lambda_{\min} = 0.025$. We can see on figure 12.3 that we obtain a good lower bound even with $d = 1$. Figure 12.4 shows a zoom of the convergence curve for $d = 5$.

The problem for obtaining upper bounds with the Gauss–Radau rule is to have an estimated value of $\lambda_{\min}$, the smallest eigenvalue of A. During the first CG iterations an estimate of $\lambda_{\min}$ can be obtained by computing the smallest eigenvalue of J_k. In the experiment of figure 12.5 the smallest eigenvalue is computed by inverse iteration solving tridiagonal systems. The computation is started with $a = 10^{-10}$ and we switch when the smallest eigenvalue of J_k has converged. Therefore at the beginning the upper bound is a large overestimate of the A-norm but, after the switch, we obtain a good upper bound.

Example CG1

This small example of order 48 was chosen to show the relation between the oscillations of the Gauss lower bound and the oscillations of the residual norm when

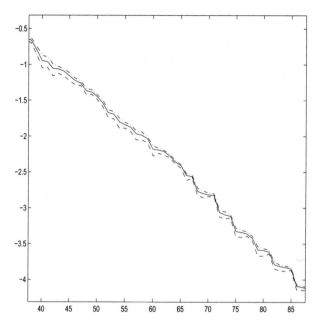

Figure 12.2 Example F3, $d = 5$, zoom of $\log_{10}$ of the A-norm of the error (solid), Gauss (dashed) and Gauss–Radau (dot-dashed)

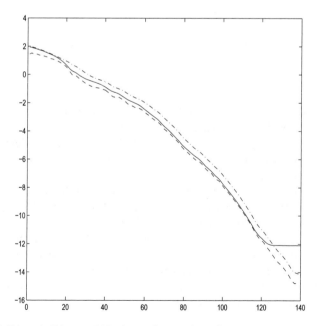

Figure 12.3 Example F4, $n = 900$, $d = 1$, $\log_{10}$ of the A-norm of the error (solid), Gauss (dashed) and Gauss–Radau (dot–dashed)

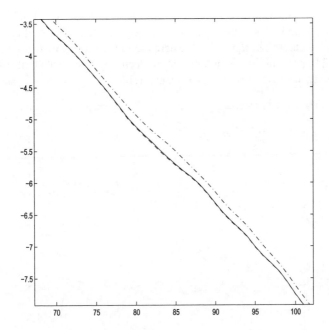

Figure 12.4 Example F4, $n = 900$, $d = 5$, zoom of $\log_{10}$ of the A-norm of the error (solid), Gauss (dashed) and Gauss–Radau (dot-dashed)

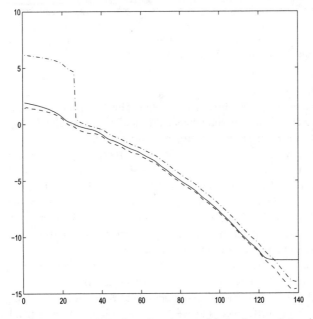

Figure 12.5 Example F4, $n = 900$, $d = 1$, estimate of $\lambda_{\min}$, $\log_{10}$ of the A-norm of the error (solid), Gauss (dashed) and Gauss–Radau (dot-dashed)

the delay d is small. In figure 12.6 the delay is $d = 1$. There are large oscillations of the lower bound. They are, of course, closely linked to the oscillations of the residual as we can see in figure 12.7. There are much few oscillations with $d = 5$ in figure 12.8, particularly when the A-norm decreases fast. Note that $n = 48$ but we need many more iterations to have a small error.

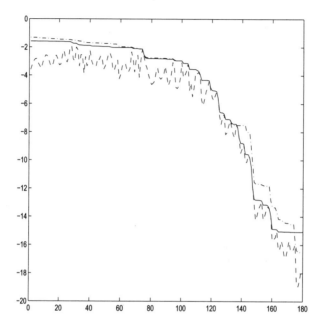

Figure 12.6 Example CG1, $d = 1$, $\log_{10}$ of the A-norm of the error (solid), Gauss (dashed) and Gauss–Radau (dot-dashed)

Example CG2

For the linear system we first use $n = 900$. The results are shown in figure 12.9 for which we choose $a = 0.002$ whence the smallest eigenvalue is $\lambda_{\min} = 0.0025$. Figure 12.10 is a zoom of the convergence curve with $d = 5$.

Then, we solve a larger problem with $n = 10000$. The approximation of the smallest eigenvalue is $a = 10^{-4}$ whence the exact value is $\lambda_{\min} = 2.3216 \ 10^{-4}$. Results with $d = 1$ are given in figure 12.11. With this problem size we need a large number of iterations. Figure 12.12 shows the results of the preconditioned conjugate gradient algorithm with an incomplete Cholesky factorization IC(0) as preconditioner; see for instance [237]. The convergence is much faster and the Gauss lower bound is closer to the A-norm of the error.

Let us consider solving a PDE problem of which we know the exact solution to demonstrate the usefulness of the stopping criterion developed by Arioli and his coauthors [9], [10]. We choose Example CG2. Since we are solving a problem arising from finite difference and we have multiplied the right-hand side of the linear system by h^2, we modify the Arioli criterion to

$$\text{If } \zeta_k \leq 0.1 * (1/n)^2((x^k)^T r^0 + c^T x^0) \text{ then stop.} \tag{12.5}$$

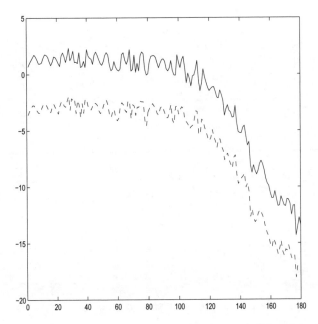

Figure 12.7 Example CG1, $d = 1$, $\log_{10}$ of the residual norm (solid) and the Gauss bound (dashed)

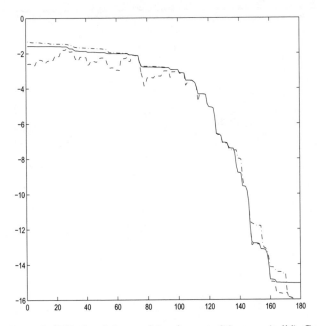

Figure 12.8 Example CG1, $d = 5$, $\log_{10}$ of the A-norm of the error (solid), Gauss (dashed) and Gauss–Radau (dot-dashed)

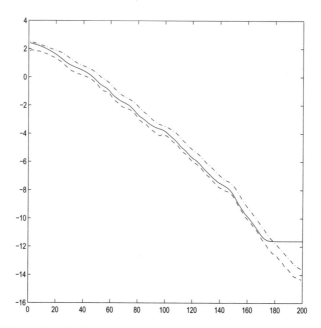

Figure 12.9 Example CG2, $d = 1$, $n = 900$, $\log_{10}$ of the A-norm of the error (solid), Gauss (dashed) and Gauss–Radau (dot-dashed)

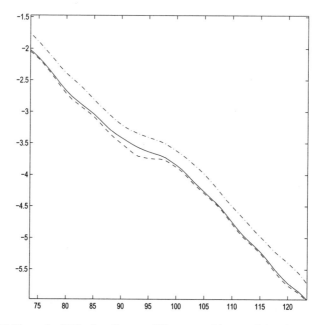

Figure 12.10 Example CG2, $d = 5$, $n = 900$, zoom of $\log_{10}$ of the A-norm of the error (solid), Gauss (dashed) and Gauss–Radau (dot-dashed)

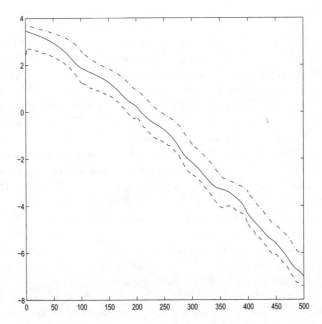

Figure 12.11 Example CG2, $d = 1$, $n = 10000$, $\log_{10}$ of the A-norm of the error (solid), Gauss (dashed) and Gauss–Radau (dot-dashed)

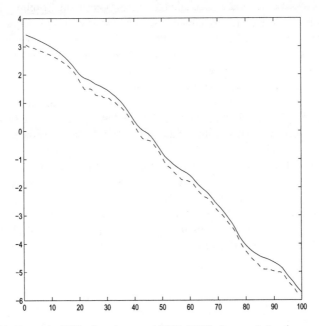

Figure 12.12 Example CG2, $d = 1$, $n = 10000$, IC(0), $\log_{10}$ of the A-norm of the error (solid) and Gauss (dashed)

where ζ_k is an estimate of $\|\epsilon^k\|_A^2$ at iteration k (obtained with iterations up to $k+d$) and c is the right-hand side of the linear system of order n. The coefficient 0.1 is somewhat arbitrary. The right-hand side of the PDE is computed such that the exact solution is $u(x, y) = \sin(\pi x) \sin(\pi y)$. When using $n = 900$, the A-norm of the difference between the "exact" solution of the linear system (obtained by Gaussian elimination) and the discretization of u is $n_u = 5.9468 \ 10^{-4}$. Using the stopping criterion of equation (12.5) we do 53 iterations and the A-norm of the difference between u and the CG approximate solution is $n_x = 7.1570 \ 10^{-4}$, which is of the same order of magnitude as the true error norm.

With $n = 10000$, we obtain $n_u = 5.6033 \ 10^{-5}$. We do 226 iterations and we have $n_x = 9.5473 \ 10^{-5}$. Using the incomplete Cholesky factorization IC(0) as a preconditioner we do 47 iterations and obtain $n_x = 5.6033 \ 10^{-5}$. Hence, this stopping criterion is working fine even for this difficult problem.

Anti-Gauss quadrature rules can also be used to obtain estimates of the A-norm of the error. For Example F4, the results with $d = 1$ are given in figure 12.13. We obtain an upper bound with the anti-Gauss rule. A zoom of the convergence curve with $d = 5$ is given in figure 12.14. Use of the anti-Gauss rule is interesting since it does not need any estimate of the smallest eigenvalue of A. However, the anti-Gauss estimate may fail since sometimes we have to take square roots of negative values. Hence, for figure 12.15 we use the generalized anti-Gauss rule with a parameter γ. We started from a value γ_0 (taken to be 1 and 0.7 for the figure). When at some iteration we find a value $\delta_k < 0$ when computing the Cholesky factorization of the Jacobi matrix, we decrease the value of γ until we find a positive definite matrix. At most we will find $\gamma = 0$ and recover the Gauss rule. Figure 12.15 shows that this strategy works fine. We can also see that using values of γ less than 1 may give better results. But, of course, with the anti-Gauss rule we may not always get an upper bound. One can also use an average between the Gauss and the anti-Gauss rules. This is shown in figure 12.16. It may give an estimate that is very close to the exact error norm but not necessarily a bound.

The fact that the anti-Gauss rule may not always give an upper bound is illustrated in figure 12.17 for Example CG1. We see that in this case the Gauss–Radau rule gives a much better result than the anti-Gauss rule but it needs an estimate of the smallest eigenvalue of A.

We now use the formula given by Brezinski [37] for estimating the A-norm of the error. Figure 12.18 displays the result of this estimate. Generally Brezinski's estimates give better results for the l_2 norm than for the A-norm.

As we have seen before, the l_2 norm of the error can also be estimated by quadrature rules. Details of the algorithm are given in [239]. A numerical experiment for Example F4 and $n = 900$ is given in figure 12.19 with a delay $d = 1$.

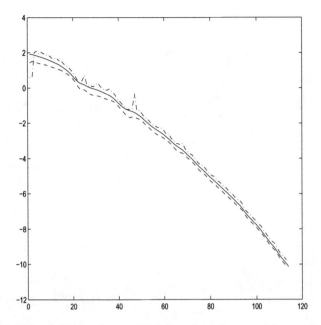

Figure 12.13 Example F4, $d = 1$, $n = 900$, $\log_{10}$ of the A-norm of the error (solid), Gauss (dashed) and anti-Gauss (dot-dashed)

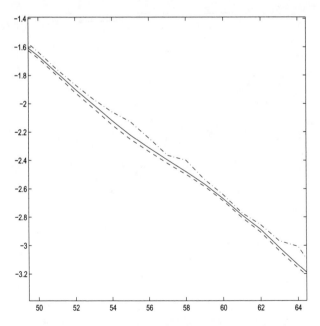

Figure 12.14 Example F4, $d = 5$, $n = 900$, zoom of $\log_{10}$ of the A-norm of the error (solid), Gauss (dashed) and anti-Gauss (dot-dashed)

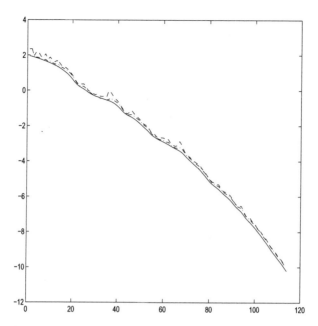

Figure 12.15 Example F4, $d = 1$, $n = 900$, $\log_{10}$ of the A-norm of the error (solid), anti-Gauss $\gamma_0 = 1$ (dashed) and $\gamma_0 = 0.7$ (dot-dashed)

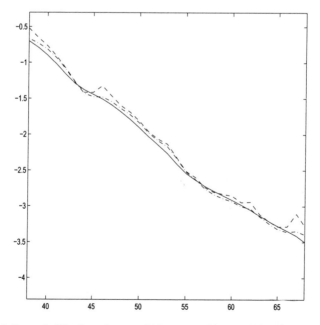

Figure 12.16 Example F4, $d = 1$, $n = 900$, zoom of $\log_{10}$ of the A-norm of the error (solid), averaged anti-Gauss $\gamma_0 = 1$ (dashed) and $\gamma_0 = 0.7$ (dot-dashed)

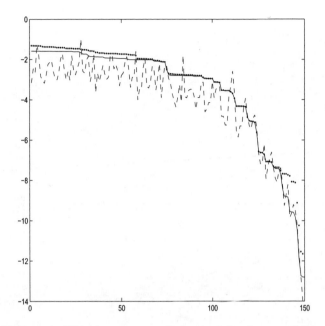

Figure 12.17 Example CG1, $d = 1$, $n = 48$, $\log_{10}$ of the A-norm of the error (solid), anti-Gauss (dashed) and Gauss–Radau (dotted)

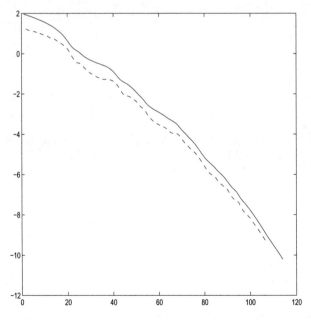

Figure 12.18 Example F4, $n = 900$, $\log_{10}$ of the A-norm of the error (solid) and Brezinski's estimate (dashed),

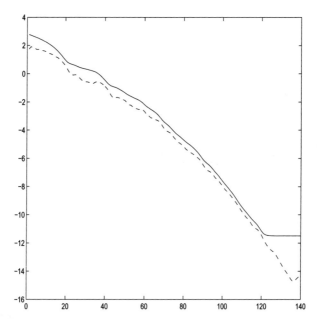

Figure 12.19 Example F4, $d = 1$, $n = 900$, $\log_{10}$ of the l_2 norm of the error (solid) and Gauss (dashed)

Chapter Thirteen

Least Squares Problems

In this chapter we are concerned with the application of the techniques we developed to estimate bilinear forms related to the solution of least squares problems. First we give a brief introduction to least squares problems. For more details see the book by Björck [30].

13.1 Introduction to Least Squares

13.1.1 Weighted Least Squares

Assume we have a data matrix A of dimension $m \times n$ with $m \geq n$ and a vector c of observations and we want to solve the linear system $Ax \approx c$ in a certain sense. If c is not in the range of A, a common way to compute a solution is to solve a least squares (LS) problem, seeking for the minimum of

$$\|W(c - Ax)\|, \tag{13.1}$$

where W is a diagonal matrix of weights (in many instances $W = I$, the identity matrix) and the norm is the l_2 (or Euclidean) norm; see, for instance, Golub and Van Loan [154] or Björck [30]. Least squares approximations were first introduced in the 19th century by Adrien-Marie Legendre and Carl Friedrich Gauss to solve some problems arising from astronomy. Of course, at that time their methods were not formulated in matrix terms. It is well known that the mathematical solution of the LS problem (13.1) satisfies the normal equations

$$A^T W^2 A x = A^T W^2 c.$$

This is obtained by computing the gradient of the functional in equation (13.1). The solution is unique if A has full rank n. If we do not use weighting for simplicity ($W = I$) we see easily that the residual norm $\|r_{LS}\|$ at the solution x_{LS} is given by

$$\|r_{LS}\|^2 = c^T c - c^T A (A^T A)^{-1} A^T c$$

and

$$\|x_{LS}\|^2 = c^T A (A^T A)^{-2} A^T c.$$

We have quadratic forms with a symmetric matrix in the right-hand sides.

Since the normal equation matrix $A^T A$ can be badly conditioned, the LS problem is often solved using orthogonal transformations, namely Givens rotations or Householder transformations; see Björck [30]. However, the solution can also be written with the help of the singular value decomposition (SVD). Assume $W = I$, using the SVD of A we have three matrices U, V, Σ such that $U^T A V = \Sigma$, where U of order m and V of order n are orthonormal and Σ is a diagonal matrix with r positive diagonal elements (the singular values), where r is the rank of A (which is the number of nonzero singular values). Let u^i, $i = 1, \ldots, m$ be the columns of U and v^i, $i = 1, \ldots, n$ be the columns of V. Then, the solution of the LS problem is

$$x_{LS} = \sum_{i=1}^{r} \frac{(u^i)^T c}{\sigma_i} v^i. \tag{13.2}$$

It gives the minimum residual norm $\|c - Ax\|$ and the smallest l_2 norm between the minimizers when the solution is not unique.

The least squares problem given in equation (13.1) can also be rephrased as

$$\text{minimize } \|Wr\| \tag{13.3}$$

subject to $Ax = c + r$. Thus the LS problem corresponds to perturbing the vector of observations c by the minimum amount r such that the right-hand side $c + r$ is in the range of A.

13.1.2 Backward Error for Least Squares Problems

When solving a linear system $Ax = c$ with a square matrix, it is of interest to know the effects of perturbations of the data on the solution. If one only perturbs the matrix A, we consider the problem

$$(A + \Delta A)y = c,$$

with a perturbation satisfying $\|\Delta A\| \leq \omega$. The normwise backward error η measures the minimal distance (in the l_2 norm) to a perturbed problem which is solved exactly by the computed solution y,

$$\eta = \inf\{\omega|, \ \omega \geq 0, \ \|\Delta A\| \leq \omega, \ (A + \Delta A)y = c\}.$$

The normwise backward error for square linear systems has been characterized by Rigal and Gaches [284]. Let $r = c - Ay$ be the residual, then

$$\eta = \frac{\|r\|}{\|y\|}.$$

A rule of thumb says that the forward relative error $\|y - x\|/\|x\|$ is approximately the product of the condition number of A and the backward error.

When solving a linear problem in the least squares sense with an $m \times n$ matrix A, we seek for the solution of

$$\min_x \|c - Ax\|.$$

The perturbation analysis of this problem is summarized in Higham's book [188]. The backward error is now defined in the Frobenius norm as

$$\mu(x) = \min \| \, (\Delta A \quad \theta \Delta c) \, \|_F,$$

subject to $(A + \Delta A)x = c + \Delta c$ where x is the computed solution. Here θ is a real parameter. We will see in chapter 14 that this is quite similar to the total least squares (TLS) problem except that here we have the computed solution x.

The backward error has been characterized by Walden, Karlson and Sun [343]. It is given by

$$\mu(x) = \left(\frac{\|r\|^2}{\|x\|^2} \nu + \min \left\{ 0, \lambda_{\min} \left(AA^T - \nu \frac{rr^T}{\|x\|^2} \right) \right\} \right)^{1/2}, \quad \nu = \frac{\theta^2 \|x\|^2}{1 + \theta^2 \|x\|^2},$$

where $r = c - Ax$ is the residual vector corresponding to the computed solution. Assuming the right-hand side c is known exactly (data least squares abbreviated as DLS), letting $\theta \to \infty$ which gives $\Delta c = 0$, the computed solution x is the exact solution for a matrix $A + \Delta A$. The backward error is then

$$\mu(x) = \min \|\Delta A\|_F$$

and

$$\mu(x) = \left(\frac{\|r\|^2}{\|x\|^2} + \min \left\{ 0, \lambda_{\min} \left(AA^T - \frac{rr^T}{\|x\|^2} \right) \right\} \right)^{1/2}.$$

Let $A = U\Sigma V^T$ be the SVD of A, where U and V (respectively of order m and n) are orthonormal matrices. If $m > n$, Σ of dimension $m \times n$ is

$$\Sigma = \begin{pmatrix} D \\ 0 \end{pmatrix},$$

D being a diagonal matrix of order n. Then, we have

$$AA^T = U \begin{pmatrix} D^2 & 0 \\ 0 & 0 \end{pmatrix} U^T.$$

Let $d = \sqrt{\nu} r / \|x\|$, the matrix to consider is $AA^T - dd^T$. If we denote $f = U^T d$, we have

$$AA^T - dd^T = U \left[\begin{pmatrix} D^2 & 0 \\ 0 & 0 \end{pmatrix} - ff^T \right] U^T.$$

The eigenvalues of $AA^T - dd^T$ are those of the matrix within brackets. Clearly some of the diagonal entries are $-f_i^2 \leq 0$. Therefore this matrix cannot be positive definite and it is likely (if there are components $f_i \neq 0$, $i = n+1, \ldots, m$) that there are negative eigenvalues. To compute the smallest negative eigenvalue of $AA^T - dd^T$ we have to solve the secular equation

$$1 - d^T (AA^T - \mu I)^{-1} d = 0.$$

Using the SVD of A this is written as

$$1 - \sum_{i=1}^{n} \frac{f_i^2}{\sigma_i^2 - \mu} + \sum_{i=n+1}^{m} \frac{f_i^2}{\mu} = 0.$$

We have a pole at 0. If $\mu \to 0$ by negative values, the left-hand side $\to -\infty$. When $\mu \to -\infty$ the left-hand side goes to 1. Hence, there is one negative root for the secular equation.

If $m \leq n$ D is of order m and we have $AA^T = UD^2U^T$, the secular equation is

$$1 - \sum_{i=1}^{n} \frac{f_i^2}{\sigma_i^2 - \mu} = 0.$$

The left-hand side is a decreasing function between poles. The limit value when $\mu \to -\infty$ is 1. Therefore there is a zero left to σ_1^2. It may or may not be negative.

The value $\mu(x)$, requiring the knowledge of the smallest eigenvalue of a rank-one perturbation to AA^T, may be expensive to compute and can lead to computational difficulties because of cancellation. It has been suggested to use

$$\tilde{\mu}(x) = \|(\|x\|^2 A^T A + \|r\|^2 I)^{-1/2} A^T r\|,$$

as an estimate of $\mu(x)$. If $F(A, x) = A^T (c - Ax)$ and D_A is the matrix of the partial derivatives related to A, then (see Grcar [165])

$$D_A D_A^T = \|r\|^2 I + \|x\|^2 A^T A.$$

The estimate is $\|D_A^\dagger F(A, x)\| = \|(D_A D_A^T)^{-1/2} A^T r\|$, where the $\dagger$ sign denotes the pseudoinverse. Grcar proved that

$$\lim_{x \to x^*} \frac{\tilde{\mu}(x)}{\mu(x)} = 1,$$

where x^* is the exact solution of the original problem. Methods to compute $\tilde{\mu}(x)$ were considered by Grcar, Saunders and Su; see also the Ph.D. thesis of Zheng Su [322]. If one knows the singular value decomposition of A, $A = U \Sigma V^T$, then

$$\|x\| \tilde{\mu}(x) = \|(\Sigma^2 + \eta^2 I)^{-1/2} \Sigma U^T r\|,$$

where $\eta = \|r\|/\|x\|$. Of course, this is quite expensive and not practical when A is large and sparse. Another possibility is to use a QR factorization of the matrix

$$K = \begin{pmatrix} A \\ \frac{\|r\|}{\|x\|} I \end{pmatrix},$$

see Zheng Su [322]. However, computing $\hat{\mu}(x) = \|x\|^2 [\tilde{\mu}(x)]^2$ fits quite well into our approach for estimating quadratic forms since

$$\hat{\mu}(x) = y^T (A^T A + \eta^2 I)^{-1} y, \quad y = A^T r.$$

Then, we can use almost exactly the method of Section 15.1.6 for solving the L-curve problem that is using the (Golub–Kahan) Lanczos bidiagonalization algorithm starting with $y/\|y\|$. This computes bidiagonal matrices B_k. To obtain the estimates we have to solve (small) problems similar to equation (15.6). One can also use the Gauss–Radau modification of B_k; see chapter 4 of Su's thesis [322].

13.2 Least Squares Data Fitting

13.2.1 Solution Using Orthogonal Polynomials

We consider the following approximation problem of fitting given data by polynomials. Giving a discrete inner product

$$\langle f, g \rangle_m = \sum_{j=1}^{m} f(t_j) g(t_j) w_j^2, \tag{13.4}$$

where the nodes t_j and the weights w_j^2, $j = 1, \ldots, m$ are given, and a set of given values y_j, $j = 1, \ldots, m$ find a polynomial q of degree $n < m$ such that the weighted sum

$$\sum_{j=1}^{m} (y_j - q(t_j))^2 \, w_j^2, \tag{13.5}$$

is minimized. The values y_j may come, for instance, from the pointwise values of a function at the nodes t_j or from experimental data. The solution of this least squares problem can be obtained by using the polynomials p_k orthogonal with respect to the inner product (13.4). This was considered by G. Forsythe in 1957 [113]. The inner product (13.4) can be seen as a Gauss quadrature formula. Let J_m be the corresponding Jacobi matrix with eigenvalues $\theta_i^{(m)} = t_i$. The values w_i are the first elements of the normalized eigenvectors of J_m. We know that

$$p^m(\theta_i^{(m)}) = \left(p_0(\theta_i^{(m)}) \quad \cdots \quad p_{m-1}(\theta_i^{(m)}) \right)^T$$

is an (unnormalized) eigenvector corresponding to $\theta_i^{(m)}$. Then, if we denote

$$P_m = \left(p^m(\theta_1^{(m)}) \quad \cdots \quad p^m(\theta_m^{(m)}) \right),$$

we have $J_m P_m = P_m \Theta_m$ where Θ_m is the diagonal matrix of the eigenvalues $\theta_i^{(m)}$. Let D_m be the diagonal matrix with diagonal elements

$$\nu_i = [p^m(\theta_i^{(m)})^T p^m(\theta_i^{(m)})]^{-1/2}.$$

We have $P_m^T P_m = D_m^{-2}$ and $P_m D_m$ is the orthonormal matrix whose columns are the normalized eigenvectors of J_m. The first elements of the columns of $P_m D_m$ are the values w_j, $j = 1, \ldots, m$. Let y^m be the vector with components y_j and d^m be a vector of coefficients d_j. Then, a polynomial q of degree $m - 1$ can be written as $q(x) = (p^m(x))^T d^m$ and if t^m is the vector of the values $t_j = \theta_j^{(m)}$, we have $q(t^m) = P_m^T d^m$. Then, the weighted sum in equation (13.5) is written as

$$\sum_{j=1}^{m} (y_j - q(t_j))^2 \, w_j^2 = \| D_m(y^m - q(t^m)) \|^2$$

$$= \| D_m(y^m - P_m^T d^m) \|^2$$

$$= \| P_m D_m^2 y^m - d^m \|^2.$$

The last expression is obtained because $P_m D_m$ is orthonormal and therefore we have $P_m D_m^2 P_m^T = I$. Clearly, $d^m = P_m D_m^2 y^m$ is the solution of the interpolation problem at nodes t_j expressed in terms of orthogonal polynomials. Taking a vector d^n with only the first n components of d^m gives the solution of the least squares fit with polynomials of degree $n < m - 1$ written as

$$d^n = P_{n,m} D_m^2 y^m,$$

where $P_{n,m}$ is the matrix of the first n rows of P_m.

Therefore, given the nodes t_j and weights w_j^2, $j = 1, \ldots, m$, we have to solve an inverse eigenvalue problem to obtain the Jacobi matrix J_m representing the orthogonal polynomials. We have already studied such problems in chapter 6. We will see in the next sections how to apply the algorithms of chapter 6 to our particular problem.

The paper [100] by Elhay, Golub and Kautsky considers the slightly different problem of updating and downdating the least squares fit. Assuming the knowledge of the solution for dimension m, adding a new triplet of data $\{t_{m+1}, w_{m+1}, y_{m+1}\}$ and computing the new solution is called updating. Removing a triplet of data is called downdating. This is particularly useful when the data come from experiments.

13.2.2 Updating the LS Solution

We assume that we know the Jacobi matrix J_n and the vector d^n with $n \leq m$. We want to compute the Jacobi matrix $J_{\tilde{n}}$ and vector $d_{\tilde{n}}$ with $\tilde{n} = n$ or $n + 1$ for the data $\{t_j, w_j, y_j\}$, $j = 1, \ldots, m + 1$. A solution always exists for a partial solution with $\tilde{n} = n$. The solution for $\tilde{n} = n + 1$ requires that $m = n$ and therefore the points t_j, $j = 1, \ldots, m$ are the eigenvalues of J_n. The update is possible only if the new point t_{m+1} is not one of the eigenvalues of J_n.

In the case $n = m$, $\tilde{n} = m + 1$, the problem is to expand J_m to a tridiagonal matrix J_{m+1} having the spectral data $\{t_j, w_j\}_{j=1,\ldots,m+1}$. The solution is given by the following theorem, see [100].

THEOREM 13.1 *Assume we know J_m. Let $\sigma_m = (w_1^2 + \cdots + w_m^2)^{1/2}$. The solution of the problem for $\{t_j, w_j, y_j\}$, $j = 1, \ldots, m + 1$ is given by*

$$J_{m+1} = Q \begin{pmatrix} J_m & 0 \\ 0 & t_{m+1} \end{pmatrix} Q^T,$$

$$\sigma_{m+1} = (\sigma_m^2 + w_{m+1}^2)^{1/2},$$

$$d^{m+1} = Q \begin{pmatrix} d^m \\ w_{m+1} y_{m+1} \end{pmatrix},$$

where the orthogonal matrix Q is uniquely determined by requiring J_{m+1} to be tridiagonal and Q to be such that

$$Q(\sigma_m e^1 + w_{m+1} e^{m+1}) = \sigma_{m+1} e^1.$$

Proof. Let $Q_m = P_m D_m$. Then, Q_m is the matrix of the eigenvectors of $J_m = Q_m \Theta_m Q_m^T$ and

$$\sigma_m (e^1)^T Q_m = (w^m)^T,$$

where w^m is the vector with components w_j, $j = 1, \ldots, m$. Similarly, we have $J_{m+1} = Q_{m+1} \Theta_{m+1} Q_{m+1}^T$ and $\sigma_{m+1}(e^1)^T Q_{m+1} = (w_{m+1})^T$ with $(w^{m+1})^T = ((w^m)^T \; w_{m+1})$. The matrix

$$Q = Q_{m+1} \begin{pmatrix} Q_m^T & 0 \\ 0 & 1 \end{pmatrix}$$

satisfies the first relation of the theorem. Moreover, one can check that $Q(\sigma_m e^1 + w_{m+1} e^{m+1}) = \sigma_{m+1} e^1$ is verified. Taking the norms of both sides of this relation gives the second relation of the theorem. Now, we have $d^m = Q_m D_m y^m$ and $d^{m+1} = Q_{m+1} D_{m+1} y^{m+1}$ with $(y^{m+1})^T = ((y^m)^T \ y_{m+1})$. Hence,

$$d^{m+1} = Q_{m+1} \begin{pmatrix} D_m y^m \\ w_{m+1} y_{m+1} \end{pmatrix} = Q_{m+1} \begin{pmatrix} Q_m^T d^m \\ w_{m+1} y_{m+1} \end{pmatrix} = Q \begin{pmatrix} d^m \\ w_{m+1} y_{m+1} \end{pmatrix},$$

which proves the third assertion. □

This theorem is not constructive since even if we know Q_m we do not know Q_{m+1}. The matrix Q of order $m+1$ can be constructed as a product of elementary Givens rotations

$$Q = R_m R_{m-1} \cdots R_1,$$

where R_j is a rotation between rows j and $m+1$. The first rotation R_1 is computed to achieve the relation $Q(\sigma_m e^1 + w_{m+1} e^{m+1}) = \sigma_{m+1} e^1$. The vector $\sigma_m e^1 + w_{m+1} e^{m+1}$ has nonzero components only in positions 1 and $m+1$. Therefore the bottom entry can be zeroed by a rotation R_1 between the first and last rows. Applying symmetrically this rotation to the matrix

$$\begin{pmatrix} J_m & 0 \\ 0 & t_{m+1} \end{pmatrix}$$

creates two nonzero entries in the last row in positions 1 and 2 and, symmetrically two nonzeros in the last column. The next rotations $R_2, R_3, \ldots$ are constructed to chase these nonzero elements to the right of the last row and to the bottom of the last column. Hence, R_2 is a rotation between rows 2 and $m+1$ to zero the entry in position $(m+1, 1)$. It modifies the element $(m+1, 2)$ and creates a nonzero element in position $(m+1, 3)$. Rotation R_3 zero the $(m+1, 2)$ entry and creates a nonzero in position $(m+1, 4)$ and so on. At step k the first $k-1$ elements of the last row and of the last column of

$$K_k = R_k R_{k-1} \cdots R_1 \begin{pmatrix} J_m & 0 \\ 0 & t_{m+1} \end{pmatrix} R_1^T \cdots R_{m-1}^T R_m^T,$$

vanish. The matrix K_k is tridiagonal up to the last row and column which have nonzero elements in position $k, k+1$ and $m+1$. Hence, K_m is tridiagonal. This updating method named RHR should be stable since it uses only orthogonal rotation matrices.

Note that we can also use this method to solve the inverse eigenvalue problem of chapter 5 starting with a pair of one node and one weight, adding a new pair at each step. The data $y_j, j = 1, \ldots, m+1$ is only involved in the computation of d^m when we know Q.

Elhay, Golub and Kautsky [100] studied also some Lanczos-type methods in the same spirit as the methods proposed in Kautsky and Golub [202] that we have considered in chapter 5 in the section devoted to the modification of the weight function by multiplication with a polynomial. They were looking for monic polynomials $\tilde{p}_k$ orthogonal with respect to the inner product $\langle . , . \rangle_{m+1}$ assuming the knowledge of the orthogonal polynomials p_k for $\langle . , . \rangle_m$. We have

$$\tilde{p}_{m+1}(t) = (t - t_{m+1}) p_m(t).$$

Both sets of polynomials satisfy

$$tp^m(t) = J_m p^m(t) + \beta_m e^m p_m(t),$$

$$t\tilde{p}^m(t) = \tilde{J}_{m+1}\tilde{p}^m(t) + \tilde{\beta}_{m+1}e^{m+1}\tilde{p}_{m+1}(t).$$

Denoting $(p^{m+1})^T = ((p^m)^T \ p_m)$, we have

$$tp^{m+1}(t) = J_{m+1}p^{m+1}(t) + e^{m+1}\tilde{p}_{m+1}(t),$$

where

$$J_{m+1} = \begin{pmatrix} J_m & \beta_m e^m \\ 0 & t_{m+1} \end{pmatrix}.$$

There exists a nonsingular lower triangular matrix L such that

$$p^{m+1} = L\tilde{p}^{m+1}.$$

Comparing the previous relations, we obtain

$$J_{m+1}L = L\tilde{J}_{m+1}, \quad \tilde{\beta}_{m+1}(e^{m+1})^T Le^{m+1} = 1.$$

Then, denoting by l^j the columns of L,

$$(1 - \delta_{j,m+1})\tilde{\beta}_j l^{j+1} + \tilde{\alpha}_j l^j + (1 - \delta_{j,1})\tilde{\beta}_{j-1}l^{j-1} = J_{m+1}l^j, \ j = 1, \ldots, m+1,$$

where $\delta_{j,i}$ is the Kronecker symbol. These relations can be used to evaluate alternately the elements of $\tilde{J}_{m+1}$ and the columns of L from the knowledge of l^1. This is clearly analogous to the Lanczos (or Stieltjes) algorithm. This algorithm exploits the special form of L. It remains to show how to compute the first column $l^1 = Le^1$. Using the relation $p^{m+1} = L\tilde{p}^{m+1}$ and the fact that the polynomials $\tilde{p}_j$ are orthonormal we have that the matrix LL^T is equal to a matrix whose elements (i, j) are the inner products $\langle p_i, p_j \rangle_{m+1}$. We remark that since L is lower triangular, $L^T e^1 = ((e^1)^T Le^1)e^1$. Therefore,

$$LL^T e^1 = ((e^1)^T Le^1)l^1 = \langle p^{m+1}, p_0 \rangle = e^1 + w_{m+1}^2 p_0 \, p^{m+1}(t_{m+1}).$$

The last relation allows us to compute the first column l^1. It turns out that for constructing the Jacobi matrix we need only the diagonal and subdiagonal elements of L. Let the jth column of L be denoted as

$$l^j = \begin{pmatrix} 0 & \ldots & 0 & \rho_j & \tau_j & \ldots \end{pmatrix}^T,$$

where ρ_j and τ_j are the elements we are interested in. We have the following relations

$$\tilde{\beta}_{j-1}\rho_{j-1} = (e^{j-1})^T \tilde{J}_{m+1}l^j = \beta_{j-1}\rho_j, \ j = 2, \ldots, m+1,$$

$$\tilde{\alpha}_j\rho_j + \tilde{\beta}_{j-1}\tau_{j-1} = (e^j)^T J_{m+1}l^j = \begin{cases} \alpha_j\rho_j + \beta_j\tau_j, & 1 \le j \le m \\ t\rho_{m+1}, & j = m+1 \end{cases}$$

$$\tilde{\beta}_{m+1} = 1/\rho_{m+1}.$$

It remains to compute ρ_j and τ_j, $j = 1, \ldots, m+1$. Denoting $M = LL^T$, we have

$$M = \text{diag}(I, 0) + u^{m+1}(u^{m+1})^T,$$

where $u^{m+1} = w_{m+1}p^{m+1}(t_{m+1})$. Thus L is the Cholesky factor of a rank-one modification of a diagonal matrix. We denote the elements of the vector u^{m+1} by $\psi_j = w_{m+1}p_{j-1}(t_{m+1})$. For $j < k$ we have

$$(e^k)^T L e^j = \psi_k q_j,$$

where the diagonal elements ρ_j and q_j satisfy

$$\rho_j^2 = 1 - \delta_{j,m+1} + \psi_j^2 (1 - q_1^2 - \cdots - q_{j-1}^2),$$

$$\rho_j q_j = \psi_j (1 - q_1^2 - \cdots - q_{j-1}^2).$$

Noticing that $\tau_j = (e^{j-1})^T L e^j$, we can evaluate ρ_j and the ratio τ_j/ρ_j which is

$$\frac{\tau_j}{\rho_j} = \frac{\psi_{j+1}\psi_j}{\psi_j^2 + 1/(1 - q_1^2 - \cdots - q_{j-1}^2)}.$$

The solution of the least squares problem can be computed recursively as

$$d_j^{m+1} = \frac{d_j^m + w_{m+1}y_{m+1}\psi_j - \psi_j \sum_{k=1}^{j-1} d_k^{m+1} q_k}{\rho_j}.$$

This method is called TLS (!) in [100]. To avoid confusion with total least squares (see chapter 14) we will denote this method as TLUDSFUS, UD for update and downdate, SFUS for special form and unscaled. In the paper [100] another method was described using determinants. We will denote it by TLUDUS.

13.2.3 Downdating the LS Solution

Before considering the downdating of the least squares fitting solution, we show how to change the weight of an existing point. This can be done by adding a new point that is the same as one of the points t_j, $j = 1, \ldots, m$, say the last one. The solution to this problem is given in the following theorem, see Elhay, Golub and Kautsky [100].

THEOREM 13.2 *Given σ_m, J_m and d^m and new data $\{t_{m+1}, w_{m+1}, y_{m+1}\}$ with $t_{m+1} = t_m$, the solution is given by*

$$\tilde{J}_m = Q J_m Q^T,$$

$$\tilde{\sigma}_m = (\sigma_m^2 + w_{m+1}^2)^{1/2},$$

$$\tilde{d}^m = Q(d^m + (\tilde{w}_m \tilde{y}_m - w_m y_m)q^m),$$

where $\tilde{w}_m = (w_m^2 + w_{m+1}^2)^{1/2}$, $\tilde{y}_m = (w_m^2 y_m + w_{m+1}^2 y_{m+1})/(w_m^2 + w_{m+1}^2)$ and the orthonormal matrix Q is uniquely determined by requiring $\tilde{J}_m$ to be tridiagonal and such that

$$Q(\sigma_m e^1 + (\tilde{w}_m - w_m)q^m) = \tilde{\sigma}_m e^1.$$

The vector q^m is the normalized eigenvector of J_m corresponding to the eigenvalue t_m scaled to have a positive first element.

Proof. The proof is very similar to the proof of theorem 13.1 (see [100]).   □

This theorem shows that one can modify the weight of a chosen node by orthogonal transformations. However, an eigenvector like q^m has to be computed. We remark that the updating methods using rotations discussed previously can be used in this case. The difference with the previous situation is that the Jacobi matrix that is obtained cannot be unreduced because it must have two equal eigenvalues. Hence, one of the subdiagonal coefficients must be zero. Note also that theorem 13.2 corresponds to an increase of the weight since $w_{m+1}^2 > 0$ corresponds to $\tilde{w}_m > w_m$.

We now turn to the downdating problem. This is similar to zeroing a weight, say, having $\tilde{w}_m = 0$. A way to achieve this is to use the same method as when modifying a weight but replacing w_{m+1} by $\imath w_{m+1}$ where $\imath^2 = -1$. The methods that have been described before in this section can be coded such that the complex quantities that are involved remain purely imaginary throughout the calculation and thus everything can be done conveniently in real arithmetic. In the method using rotations the similarity matrices which are involved are

$$\begin{pmatrix} c & \imath s \\ -\imath s & c \end{pmatrix},$$

with $c^2 - s^2 = 1$, embedded in an identity matrix. These matrices are called hyperrotations and they are complex orthogonal. Contrary to plane rotations they can be badly conditioned.

Another possibility is, so to speak, to "reverse" the rotation method. The solution is given in the following theorem.

THEOREM 13.3 *Given σ_{m+1}, J_{m+1} and d^{m+1}, the solution when removing the triplet $\{t_{m+1}, w_{m+1}, y_{m+1}\}$ from the data is given by*

$$\begin{pmatrix} \tilde{J}_m & 0 \\ 0 & t_{m+1} \end{pmatrix} = Q J_{m+1} Q^T,$$

$$\tilde{\sigma}_m = (\sigma_m^2 - w_{m+1}^2)^{1/2},$$

$$\begin{pmatrix} \tilde{d}^m \\ w_{m+1} y_{m+1} \end{pmatrix} = Q d^{m+1},$$

where the orthogonal matrix Q is uniquely determined by requiring $\tilde{J}_m$ to be tridiagonal and such that

$$Q(\sigma_{m+1} e^1 - w_{m+1} q^{m+1}) = \tilde{\sigma}_m e^1,$$

q^{m+1} being the normalized eigenvector of J_{m+1} corresponding to t_{m+1} scaled to have a positive first element.

Proof. See Elhay, Golub and Kautsky [100].   □

The method based on theorem 13.3 is called REV. The paper [100] also shows how to downdate a partial matrix J_n with $n < m + 1$. The existence of a solution is equivalent to the existence of a matrix $J_{\hat{n}}$ with $\hat{n} > n$ such that J_n is a submatrix of $J_{\hat{n}}$ and the node to be removed is an eigenvalue of $J_{\hat{n}}$. Note that obtaining this matrix is quite similar to the problem we solved for the Gauss–Radau quadrature rule.

13.3 Numerical Experiments

13.3.1 LS Test Problems

In this section we describe the results of numerical experiments for updating and downdating a least squares fit with the methods we have described above and we show how to improve some of them. Let us consider the following problems,

Example LS1

The nodes, weights and data are given by

$$t_m = -1+2(m-1)/(n-1), \ w_m = 1/\sqrt{n}, \ y_m = 1.5+\sin(4t_m), \ m = 1,\ldots,n.$$

Example LS2

$$\rho = 0.9, \ t_1 = 0.01, \ t_n = 100,$$

$$t_m = t_1 + ((m-1)/(n-1))(t_n - t_1)\rho^{n-m}, \ m = 2,\ldots,n-1,$$

$$w_m = 1/\sqrt{n}, \ y_m = 1.5 + \sin(4t_m).$$

Example LS1 is close to one in Elhay, Golub and Kautsky [100]. It has regularly distributed nodes. Example LS2 corresponds to the Strakoš Example F3; see chapter 11.

13.3.2 Solutions Using Rotations

We first consider the following experiment. Starting from

$$\sigma = |w_1|, \ \alpha_1 = t_1, \ d_1 = y_1\sigma,$$

we recursively build up the symmetric tridiagonal matrices J_m, $m = 2,\ldots,n$ whose diagonal elements are α_i and subdiagonal elements are β_i adding a point t_m at each step. Having obtained the final J_n, we downdate the solution by successively removing $t_m, m = n,\ldots,2$. In the updating phase we check the eigenvalues of J_m. They must be equal to $t_1,\ldots,t_m$. When downdating we compare the Jacobi matrices we obtain with the ones that have been computed in the updating phase.

We first do the updating using RHR and the downdating using hyperbolic rotations. Let us start with Example LS1 using a small number of points $n = 8$. The absolute values of the differences between the points t_m and the eigenvalues of J_n are at the roundoff level. This means that the updating process using rotations works quite well, at least for this small problem. Now, we downdate by removing one point at a time. The $\log_{10}$ of the relative differences in the elements of the matrix J_m and the Fourier coefficients computed as

$$\frac{\|\alpha - \tilde{\alpha}\|}{\|\tilde{\alpha}\|},$$

where α (resp. $\tilde{\alpha}$) is the diagonal of J_m obtained when downdating (resp. updating) are given in figure 13.1. In other words we compare the solution we obtain when downdating to what we had previously when updating. Therefore, the figure is to be read from right to left. The results when downdating the last point $t_n = 1$ are given

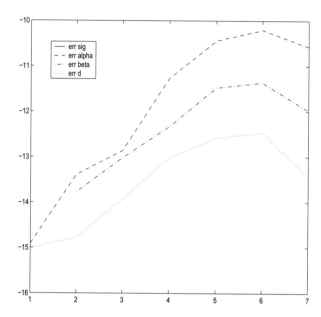

Figure 13.1 Example LS1, $\log_{10}$ of relative errors when downdating, $n = 8$, RHR and hyperbolic rotations

for abscissa 7 and so on. We see that the relative errors are quite large, up to 10^{-10}, except for σ for which they are negligible. Moreover, the errors increase when n increases, up to the point where the algorithm is no longer able to downdate.

These downdating results can be improved a little by a careful reordering of the floating point operations (specially when computing the rotations). With this version of the algorithm and for Example LS1, the breakdown when no downdating is possible happens for $n = 13$.

Example LS2 breaks down at $n = 13$. The results for $n = 12$ which are shown in figure 13.2 are worst than for the other example. The conclusion of these experiments (and others not reported here) is that the hyperbolic rotations that are used for downdating are highly unstable and cannot be used even for moderate values of n.

As we have seen, Elhay, Golub and Kautsky [100] proposed another downdating method called REV based on an eigenvector of J_m, see theorem 13.3. In [100] its components are obtained by solving a triangular system. The results for Example LS1 are given in figure 13.3. The errors are not better than with hyperbolic rotations (with modifications). However, REV is more robust and allows us to use larger values of n. Results for $n = 19$ are given in figure 13.4. For $n = 20$, we found a value $\beta_j = 0$ and the algorithm has to stop.

When downdating the points in the reverse order as they were added for updating, one can compute the eigenvectors of the matrices J_j incrementally. Assume that we are updating for J_n to J_{n+1}. Let Z (resp. W) be the matrix of the eigenvectors

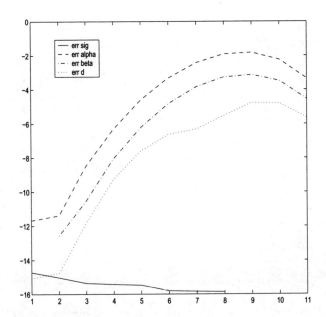

Figure 13.2 Example LS2, $\log_{10}$ of relative errors when downdating with modifications of RHR and hyperbolic rotations, $n = 12$

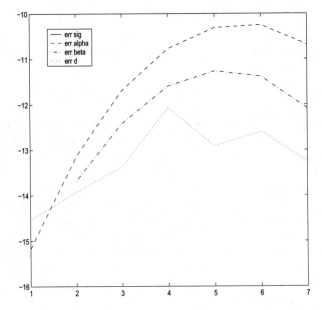

Figure 13.3 Example LS1, $\log_{10}$ of relative errors when downdating with REV, $n = 8$

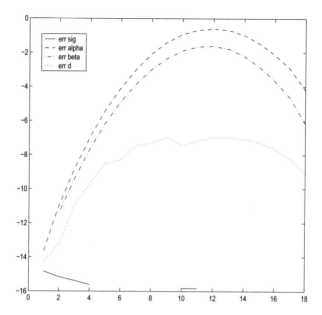

Figure 13.4 Example LS1, $\log_{10}$ of relative errors when downdating with REV, $n = 19$

of J_n (resp. J_{n+1}) and $Q = R_n \cdots R_1$ be the products of the rotations in RHR. We have the relation

$$W = Q \begin{pmatrix} Z & 0 \\ 0 & 1 \end{pmatrix} = R_n \cdots R_1 \begin{pmatrix} Z & 0 \\ 0 & 1 \end{pmatrix}.$$

Therefore, starting from $Z = 1$ for $n = 1$, we can compute the eigenvectors by applying the rotations that are obtained at each step. Reciprocally, from the eigenvectors of J_{n+1} we have

$$\begin{pmatrix} Z & 0 \\ 0 & 1 \end{pmatrix} = R_1^T \cdots R_n^T W.$$

If we know the eigenvalues of J_n that are (approximately) the points t_j, we can also recover J_n from Z (removing the last point we added) by taking the tridiagonal part of $Z \Lambda Z^T$ where Λ is the diagonal matrix of the eigenvalues (although this is only marginally better, what is important is to have good eigenvectors). Let us see how this is working in figures 13.5 and 13.6. We see that the results are good even for large data sets. However, there are several serious drawbacks with this algorithm. First, it is much more expensive than the other ones. So recomputing the solution for the downdated data set from scratch may be cheaper. If we want to remove any data point and not just the last one, we can use the REV algorithm modified to use an eigenvector computed using the rotations and not by solving a linear system. Then, we can compute the downdated eigenvectors by deflation after applying the rotations computed by REV.

We now try to reproduce the numerical results of Elhay, Golub and Kautsky [100] using a sliding window in the data. The data are generated for N points;

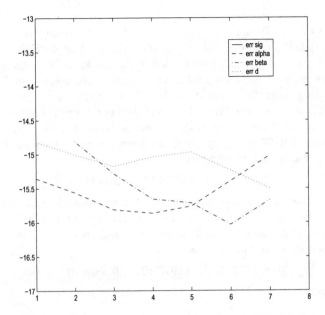

Figure 13.5 Example LS1, $\log_{10}$ of relative errors when downdating with eigenvectors, $n =$ 8

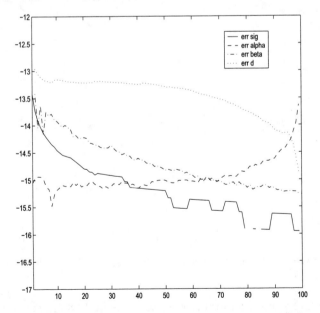

Figure 13.6 Example LS1, $\log_{10}$ of relative errors when downdating with eigenvectors, $n =$ 100

let $Y_k = \{t_j, w_j, y_j\}_{j=k}^{k+M-1}$, $k = 1, \ldots, N - M + 1$ be the sliding window. A least squares fit of dimension n ($n <= M$) is computed for the data Y_1, that is, the points from 1 to M. Then, for $k = 2, \ldots, N - M + 1$, the solution is computed for Y_k by first updating with the data $\{t_{k+M}, w_{k+M}, y_{k+M}\}$ on the right and second downdating $\{t_{k-1}, w_{k-1}, y_{k-1}\}$ on the left of the window. The authors compare the solution every n_s steps with a reference solution computed from scratch by updating. They used the values $N = 50, M = 10, n = 5, n_s = 5$. We first compare the values of the entries of J_m and the Fourier coefficients d at the end of the process. We do not use a random perturbation of the data as in [100]. The relative errors of RHRud (for updating and downdating) for Example LS1 at the end for σ, α, β and d are, respectively,

$$0, \ 2.8951 \ 10^{-7}, \ 9.0746 \ 10^{-8}, \ 2.3494 \ 10^{-10}.$$

The relative errors at each step (as a function of k, when we add the data for $k + M$ and remove the data for k) are shown in figure 13.7. The errors increase with k as the window is sliding and the quality of the solution decreases as N increases. For $N = 190$, we obtain as relative errors

$$0, \ 8.1277 \ 10^{-1}, \ 9.1918 \ 10^{-1}, \ 6.5218 \ 10^{-6},$$

with bad results for α and β.

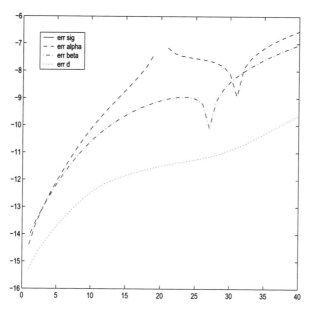

Figure 13.7 Example LS1, RHRud, $\log_{10}$ of relative errors with a sliding window, $N = 50, M = 10, n = 5$

Example LS2 gives better results as we see in figure 13.8. The relative errors at the end for σ, α, β and d and $N = 50, M = 10, n = 5$ are

$$0, \ 1.2783 \ 10^{-12}, \ 6.9897 \ 10^{-13}, \ 5.0452 \ 10^{-13}.$$

We were able to compute up to $N = 340, M = 10, n = 5$ with the following results:

$$1.6184\ 10^{-16},\ 7.3216\ 10^{-12}, 3.7765\ 10^{-12},\ 7.5073\ 10^{-13}.$$

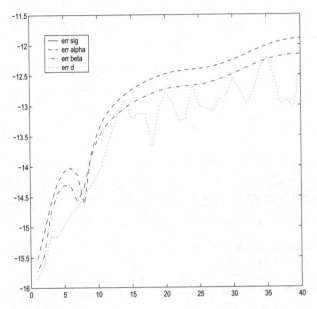

Figure 13.8 Example LS2, RHRud, $\log_{10}$ of relative errors with a sliding window, $N = 50, M = 10, n = 5$

Things are different when we update with RHR and downdate with REV; see figure 13.9. For Example LS1 the relative errors at the end for σ, α, β and d are

$$0,\ 3.6593\ 10^{-1},\ 3.4459\ 10^{-1},\ 3.2987\ 10^{-4}.$$

The results are much better for example LS2 as it can be seen in figure 13.10.

When using the methods that compute all the eigenvectors when updating and downdating, we obtain the following results for Example LS1 with $N = 50, M = 10, n = 5$:

$$0,\ 1.8244\ 10^{-14},\ 5.8681\ 10^{-14},\ 2.9810\ 10^{-13}.$$

The downdating is done using REV but with an eigenvector computed during the updating and not by solving a linear system as in [100]. This method works fine for $N = 50, M = 10, n = 5$, the relative errors being of the order 10^{-13}; see figure 13.11. For $N = 190$ we obtain

$$0,\ 2.2019\ 10^{-14},\ 2.2132\ 10^{-13},\ 7.8349\ 10^{-13}.$$

Example LS2 gives

$$0,\ 3.8045\ 10^{-15},\ 2.6309\ 10^{-15},\ 1.9726\ 10^{-14}.$$

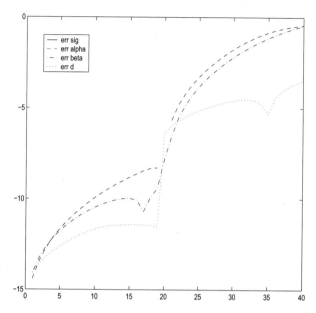

Figure 13.9 Example LS1, RHRu + REV, $\log_{10}$ of relative errors with a sliding window,
$N = 50, M = 10, n = 5$

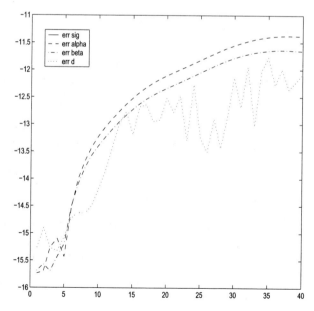

Figure 13.10 Example LS2, RHRu + REV, $\log_{10}$ of relative errors with a sliding window,
$N = 50, M = 10, n = 5$

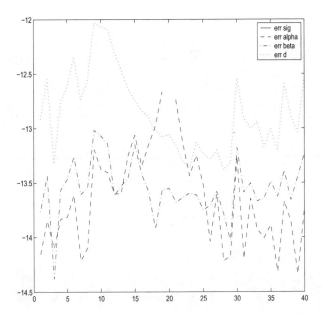

Figure 13.11 Example LS1, eigenvectors, $\log_{10}$ of relative errors with a sliding window, $N = 50, M = 10, n = 5$

Just to compare with other methods, for $N = 340$ we obtain

$$1.6184 \ 10^{-16}, \ 8.7078 \ 10^{-15}, 1.3943 \ 10^{-14}, \ 3.8488 \ 10^{-14}.$$

We note that if we just want to discard one point it is not necessary to compute all the eigenvectors. We just have to store and use the rotations. However, the computation of the Fourier coefficients is easier when using the eigenvectors.

There are many other ways to compute an eigenvector for a known eigenvalue. This problem has been considered at length during the last fifteen years by Dhillon and Parlett [84], [85], [268], based on remarks of J. Wilkinson and works of V. Fernando [105], [107]; see also [83], [86], [87]. The solution uses twisted factorizations of the Jacobi matrix J_m, see chapter 3. Making only this change to the method using eigenvectors, we obtain the following results for Example LS1, $N = 50, M = 10, n = 5$ (see figure 13.12)

$$0, \ 2.4329 \ 10^{-16}, \ 3.5398 \ 10^{-16}, \ 4.1946 \ 10^{-15}.$$

Example LS2 gives

$$0, \ 4.9989 \ 10^{-16}, \ 2.9341 \ 10^{-16}, \ 1.3637 \ 10^{-15}.$$

It is now interesting to see what we can obtain if we do the updating without using the eigenvectors and the downdating with the eigenvectors computed using the Dhillon–Parlett algorithm. So we do the updating by using RHRu and the downdating using an eigenvector. We obtain the following results for Example LS1, $N = 50, M = 10, n = 5$ (see figure 13.13)

$$0, \ 3.1013 \ 10^{-16}, \ 8.5406 \ 10^{-16}, \ 3.3256 \ 10^{-15}.$$

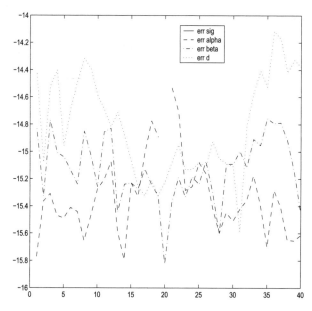

Figure 13.12 Example LS1, eigenvector Dhillon–Parlett, $\log_{10}$ of relative errors with a sliding window, $N = 50, M = 10, n = 5$

The results are almost as good as when using the eigenvectors for updating. There is no growth of errors.

Example LS2 gives

$$0, \ 4.3906 \ 10^{-16}, \ 1.0122 \ 10^{-15}, \ 1.7337 \ 10^{-15},$$

and the results are in figure 13.14 for $N = 50$. The conclusion of this experiment is that updating with rotations and downdating with a good eigenvector allows us to solve large problems in a stable way with good results.

Now that we have another method to compute the eigenvector, we return to the first experiment where we first update and then downdate the points by deleting the last one. We obtain the results of figure 13.15, which are much better than with the other methods.

The problem with using the eigenvector to downdate is that we have to know that the Jacobi matrix has the point t to be removed as an eigenvalue. This is not true when we compute a partial solution with $n < m$. A possible solution (as we used above) is to always compute the full Jacobi matrix of size m and then to use the appropriate submatrix. Moreover, this has the added advantage of providing least squares fitting of different degrees. This is not too costly if m is not much different from n as in the examples we used. Elhay, Golub and Kautsky [100] considered the problem of downdating a partial solution.

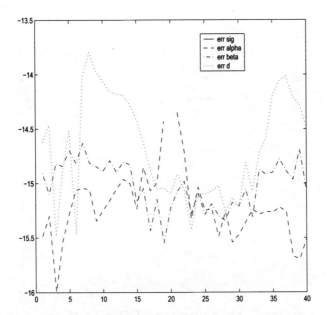

Figure 13.13 Example LS1, RHRu + eigenvector Dhillon–Parlett, $\log_{10}$ of relative errors with a sliding window, $N = 50, M = 10, n = 5$

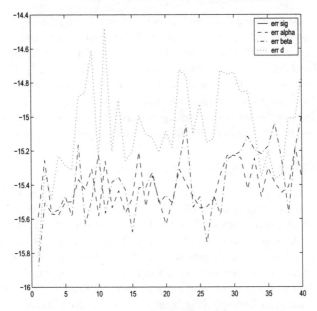

Figure 13.14 Example LS2, RHRu + eigenvector Dhillon–Parlett, $\log_{10}$ of relative errors with a sliding window, $N = 50, M = 10, n = 5$

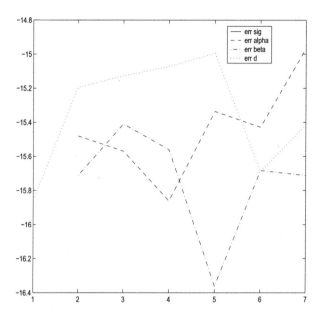

Figure 13.15 Example LS1, RHRu + eigenvector Dhillon–Parlett, $\log_{10}$ of relative errors when downdating, $n = 8$

13.3.3 Solution Using Lanczos-Like Methods

We consider the experiment where we slide the window by adding a node at the right of the interval and removing a node at the left end at each step. For Example LS1 we experiment difficulties with TLUDUS and TLUDSFUS after $N = 30$. For Example LS2 the results are much better. Figures 13.16 and 13.17 display the results of algorithm TLUDUS for $N = 50$ and $N = 200$, respectively. The results for TLUDSFUS are given in figures 13.18 and 13.19.

The general conclusion is that the algorithms which work best for downdating use orthogonal transformations and accurate eigenvectors.

13.3.4 Solution Using the Lanczos Algorithm

The problem of computing the LS solution by computing the entries of the Jacobi matrix is an inverse eigenvalue problem. As we have seen in chapter 6 this problem can be solved in exact arithmetic by the Lanczos algorithm. In finite precision arithmetic the Lanczos algorithm may suffer from a severe growth of rounding errors. Nevertheless, let us see how it works on our Examples LS1 and LS2. We check if the eigenvalues of the computed Jacobi matrix are equal to the given points t_j and how the first components of the eigenvectors are related to the weights w_j in the inner product. For Example LS1, after $N = 50$ Lanczos iterations the relative errors on the nodes and weights in l_2 norm are

$$e_t = 4.9736 \ 10^{-16}, \quad e_w = 6.4759 \ 10^{-15},$$

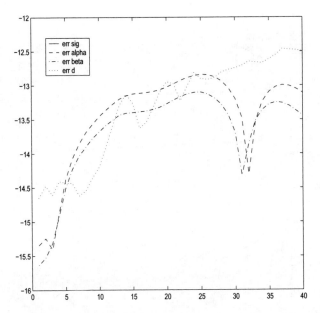

Figure 13.16 Example LS2, TLUDUS, $\log_{10}$ of relative errors with a sliding window, $N = 50, M = 10, n = 5$

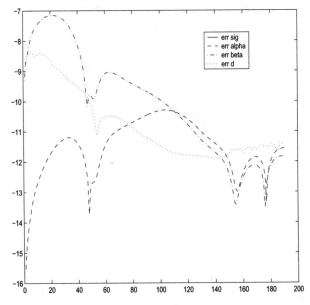

Figure 13.17 Example LS2, TLUDUS, $\log_{10}$ of relative errors with a sliding window, $N = 200, M = 10, n = 5$

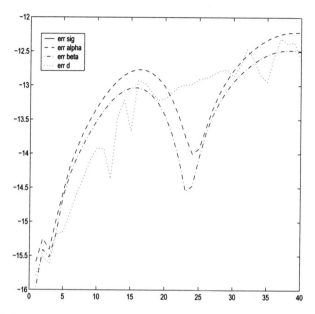

Figure 13.18 Example LS2, TLUDSFUS, $\log_{10}$ of relative errors with a sliding window, $N = 50, M = 10, n = 5$

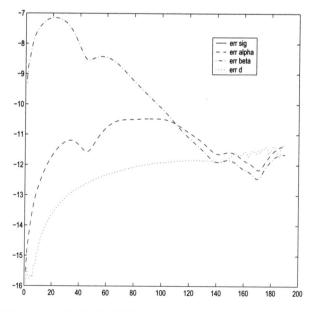

Figure 13.19 Example LS2, TLUDSFUS, $\log_{10}$ of relative errors with a sliding window, $N = 200, M = 10, n = 5$

with full double reorthogonalization at each Lanczos iteration. The level of orthogonality computed as the maximum of the absolute values of off-diagonal elements of $V_m^T V_m$ is $6.6613 \ 10^{-16}$, where V_m is the matrix of the Lanczos vectors which are orthogonal in exact arithmetic.

Without reorthogonalization, we have

$$e_t = 1.3203 \ 10^{-6}, \quad e_w = 2.6883 \ 10^{-6}.$$

The level of orthogonality is 0.0017. Of course, the results are much worse. However, this example is not the worst one for the Lanczos algorithm because the gaps between successive eigenvalues are constant and the first convergence of a Ritz value toward an eigenvalue does not occur rapidly. Let us consider Example LS2 with $N = 50$. With double reorthogonalization the level of orthogonality is $6.6613 \ 10^{-16}$ and the errors are

$$e_t = 5.2394 \ 10^{-16}, \quad e_w = 5.1508 \ 10^{-14}.$$

Without reorthogonalization, we have a level of orthogonality 0.7552 and

$$e_t = 0.6778, \quad e_w = 0.5391.$$

Therefore, the computed Jacobi matrix does not have the given nodes as eigenvalues. This example was designed by Z. Strakoš to have a fast convergence of a Ritz value and a rapid growth of the rounding errors which explains the complete loss of orthogonality.

To compute the vector of Fourier coefficients d^n of the solution from the Lanczos algorithm we note that we have

$$d^m = P_m D_m^2 y^m,$$

where $P_m D_m$ is the matrix of the normalized eigenvectors and d^n with $n \leq m$ is obtained by taking the n first rows of the previous expression. When applying the Lanczos algorithm to a diagonal matrix, the matrix of eigenvectors is the identity matrix I. At the end, we have the computed eigenvectors which are $V_m Z_m$ where Z_m is the matrix of the eigenvectors of J_m. Therefore, we have $V_m Z_m = I$ which gives $Z_m = V_m^T$. The solution is given by

$$d^m = V_m^T D_m y^m. \tag{13.6}$$

The diagonal elements of D_m are given by

$$(D_m)_{i,i} = \left(\sum_{j=0}^{m-1} p_j(\theta_i^{(m)})^2 \right)^{\frac{1}{2}}.$$

The polynomials p_j can be evaluated at $\theta_i^{(m)}$ using J_m. In exact arithmetic the Ritz values at the last iteration are equal to the nodes, $\theta_i^{(m)} = t_i$. The least squares solution is obtained by taking the first n rows of the left-hand side of equation (13.6). Note that we have to keep the Lanczos vectors to compute the solution.

It is not so obvious how to update the Jacobi matrix when using the Lanczos algorithm. It amounts to computing the Lanczos coefficients when going from a diagonal matrix of order m to a matrix of order $m + 1$ by adding the new point

t_{m+1} and also adding a component w_{m+1} to the initial vector. In exact arithmetic, this can be solved by using the relation between the Lanczos algorithm and the QR factorization. We have seen that $\Lambda = V_m R_m$ and that V_m is the matrix of the Lanczos vectors. The coefficients of the Jacobi matrix can be obtained from the two main diagonals of R_m. Then, we can use methods for updating the QR factorization. The downdating problem can be be seen as zeroing a component (say the last one) of the initial vector in the Lanczos algorithm. But it is not clear how one can exploit this.

13.3.5 Solution Using the Stieltjes Algorithm

The Stieltjes algorithm was described in chapter 6. We use the Matlab routine Stieltjes.m from the package OPQ by Gautschi [132] and compute the Jacobi matrix. For Example LS1, after $N = 50$ iterations the relative errors in the l_2 norm are

$$e_t = 3.3769 \ 10^{-7}, \quad e_w = 7.1688 \ 10^{-7}.$$

Now consider Example LS2 with $N = 50$. The errors are

$$e_t = 0.6815, \quad e_w = 0.5371.$$

Therefore, the behavior of the Stieltjes algorithm is almost the same as it is for the Lanczos algorithm without reorthogonalization.

13.3.6 The Gragg and Harrod algorithm

We use the implementation provided in Gautschi's package OPQ [132]. We check the eigenvalues (using the Matlab QR algorithm) and the first components of the eigenvectors of the computed Jacobi matrix. For Example LS1, for $N = 50$ the relative errors in the l_2 norm are

$$e_t = 4.6782 \ 10^{-16}, \quad e_w = 6.0235 \ 10^{-15}.$$

Example LS2 gives

$$e_t = 4.0544 \ 10^{-16}, \quad e_w = 8.7212 \ 10^{-14}.$$

The method proposed by Gragg and Harrod [164] constructs the solution incrementally; see chapter 5. Therefore, it can be used to update the solution. The problem of downdating was not considered in this paper.

13.3.7 The Laurie QD Algorithm

In [220] Laurie proposed to use variants of the QD algorithm to recover the Jacobi matrix from nodes and (positive) weights. We use the algorithm pftoqd. For Example LS1, the smallest nodes are negative. Thus we have to use the trick mentioned in [220] to compute the Jacobi matrix by omitting the last step and shifting. The errors for $N = 50$ are

$$e_t = 4.4347 \ 10^{-16}, \quad e_w = 5.3900 \ 10^{-15}.$$

For Example LS2, we obtain the following errors:

$$e_t = 3.3720 \ 10^{-16}, \quad e_w = 2.4849 \ 10^{-14}.$$

13.4 Numerical Experiments for the Backward Error

Let us consider a small problem of dimension 20×10 (Example BK1). We use an example derived from those of U. von Matt for solving ill-posed problems; see chapter 15. Let A be an $m \times n$ matrix such that

$$A = U_s \Sigma_s V_s^T, \quad U_s = I - 2\frac{u_s u_s^T}{\|u_s\|^2}, \quad V_s = I - 2\frac{v_s v_s^T}{\|v_s\|^2},$$

where u_s and v_s are vectors whose components are $\sin(2\pi i/(l+1))$ where l is the length of the vector. Σ_s is an $m \times n$ diagonal matrix with diagonal elements $[1, \cdots, \sqrt{n}]$. Then the singular values are perturbed by 10^{-3} times a random number. Let x_s be a vector whose ith component is $1/i$ and $c_s = Ax_s$. The right-hand side is

$$c = c_s + \xi \, \text{randn}(m, 1).$$

Let x be an approximate solution of the least squares problem and $r = c - Ax$ be the residual. We first compute the smallest eigenvalue

$$\lambda_{\min} \left(AA^T - \nu \frac{rr^T}{\|x\|^2} \right),$$

using the Matlab 6 eigenvalue solver. Since we know that the smallest eigenvalue is negative, we then try to compute $\lambda_{\min}$ using the SVD of A and solving for the smallest solution of a secular equation. If $d = \sqrt{\nu} r/\|x\|$ and if we denote the SVD as $A = U\Sigma V^T$ we have to solve

$$1 - \sum_{i=1}^{n} \frac{f_i^2}{\sigma_i^2 - \mu} + \sum_{i=n+1}^{m} \frac{f_i^2}{\mu} = 0,$$

where $f = U^T d$ and σ_i are the singular values of A. We use the algorithm BNS1 of chapter 9. The function is interpolated by a rational function

$$1 + \frac{p}{\mu} - r - \frac{s}{\delta - \mu}$$

with $p = \sum_{i=n+1}^{m} f_i^2$, δ is the smallest singular value squared and r and s are determined by interpolation of the function and its derivative. Since these secular equations are difficult to solve because the function is often flat around the zero, if the value of the function at the approximate zero is not small enough, we refine the zero by bisection.

From the computations of the smallest eigenvalue we compute the backward error

$$\mu(x) = \left(\frac{\|r\|^2}{\|x\|^2} \nu + \min \left\{ 0, \lambda_{\min} \left(AA^T - \nu \frac{rr^T}{\|x\|^2} \right) \right\} \right)^{1/2}, \quad \nu = \frac{\|x\|^2}{1 + \|x\|^2}.$$

We also compute the estimate of the backward error

$$\tilde{\mu}(x) = \|(\|x\|^2 A^T A + \|r\|^2 I)^{-1/2} A^T r\|,$$

using the SVD of A.

Let $x^0 = (A^T A)^{-1} A^T c$. As the approximate solution we use $x = x^0 + \epsilon w$ where w is a random vector. In table 13.1 $\lambda_{\min}$ is the eigenvalue computed directly, $\lambda_{\sec}$ is the zero obtained by the secular equation solver and λ_{bis} the refined solution obtained by bisection. The line below gives the corresponding values of the secular function (which must be zero at the solution). Then μ and μ_{bis} are the backward errors obtained from $\lambda_{\min}$ and λ_{bis}, respectively. Finally $\tilde{\mu}$ is the estimate of μ. We give the results for several values of ϵ starting from the solution obtained by solving the normal equations.

For $\epsilon = 0$, both eigenvalues are approximately the same and we do not need to refine with bisection but the value of the secular function is smaller with $\lambda_{\sec}$. The backward errors are zero and the estimate is at the roundoff level. When ϵ increases the difference between $\lambda_{\min}$ and $\lambda_{\sec}$ increases too. This is because the secular equation is more difficult to solve. However, the refined solution λ_{bis} gives smaller values of the secular function. We see that with a large perturbation $\lambda_{\min}$ is not accurate at all, $\lambda_{\sec}$ is better but not really good and we have to rely on λ_{bis}. The value $\tilde{\mu}$ always gives a good estimate of the backward error. Hence it is worth looking at computing this estimate for large problems for which it is not feasible to compute the SVD of A.

Table 13.1 Example BK1, $m = 20, n = 10$

ϵ	$\lambda_{\min}, f$	$\lambda_{\sec}, f$	$\lambda_{\mathrm{bis}}, f$
0	$-4.4434299056\ 10^{-6}$	$-4.4434299053\ 10^{-6}$	-
	$6.5650\ 10^{-11}$	$-5.3129\ 10^{-32}$	
	μ	μ_{bis}	$\tilde{\mu}$
	0	0	$7.41124\ 10^{-16}$
ϵ	$\lambda_{\min}, f$	$\lambda_{\sec}, f$	$\lambda_{\mathrm{bis}}, f$
10^{-5}	$-4.44363798\ 10^{-6}$	$-4.44356566\ 10^{-6}$	$-4.443363799\ 10^{-6}$
	$-2.1870\ 10^{-10}$	$4.5428\ 10^{-5}$	$1.4646\ 10^{-11}$
	μ	μ_{bis}	$\tilde{\mu}$
	$2.98008\ 10^{-5}$	$2.98007\ 10^{-5}$	$3.82161\ 10^{-5}$
ϵ	$\lambda_{\min}, f$	$\lambda_{\sec}, f$	$\lambda_{\mathrm{bis}}, f$
10^{-3}	$-4.436828\ 10^{-6}$	$-6.452511\ 10^{-6}$	$-4.436837\ 10^{-6}$
	$-1.9327\ 10^{-6}$	$3.1238\ 10^{-1}$	$3.1330\ 10^{-11}$
	μ	μ_{bis}	$\tilde{\mu}$
	$2.977886\ 10^{-3}$	$2.977884\ 10^{-3}$	$3.81698\ 10^{-3}$
ϵ	$\lambda_{\min}, f$	$\lambda_{\sec}, f$	$\lambda_{\mathrm{bis}}, f$
10^{-1}	$-3.86786\ 10^{-6}$	$-1.77244\ 10^{-2}$	$-3.93425\ 10^{-6}$
	$-1.6621\ 10^{-2}$	$9.6995\ 10^{-1}$	$-8.8366\ 10^{-12}$
	μ	μ_{bis}	$\tilde{\mu}$
	$2.761580\ 10^{-1}$	$2.761579\ 10^{-1}$	$3.356922\ 10^{-1}$
ϵ	$\lambda_{\min}, f$	$\lambda_{\sec}, f$	$\lambda_{\mathrm{bis}}, f$
10	$-2.79253\ 10^{-7}$	-3.14999	-2.09544
	-1.1168	0.1862	$1.6259\ 10^{-11}$
	μ	μ_{bis}	$\tilde{\mu}$
	2.26368	1.74035	1.70267

We wish to compute an estimate of

$$\hat{\mu} = y^T (A^T A + \eta^2 I)^{-1} y, \quad y = A^T r, \quad \eta = \frac{\|r\|}{\|x\|},$$

and then $\tilde{\mu} = \sqrt{\hat{\mu}}/\|x\|$. To obtain a lower bound corresponding to Gauss quadrature we use the Golub–Kahan bidiagonalization algorithm with r as a starting vector. At each iteration k we obtain a bidiagonal matrix C_k of dimension $(k + 1) \times k$. The approximate value of $\hat{\mu}$ is

$$\|y\|^2 (e^1)^T C_k (C_k^T C_k + \eta^2 I)^{-1} C_k^T e^1.$$

We compute this using the SVD of $C_k = U_k S_k V_k^T$. Let S_k be the diagonal matrix of order k of the singular values s_i, $u^k = U_k^T e^1$ and $f = (u^k)_{1:k}$ the vector of the first k components of u^k. The approximation is

$$\|y\|^2 \sum_{i=1}^{k} \frac{f_i^2}{s_i^2 + \eta^2}.$$

If it is needed one can also compute upper bounds by suitably modifying C_k to obtain the Gauss–Radau quadrature rule.

On the small Example BK1 with $\epsilon = 10^{-3}$ after nine iterations we obtain a value $3.8169799 \; 10^{-3}$ but we have already $3.7960733 \; 10^{-3}$ after three iterations. If we use a similar example but of dimension 2000×1000, the value of μ is 0.2535201 and $\tilde{\mu}$ is 0.3210393. After 25 iterations we obtain 0.3203181. If $\epsilon = 0.1$ we have $\tilde{\mu} = 14.0104382$ and after seven iterations we obtain 14.0100725. Hence, this technique works nicely and we have only to compute the SVD of small matrices.

Chapter Fourteen

Total Least Squares

14.1 Introduction to Total Least Squares

In least squares (LS) we have only a perturbation of the right-hand side as in equation (13.3) whereas total least squares (TLS) considers perturbations of the vector of observations c and of the $m \times n$ data matrix A. Given two nonsingular diagonal weighting matrices W_L of order m and W_R of order $n+1$, we consider the problem

$$\min_{E, r} \; \|W_L (E \quad r) W_R\|_F, \tag{14.1}$$

subject to the constraint $(A + E)x = c + r$, which means finding the smallest perturbations E and r such that $c + r$ is in the range of $A + E$. The norm $\| \cdot \|_F$ is the Frobenius norm, which is the square root of the sum of the squares of all the entries of the given matrix. The matrix E is $m \times n$ and r is a vector with m components. This type of minimization problem has been considered by statisticians since the beginning of the 20th century. For examples of applications in different areas of scientific computing, see for instance Arun [12], Fierro, Golub, Hansen and O'Leary [108], Mühlich and Mester [245], Pintelon, Guillaume, Vandersteen and Rolain [275], Sima, Van Huffel and Golub [302] and Xia, Saber, Sharma and Murat Tekalp [352]. Here we follow the exposition of Golub and Van Loan [153]; see also [154]. A detailed treatment of the TLS problem is given in the book by Van Huffel and Vandewalle [333]. We will also rely on the recent results of Paige and Strakoš [258], [259], [260], [261] and [262].

As pointed out in [153], TLS problems may fail to have a solution. The solution of the TLS problem (when it exists) is given in the following theorem.

THEOREM 14.1 *Let $C = W_L (A \quad c) W_R$ and $U^T C V = \Sigma$ be its SVD. Assume that the singular values of C are such that*

$$\sigma_1 \geq \cdots \geq \sigma_k > \sigma_{k+1} = \cdots \sigma_{n+1}.$$

Then the solution of the TLS problem (14.1) is given by

$$\min \|W_L (E \quad r) W_R\|_F = \sigma_{n+1},$$

and

$$x_{TLS} = -\frac{W_R^1 y}{\alpha w_{n+1}^R},$$

where the vector $(y \quad \alpha)^T$ of norm 1 with $\alpha \neq 0$ is in the subspace S_k spanned by the right singular vectors $\{v^{k+1}, \ldots, v^{n+1}\}$ of V and W_R^1 is a diagonal matrix of *order n whose diagonal elements are the first n diagonal elements w_j^R of W_R and w_{n+1}^R is the last element which is omitted. If there is no such vector with $\alpha \neq 0$, the TLS problem has no solution.*

Proof. We follow [153]. Since $c + r$ is in the range of $A + E$, there is an x such that $(A + E)x = c + r$. This compatibility condition can be written as

$$[W_L (A \quad c) W_R + W_L (E \quad r) W_R] W_R^{-1} \begin{pmatrix} x \\ -1 \end{pmatrix} = 0. \qquad (14.2)$$

The TLS problem is thus equivalent to finding a perturbation matrix Δ of dimension $m \times (n + 1)$ having minimal Frobenius norm such that the matrix $C + \Delta$ is rank deficient. It can be shown (see, for instance, Stewart [310]) that the solution of this problem is

$$\min_{\text{rank}(C + \Delta) < n + 1} \|\Delta\|_F = \sigma_{n+1}.$$

The minimum is attained by $\Delta = -Cvv^T$ where v is a vector of norm 1 in the subspace S_k. Then if $v = (y \quad \alpha)^T$, it is easy to check that x_{TLS} in the theorem satisfies equation (14.2). $\qquad \square$

Golub and Van Loan [153] gave some results about the sensitivity of the TLS problem as well as characterizations of the solution. The right singular vectors v^i are the eigenvectors of $C^T C$ and S_k is the invariant subspace associated to the smallest eigenvalue σ_{n+1}^2. Then the TLS solution x_{TLS} solves the eigenvalue problem

$$C^T C W_R^{-1} \begin{pmatrix} x \\ -1 \end{pmatrix} = \sigma_{n+1}^2 W_R^{-1} \begin{pmatrix} x \\ -1 \end{pmatrix}. \qquad (14.3)$$

Let $\hat{A} = W_L A W_R^1$, $\hat{c} = W_L c$ and $\lambda = w_{n+1}^R$. Then equation (14.3) can be written as

$$\begin{pmatrix} \hat{A}^T \hat{A} & \lambda \hat{A}\hat{c} \\ \lambda \hat{c}^T \hat{A} & \lambda^2 \hat{c}^T \hat{c} \end{pmatrix} \begin{pmatrix} (W_R^1)^{-1}x \\ -\lambda^{-1} \end{pmatrix} = \sigma_{n+1}^2 \begin{pmatrix} (W_R^1)^{-1}x \\ -\lambda^{-1} \end{pmatrix}. \qquad (14.4)$$

Let $\hat{U} \hat{A} \hat{V} = \hat{\Sigma}$ be the SVD of $\hat{A}$ with singular values $\hat{\sigma}_j$, the smallest being $\hat{\sigma}_n$. If we denote

$$K = \hat{\Sigma}^T \hat{\Sigma}, \; g = \hat{\Sigma}^T \hat{U}^T c, \; h^2 = \hat{c}^T \hat{c}, \; z = \hat{V}^T (W_R^1)^{-1}x,$$

equation (14.4) writes

$$\begin{pmatrix} K & \lambda g \\ \lambda g^T & \lambda^2 h^2 \end{pmatrix} \begin{pmatrix} z \\ -\lambda^{-1} \end{pmatrix} = \sigma_{n+1}^2 \begin{pmatrix} z \\ -\lambda^{-1} \end{pmatrix}.$$

This gives the two following equations

$$(K - \sigma_{n+1}^2)z = g, \quad \frac{\sigma_{n+1}^2}{\lambda^2} + g^T z = h^2.$$

With these notations we have the following result from Golub and Van Loan [153].

THEOREM 14.2 *If $\hat{\sigma}_n > \sigma_{n+1}$, then x_{TLS} exists and is the unique solution of the TLS problem. It is written as*

$$x_{TLS} = W_R^1(\hat{A}^T\hat{A} - \sigma_{n+1}^2 I)^{-1}\hat{A}^T\hat{c}.$$

Moreover, σ_{n+1} satisfies the secular equation

$$\sigma_{n+1}^2 \left[\frac{1}{\lambda^2} + \sum_{i=1}^{n} \frac{d_i^2}{\hat{\sigma}_i^2 - \sigma_{n+1}^2} \right] = \rho_{LS}^2,$$

with the vector $d = \hat{U}^T\hat{c}$ and $\rho_{LS}^2 = \|W_L(c - Ax_{LS})\|^2$.

For simplicity of notations we will no longer consider weighting in the rest of the chapter. Without weighting the norm of the residual $r_{TLS} = c - Ax_{TLS}$ can be written as

$$\|r_{TLS}\|^2 = c^Tc - c^TA(A^TA - \sigma_{n+1}^2 I)^{-1}A^Tc + \sigma_{n+1}^2 c^TA(A^TA - \sigma_{n+1}^2 I)^{-2}A^Tc$$

and

$$\|x_{TLS}\|^2 = c^TA(A^TA - \sigma_{n+1}^2 I)^{-2}A^Tc.$$

To prove the first relation we use the following identity which holds for $\mu \neq 0$:

$$I - A(A^TA + \mu I)^{-1}A^T = \mu(AA^T + \mu I)^{-1}.$$

It can be proved by noticing that $A(A^TA + \mu I)^{-1} = (AA^T + \mu I)^{-1}A$. By definition, we have $\|r_{TLS}\| \geq \|r_{LS}\|$ since LS gives the minimum of the residual norm over all vectors. By using the SVD of A it turns out that

$$\|r_{TLS}\|^2 = \|r_{LS}\|^2 + \sum_{i=1}^{n} d_i^2 \frac{\sigma_{n+1}^4}{\hat{\sigma}_i^2(\hat{\sigma}_i^2 - \sigma_{n+1}^2)^2}. \tag{14.5}$$

This can also be written as

$$\|r_{TLS}\|^2 = \|r_{LS}\|^2 + \sigma_{n+1}^4 c^TA(A^TA)^{-1}(A^TA - \sigma_{n+1}^2 I)^{-2}A^Tc.$$

From Van Huffel and Vandewalle [333], we have

$$\|r_{TLS}\| = \sigma_{n+1}(1 + \|x_{TLS}\|^2)^{1/2}.$$

These relations show that σ_{n+1} must satisfy the equation

$$\sigma_{n+1}^2 = c^Tc - c^TA(A^TA - \sigma_{n+1}^2 I)^{-1}A^Tc.$$

This is a secular equation for the smallest singular value σ_{n+1} of $(A \quad c)$. We have seen how to solve secular equations in chapter 9, where numerical experiments were described.

Note that the condition $\hat{\sigma}_n > \sigma_{n+1}$ in theorem 14.2 is sufficient but not necessary. If $\hat{\sigma}_n = \sigma_{n+1}$ a solution may or may not exist. In case the solution does not exist Van Huffel and Vandewalle [333] impose an additional restriction on the perturbations,

$$(E \quad r)(v^{q+1} \quad \cdots \quad v^{n+1}) = 0,$$

where q is the maximal index for which the vectors v^i have a last nonzero component. The problem with this additional condition is called nongeneric TLS. It has always a unique solution.

The characterization of the solution in theorem 14.2 is also interesting for comparison of the TLS solution with the solutions of regularized LS problems that we will study in section 15.1. Anticipating what we will see, a regularized LS solution can be written as

$$x_{RLS}(\mu) = (A^T A + \mu I)^{-1} A^T c,$$

with a parameter $\mu > 0$. We see that $x_{TLS} = x_{RLS}(-\sigma_{n+1}^2)$. Therefore, TLS appears as a "deregularizing" procedure since it corresponds to a negative μ.

If we consider only perturbations of the data matrix A without perturbations on the right-hand side, the problem is called data least squares (DLS). Without weighting, it reads

$$\min_{E} \quad \|E\|_F, \tag{14.6}$$

subject to $(A + E)x = c$. An alternative formulation is

$$\min_{x} \quad \|c - Ax\|^2/\|x\|^2.$$

A backward perturbation analysis of DLS was done by Chang, Golub and Paige [62].

14.2 Scaled Total Least Squares

All these approaches, LS, DLS and TLS, have been unified by several researchers. We consider what has been proposed by Paige and Strakoš [261], [259], [258], [260]. The scaled total least squares (STLS) problem is

$$\underset{E, r}{\text{minimize}} \quad \| (E \quad r) \|_F, \tag{14.7}$$

subject to $(A + E)x\gamma = c\gamma + r$, where γ is a parameter. The vector $x = x(\gamma)$ is the STLS solution and $x(\gamma)\gamma$ is the TLS solution of the problem (14.7) that we have already exhibited for TLS. In [261], Paige and Strakoš show that when $\gamma \to 0$, $x(\gamma)$ becomes the LS solution. On the other end, when $\gamma \to \infty$, $x(\gamma)$ becomes the DLS solution. Paige and Strakoš consider matrices with complex entries; here we consider only real matrices. An alternative formulation of the STLS problem is

$$\min_{r, x} \quad \|r\|^2 + \|c\gamma + r - Ax\|^2/\|x\|^2.$$

Another formulation is

$$\min_{x} \quad \|c\gamma - Ax\|^2/(1 + \|x\|^2).$$

By interlacing properties, one has

$$\sigma_{\min}[(A \quad c\gamma)] \leq \sigma_{\min}[A].$$

When we have a strict inequality, the STLS norm for the solution is $\sigma_{\min}[(\,A \quad c\gamma\,)]$ and the STLS problem can be solved with the SVD, the dimension of the problem permitting. However, the theory of [261] is not based on this inequality on singular values which is a sufficient condition. Let $\mathcal{U}_{\min}$ be the left singular subspace of A corresponding to $\sigma_{\min}[A]$, the condition used in [261] is

$$\text{the } m \times n \text{ matrix } A \text{ has rank } n \text{ and } c \not\perp \mathcal{U}_{\min}. \tag{14.8}$$

The rationale behind this is given in the following theorem, see [261].

THEOREM 14.3 *If σ_i^A, $i = 1, \ldots, n$ are the singular values of A in descending order and $u^1, \ldots, u^n$ the corresponding left singular vectors, then*

$$\sigma_{\min}[(\,A \quad c\gamma\,)] = \sigma_{\min}[A],$$

if and only if

$$d_i = (u^i)^T c = 0, \; i = k+1, \ldots, n,$$

where $\sigma_{k+1}^A = \cdots = \sigma_n^A$ and

$$\psi_k(\sigma, \gamma) = \gamma^2 \|r\|^2 - \sigma^2 - \gamma^2 \sigma^2 \sum_{i=1}^{k} \frac{d_i^2}{(\sigma_i^A)^2 - \sigma^2} \geq 0.$$

Proof. See [261]. $\qquad\qquad\qquad\qquad\qquad\qquad\qquad\qquad\qquad\qquad\qquad\qquad\square$

Theorem 14.3 shows that condition (14.8) implies that $\sigma_{\min}[(\,A \quad c\gamma\,)] < \sigma_{\min}[A]$. As we know from theorem 14.2, the solution of the STLS problem satisfies a secular equation.

PROPOSITION 14.4 *If condition (14.8) holds then the solution $\sigma_{\min}[(\,A \quad c\gamma\,)]$ is the smallest nonnegative scalar σ satisfying*

$$\gamma^2 \|r\|^2 - \sigma^2 - \gamma^2 \sigma^2 \sum_{i=1}^{k} \frac{d_i^2}{(\sigma_i^A)^2 - \sigma^2} = 0,$$

$\|r\|$ being the LS distance.

Proof. See [261]. $\qquad\qquad\qquad\qquad\qquad\qquad\qquad\qquad\qquad\qquad\qquad\qquad\square$

The choice of condition (14.8) was made by Paige and Strakoš because any scaled total least squares problem can be reduced by orthogonal transformations to what they called a "core" problem satisfying this condition. The problem with matrix A and right-hand side c can be transformed as

$$P^T (\,c \quad AQ\,) = \begin{pmatrix} c_1 & A_{1,1} & 0 \\ 0 & 0 & A_{2,2} \end{pmatrix},$$

where P and Q are orthonormal matrices. The original problem reduces to solving

$$A_{1,1} x_1 \approx c_1, \quad A_{2,2} x_2 \approx 0, \quad x = Q \begin{pmatrix} x_1 \\ x_2 \end{pmatrix}.$$

The problem $A_{1,1} x_1 \approx c_1$ was called a "core" problem by Paige and Strakoš [262] if the matrix $A_{2,2}$ is of maximal dimension. Generally, there is no reason for not taking $x_2 = 0$.

The paper [262] proposed to compute the "core" problem by reducing $(c \quad A)$ to upper bidiagonal form by orthogonal transformations P and Q. They are partitioned as $P = (P_1 \quad P_2)$ and $Q = (Q_1 \quad Q_2)$, giving $(c_1 \quad A_{1,1}) = P_1^T (c \quad AQ_1)$ with either

$$
(c_1 \quad A_{1,1}) = \begin{pmatrix} \beta_1 & \alpha_1 & & \\ & \beta_2 & \alpha_2 & \\ & & \ddots & \ddots \\ & & & \beta_p & \alpha_p \end{pmatrix}, \quad \beta_i \alpha_i \neq 0, \ i = 1, \dots, p
$$

if $\beta_{p+1} = 0$ or $p = m$; or

$$
(c_1 \quad A_{1,1}) = \begin{pmatrix} \beta_1 & \alpha_1 & & \\ & \beta_2 & \alpha_2 & \\ & & \ddots & \ddots \\ & & & \beta_p & \alpha_p \\ & & & & \beta_{p+1} \end{pmatrix}, \quad \beta_i \alpha_i \neq 0, \ i = 1, \dots, p, \quad \beta_{p+1} \neq 0
$$

if $\alpha_{p+1} = 0$ or $p = n$. The proofs that these matrices correspond to "core" problems are given in [262]. In particular, if $c \not\perp$ range of A then $A_{1,1}$ has no zero or multiple singular values. The important points are the following:

- The matrix $A_{1,1}$ has no zero or multiple singular values,

- $A_{1,1}$ has minimal dimensions and $A_{2,2}$ maximal dimensions,

- All components of c_1 in the left singular vector subspaces of $A_{1,1}$ are nonzero. Then we can solve the TLS problem $A_{1,1} x_1 \simeq c_1$.

This approach gives the TLS solution determined by Golub and Van Loan [153] if it exists. In the other case we obtain the nongeneric minimum norm TLS solution of Van Huffel and Vandewalle [333].

"Core" formulations for ill-posed problems are used in the paper [301] by Sima and Van Huffel. Hnětynkovà and Strakoš [190] show how to obtain the "core" formulation from the Golub–Kahan bidiagonalization algorithm, see chapter 4.

14.3 Total Least Squares Secular Equation Solvers

In this section we are interested in solving the secular equations giving σ_{n+1} or approximations of these equations using the techniques we developed for computing estimates of quadratic forms. Let us first recall the secular equations we have to consider. This methodology can be applied to the TLS problem if the solution exists or to the problem which is given by the "core" formulation.

14.3.1 TLS and DLS Secular Equations

Let r and E be, respectively, the perturbations of the right-hand side c and the matrix A. Let $C = (A \ c)$ and $(\sigma_{n+1}^C)^2$ be the smallest eigenvalue of $C^T C$. Then the TLS solution x_{TLS} solves the eigenvalue problem

$$C^T C \begin{pmatrix} x \\ -1 \end{pmatrix} = (\sigma_{n+1}^C)^2 \begin{pmatrix} x \\ -1 \end{pmatrix}. \tag{14.9}$$

The equation (14.9) can be written as

$$\begin{pmatrix} A^T A & Ac \\ c^T A & c^T c \end{pmatrix} \begin{pmatrix} x \\ -1 \end{pmatrix} = (\sigma_{n+1}^C)^2 \begin{pmatrix} x \\ -1 \end{pmatrix}. \tag{14.10}$$

By eliminating x from equation (14.10) we find that $\sigma^2 = (\sigma_{n+1}^C)^2$ satisfies the secular equation

$$c^T c - c^T A (A^T A - \sigma^2 I)^{-1} A^T c = \sigma^2. \tag{14.11}$$

We are looking for the smallest σ^2 satisfying this equation with $\sigma < \sigma_{\min}(A)$, the smallest singular value of A.

For data least squares (DLS) when only the matrix is perturbed, the secular equation is

$$c^T c - c^T A (A^T A - \sigma^2 I)^{-1} A^T c = 0. \tag{14.12}$$

To simplify these secular equations we can use the relation

$$I - A(A^T A - \sigma^2 I)^{-1} A^T = -\sigma^2 (AA^T - \sigma^2 I)^{-1},$$

which is valid if $\sigma^2 \neq 0$. Therefore, the secular equation (14.11) reduces to

$$c^T (AA^T - \sigma^2 I)^{-1} c + 1 = 0. \tag{14.13}$$

The DLS secular equation (14.12) becomes

$$c^T (AA^T - \sigma^2 I)^{-1} c = 0. \tag{14.14}$$

Assuming that the matrix A of dimension $m \times n, m \geq n$ is of full rank, the secular equations can be written using the SVD of $A = U\Sigma V^T$, where U and V are orthogonal matrices and Σ is a rectangular diagonal matrix of the same size as A which can be written as

$$\Sigma = \begin{pmatrix} D \\ 0 \end{pmatrix},$$

where D is a diagonal matrix of order n with the singular values on the diagonal. The matrix $A^T A$ is of order n and

$$A^T A = V\Sigma^T U^T U\Sigma V^T = V\Sigma^T \Sigma V^T = VD^2 V^T.$$

The matrix $VD^2 V^T$ is the spectral decomposition of $A^T A$. On the other hand, the matrix AA^T is of order m and

$$AA^T = U\Sigma V^T V\Sigma^T U^T = U\Sigma\Sigma^T U^T = U \begin{pmatrix} D^2 & 0 \\ 0 & 0 \end{pmatrix} U^T.$$

The matrix AA^T of order m is singular and of rank n. The secular equations can be written using the SVD of A. Since

$$(A^T A - \sigma^2 I)^{-1} = [V(D^2 - \sigma^2 I)V^T]^{-1} = V(D^2 - \sigma^2 I)^{-1} V^T,$$

equation (14.11) is

$$c^T c - c^T U \Sigma (D^2 - \sigma^2 I)^{-1} \Sigma^T U^T c = \sigma^2.$$

Let $\xi = U^T c$; the TLS secular equation (14.11) is

$$c^T c - \sum_{i=1}^{n} \frac{\xi_i^2 \sigma_i^2}{\sigma_i^2 - \sigma^2} = \sigma^2. \tag{14.15}$$

Similarly, the other form of the TLS secular equation is written as

$$\sum_{i=1}^{n} \frac{\xi_i^2}{\sigma_i^2 - \sigma^2} - \sum_{i=n+1}^{m} \frac{\xi_i^2}{\sigma^2} + 1 = 0. \tag{14.16}$$

Are they equivalent (as they must be)? In equation (14.15) the term $c^T c$ is nothing other than

$$c^T c = c^T U U^T c = \sum_{i=1}^{m} \xi_i^2.$$

Therefore, equation (14.15) is

$$\sum_{i=1}^{n} \xi_i^2 \left[1 - \frac{\sigma_i^2}{\sigma_i^2 - \sigma^2} \right] + \sum_{i=n+1}^{m} \xi_i^2 = \sigma^2.$$

This implies

$$\sum_{i=n+1}^{m} \xi_i^2 - \sum_{i=1}^{n} \frac{\xi_i^2 \sigma^2}{\sigma_i^2 - \sigma^2} = \sigma^2. \tag{14.17}$$

Clearly, if $\sigma^2 \neq 0$, we can divide by σ^2 and we recover equation (14.16). Therefore, we can use either form of the TLS equation. However, they have different properties since equation (14.16) has a pole at 0. To find the TLS solution we are interested mainly in the behavior of the secular functions for $\sigma \leq \sigma_{\min}(A)$.

An example for equation (14.17) (as a function of σ) written as

$$\sigma^2 - \sum_{i=n+1}^{m} \xi_i^2 + \sum_{i=1}^{n} \frac{\xi_i^2 \sigma^2}{\sigma_i^2 - \sigma^2} = 0 \tag{14.18}$$

is displayed in figure 14.1. If we look at the same function as a function of σ^2, we have figure 14.2. An example for equation (14.16) (as a function of σ) is given in figure 14.3. If we look at the same function as a function of σ^2, we have figure 14.4.

It is not obvious which is the best form of equation to choose and which variable, σ or σ^2. Moreover, in some practical problems the poles can be very close to each other. In the TLS problem, if $\sigma_{\min}(A)$ is very close to zero, it may be difficult to find a zero of the secular function in this interval.

14.3.2 Approximation of the TLS Secular Equation

We approximate the quadratic form in the TLS secular equation (14.11) by using the Golub–Kahan bidiagonalization algorithm with c as a starting vector. It

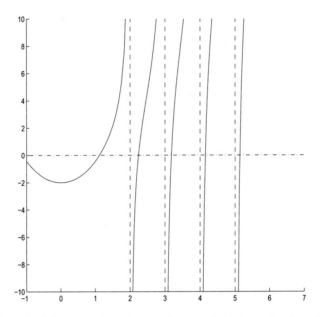

Figure 14.1 Example of TLS secular function (14.17) as a function of σ

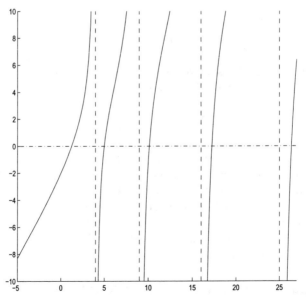

Figure 14.2 Example of TLS secular function (14.17) as a function of σ^2

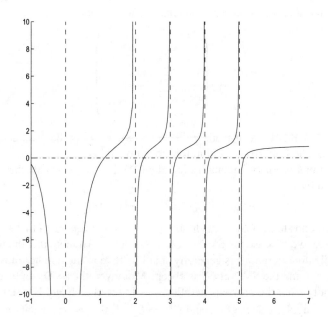

Figure 14.3 Example of TLS secular function (14.16) as a function of σ

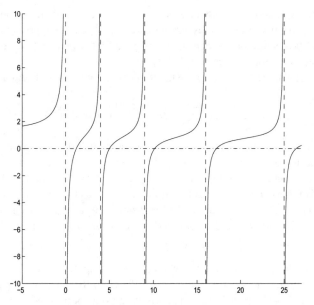

Figure 14.4 Example of TLS secular function (14.16) as a function of σ^2

is named "Lanczos bidiagonalization II" or "Bidiag 1" and has been described in chapter 4. It reduces A to lower bidiagonal form and generates a matrix

$$C_k = \begin{pmatrix} \gamma_1 & & & \\ \delta_1 & \ddots & & \\ & \ddots & \ddots & \\ & & \ddots & \gamma_k \\ & & & \delta_k \end{pmatrix},$$

a $(k+1) \times k$ matrix such that $C_k^T C_k = J_k$ where J_k is the tridiagonal matrix generated by the Lanczos algorithm for the matrix $A^T A$.

At iteration k of the bidiagonalization algorithm, we approximate the TLS secular equation by

$$c^T c - \|c\|^2 (e^1)^T C_k (C_k^T C_k - \sigma^2 I)^{-1} C_k^T e^1 = \sigma^2. \qquad (14.19)$$

This corresponds to the Gauss quadrature rule. To solve equation (14.19) when k is not too large, we first use the SVD of $C_k = U_k S_k V_k^T$. We note that the number of bidiagonalization iterations is generally small relative to the number of rows of A. Hence, computing the SVD of C_k is cheap. Moreover, the SVD can be computed incrementally from the previous iteration $k-1$; see Bunch and Nielsen [44]. Let $\sigma_i^{(k)}$ be the singular values of C_k and $\xi^{(k)} = U_k^T e^1$, the secular equation (14.19) is written as

$$\frac{(\xi_{k+1}^{(k)})^2}{\sigma^2} - \sum_{i=1}^{k} \frac{(\xi_i^{(k)})^2}{(\sigma_i^{(k)})^2 - \sigma^2} = \frac{1}{\|c\|^2}. \qquad (14.20)$$

To solve this equation in the interval $]0, \sigma_{\min}^{(k)}[$ we use the secular equation solvers of chapter 9. We note that as a by-product we can obtain an approximation of the smallest singular value of A when computing the SVD of C_k. So we can (approximately) check the condition $\sigma < \sigma_{\min}(A)$.

If we do not want to (or cannot) use the SVD of C_k, we can write equation (14.19) as

$$\frac{1}{\|c\|^2} + (e^1)^T (C_k C_k^T - \sigma^2)^{-1} e^1 = 0.$$

Then, we cannot use the secular equations solvers which do a partition of the secular equation in two pieces since this needs the knowledge of the poles. Moreover, we cannot compute the good starting point proposed by Melman [232], [233]; see chapter 9. But we can use rational interpolation, the function and its derivatives being computed by solving tridiagonal linear systems. When an approximate solution σ_{TLS}^2 has been computed, the corresponding solution x_{TLS} is obtained by solving

$$x_{TLS} = (A^T A - \sigma_{TLS}^2 I)^{-1} A^T c.$$

If A is not too large, one can store the vectors q^k computed during the bidiagonalization algorithm. Then, if k is the number of iterations and $J_k = C_k^T C_k$, we can solve $(J_k - \sigma_{TLS}^2 I) z = \|A^T c\| e^1$ and we obtain the solution with $x_{TLS} = Q_k z$. When A is large it is not feasible to store the vectors q^k. The bidiagonalization algorithm has to be rerun to obtain the solution.

14.3.3 The Gauss–Radau Rule

We implement the Gauss–Radau rule to approximate the quadratic form in the TLS secular equation (14.11) by using the Golub–Kahan bidiagonalization algorithm with $A^T c$ as a starting vector. It is named "Lanczos bidiagonalizaion I" or "Bidiag 2"; see chapter 4. It reduces A to upper bidiagonal form. If

$$B_k = \begin{pmatrix} \gamma_1 & \delta_1 & & \\ & \ddots & \ddots & \\ & & \gamma_{k-1} & \delta_{k-1} \\ & & & \gamma_k \end{pmatrix},$$

the matrix B_k is the Cholesky factor of the Lanczos matrix J_k and $B_k^T B_k = J_k$.

The Gauss rule approximates equation (14.11) by

$$\|c\|^2 - \|A^T c\|^2 (e^1)^T (B_k^T B_k - \sigma^2 I)^{-1} e^1 = \sigma^2.$$

To obtain the Gauss–Radau rule we must modify J_k (or B_k) in order to have a prescribed eigenvalue z. Let ω be the solution of

$$(B_k^T B_k - zI)\omega = (\gamma_{k-1} \delta_{k-1})^2 e^k,$$

where e^k is the last column of the identity matrix of order k. Then, let

$$\tilde{\omega}_k = (z + \omega_k) - \frac{(\gamma_{k-1} \delta_{k-1})^2}{\gamma_{k-1}^2} = (z + \omega_k) - \delta_{k-1}^2.$$

The modified matrix giving the Gauss–Radau rule is

$$\tilde{B}_k = \begin{pmatrix} \gamma_1 & \delta_1 & & \\ & \ddots & \ddots & \\ & & \gamma_{k-1} & \delta_{k-1} \\ & & & \tilde{\gamma}_k \end{pmatrix},$$

where $\tilde{\gamma}_k = \sqrt{\tilde{\omega}_k}$. Using $\tilde{B}_k$ we solve the secular equation

$$\|c\|^2 - \|A^T c\|^2 (e^1)^T (\tilde{B}_k^T \tilde{B}_k - \sigma^2 I)^{-1} e^1 = \sigma^2, \tag{14.21}$$

by using the SVD of $\tilde{B}_k$. This gives

$$f(t) = \alpha + \rho t + \sum_{i=1}^{k} \frac{\xi_i^2}{d_i - t} = 0,$$

where $\sigma^2 = \sigma_{\min}^2 + \rho t$ with $\rho = \|A^T c\|^2$, $\alpha = \sigma_{\min}^2 - \|c\|^2$. The variable in the denominator is $d_i = (\sigma_i^2 - \sigma_{\min}^2)/\rho$. The vector ξ is defined as $\xi = V_k^T e^1$ where $\tilde{B}_k = U_k S_k V_k^T$ is the SVD of $\tilde{B}_k$.

This equation is solved using the Newton method or preferably a rational approximation $a + b/t$ of the sum. This leads to solving a quadratic equation but we are only interested in the negative solution.

In the following numerical experiments we will also check the convergence of the smallest singular value of C_k to stop the Lanczos iterations. The matrix $C_k^T C_k$

is tridiagonal. When we want to compute the smallest eigenvalue of $J_{k+1} = C_{k+1}^T C_{k+1}$ we need to compute the smallest root of the equation

$$f(\lambda) = \lambda - \alpha_{k+1} - \eta_k^2 (e^k)^T (J_k - \lambda I)^{-1} e^k = 0.$$

We use a third-order rational interpolation such that

$$a + b\lambda + \frac{c}{\theta_1 - \lambda} = f$$

for a given λ, where θ_1 is the smallest eigenvalue of J_k. The coefficients a, b and c are found by solving the previous equation together with

$$b + \frac{c}{(\theta_1 - \lambda)^2} = f',$$

$$\frac{2c}{(\theta_1 - \lambda)^3} = f''.$$

The first and second derivatives are given by

$$f'(\lambda) = 1 + \eta_k^2 (e^k)^T (J_k - \lambda I)^{-2} e^k,$$

$$f''(\lambda) = 2\eta_k^2 (e^k)^T (J_k - \lambda I)^{-3} e^k.$$

The function and the derivatives are computed (for a given λ) by solving tridiagonal linear systems. The starting value of λ is taken as a number a little smaller than θ_1.

14.3.4 Numerical Experiments

In this section we report numerical experiments in which we solve the TLS secular equations. We start with a straightforward application of the algorithms of chapter 9. Then, we show how to improve them by trying to do a smaller number of secular solver iterations.

The Gauss Rule

We use examples derived from those of U. von Matt for solving ill-posed problems; see chapter 15. Let A_s be an $m \times n$ matrix such that

$$A_s = U_s \Sigma_s V_s^T, \quad U_s = I - 2\frac{u_s u_s^T}{\|u_s\|^2}, \quad V_s = I - 2\frac{v_s v_s^T}{\|v_s\|^2},$$

where u_s and v_s are random vectors (generated using "randn"). Σ_s is an $m \times n$ diagonal matrix with diagonal elements $[1, \ldots, \sqrt{n}]$. Let x_s be a vector whose ith component is $1/i$ and $c_s = A_s x_s$. The matrix is generated as

$$A = A_s + \xi \, \text{randn}(m, n).$$

The right-hand side is

$$c = c_s + \xi \, \text{randn}(m, 1).$$

The parameter ξ is 0.3. We begin with a small example (Example TLS1) with $m = 100$ and $n = 50$. The smallest singular value of A is 1.5565918. The smallest singular value of $[A, c]$ which is the "exact" solution of the TLS problem is 1.464891451263777. Therefore, the sufficient condition for the existence of the TLS solution is satisfied.

To attain a relative change of $\varepsilon = 10^{-6}$ for σ^2, it requires 28 bidiagonalization iterations. The total number of iterations of the secular equation solver using algorithm BNS1 (see chapter 9) is 73, the minimum number of iterations per bidiagonalization step is 1, the maximum 3 and the average 2.52. The solution computed by the Gauss rule is 1.464892131470029 with six exact decimal digits. Results for the other algorithms of chapter 9 are given in table 14.1. The column "Secul. it." gives the total number of iterations for all the Lanczos steps. The column "Av. it." is the average number of secular iterations per Lanczos step.

Table 14.1 Example TLS1, $m = 100, n = 50, \varepsilon = 10^{-6}$

Method	Lanczos it.	Secul. it.	Min it.	Max it.	Av. it.	Solution
BNS1	28	73	1	3	2.52	1.464892131470029
BNS2	28	74	1	3	2.55	1.464892131470029
MW	28	73	1	3	2.52	1.464892131470029
FW1	28	73	1	3	2.52	1.464892131470029
GR	28	74	1	3	2.55	1.464892131470028

Note that all methods give (almost) the same number of secular iterations. With a stopping criterion of 10^{-10}, the number of Lanczos iterations is 38 with a total of 108 secular iterations for BNS1. The solution is 1.464891451267460 with 12 exact decimal digits. Results are given in table 14.2. Note that these approximate solutions are upper bounds. This is because the Gauss rule gives a lower bound for the quadratic form and the graph of the approximate function is below the graph of the exact one. Since it is an increasing function between poles, the approximate root is to the right of the exact one.

Table 14.2 Example TLS1, $m = 100, n = 50, \varepsilon = 10^{-10}$

Method	Lanczos it.	Secul. it.	Min it.	Max it.	Av. it.	Solution
BNS1	38	108	1	3	2.77	1.464891451267460
BNS2	38	108	1	3	2.77	1.464891451267460
MW	38	108	1	3	2.77	1.464891451267460
FW1	38	108	1	3	2.77	1.464891451267460
GR	38	109	1	3	2.79	1.464891451267460

With the same problem let us vary the level of noise ξ. Results using BNS1 are given in table 14.3. The number of Lanczos iterations does not depend too much on ξ.

Then we increase the size of the problem to $m = 1000$ and $n = 500$ (Example TLS2). The smallest singular value of A is 5.230339. The smallest singular value of $[A, c]$ is 5.200943688079055.

To attain a stopping criterion of 10^{-6}, it takes 75 bidiagonalization iterations. The total number of iterations of the secular solver using BNS1 is 263, the min-

Table 14.3 Example TLS1, $m = 100, n = 50$, BNS1 $\varepsilon = 10^{-6}$

Noise	Lanczos it.	Secul. it.	Solution	Exact solution
$0.3\ 10^{-2}$	30	57	0.01703479103104873	0.01703478979190218
$0.3\ 10^{-1}$	26	49	0.169448388286749	0.1694483528865543
0.3	28	73	1.464892131470029	1.464891451263777
30	33	64	88.21012648624229	88.21012652906667

imum number per bidiagonalization step is 1, the maximum 5 and the average is 3.46. The Gauss rule computed solution is 5.200958880262121 with five exact decimal digits. Results are given in table 14.4. There is not much difference in the number of secular iterations except for GR.

Table 14.4 Example TLS2, $m = 1000, n = 500$, $\varepsilon = 10^{-6}$

Method	Lanczos it.	Secul. it.	Min it.	Max it.	Av. it.	Solution
BNS1	75	263	1	5	3.46	5.200958880262121
BNS2	75	257	1	5	3.38	5.200958880262426
MW	75	263	1	5	3.46	5.200958880262122
FW1	75	263	1	5	3.46	5.200958880262118
GR	75	244	1	4	3.21	5.200958880262104

With a criterion of 10^{-10}, the number of iterations is 101 with a total of secular iterations of 431. The solution is 5.200943723128725 with seven exact decimal digits. Results are given in table 14.5.

Table 14.5 Example TLS2, $m = 1000, n = 500$, $\varepsilon = 10^{-10}$

Method	Lanczos it.	Secul. it.	Min it.	Max it.	Av. it.	Solution
BNS1	101	431	1	5	4.23	5.200943723128725
BNS2	101	441	1	5	4.32	5.200943723128731
MW	101	431	1	5	4.23	5.200943723128735
FW1	101	431	1	5	4.23	5.200943723128736
GR	101	440	1	5	4.31	5.200943723128733

For larger problems we were not able to store A which is a dense matrix in Matlab. We use the vectors u_s and v_s to do matrix multiplies with A_s or A_s^T. Therefore, since we do not store A_s, we cannot perturb the matrix by adding a random perturbation matrix. After computing the right-hand side, we perturb the singular values in the same way as the right-hand side. Negative singular values after perturbations are replaced by 10^{-2} times a random number.

First, we use the same initial singular value distribution as in the previous examples. To see the differences with the previous examples, we use a small problem with $m = 100$, $n = 50$ and the same noise as before. The smallest singular value of A is 1.119018116118189 and the TLS parameter is 0.9286027768624542. Note that the solution is much smaller than the minimum singular value of A. We did 28 Lanczos bidiagonalization iterations with a stopping criterion of 10^{-6}. The solution given by the Gauss rule is 0.9286027937283224 obtained with 64 secular equation solver iterations using BNS1. We observe the same behavior as with Example TLS1.

Then we solve a larger problem with $m = 10000$, $n = 5000$ (Example TLS3). The smallest singular value of A is 1.418961206071727. For this problem, it is not feasible to compute the "exact" TLS parameter using the SVD of $[A, c]$. Since all methods give more or less the same results, we only give the results for BNS1 in table 14.6.

Table 14.6 Example TLS3, $m = 10000, n = 5000$, noise = 0.3

ε	Method	Lanc. it.	Secul. it.	Min it.	Max it.	Av. it.	Solution
10^{-6}	BNS1	250	273	1	2	1.09	1.418582932414374
10^{-10}	BNS1	328	660	1	3	2.01	1.418576233569240

Since we now have a good way to compute the solution of the TLS secular equation, let us try to optimize the algorithm. To save computing time we may start solving the secular equation only after some Lanczos bidiagonalization iterations since the solution obtained at the beginning is not really meaningful. It is an open problem to decide how many Lanczos iterations have to be done before starting to solve the secular equation. Another possibility which is explored in table 14.7 is to compute the SVD of B_k and to solve the secular equation only every ν Lanczos iterations. We use a frequency of 10 or 50 iterations.

Table 14.7 Example TLS3, $m = 10000, n = 5000$, noise = 0.3, $\varepsilon = 10^{-6}$

ν	Method	Lanc. it.	Secul. it.	Min it.	Max it.	Av. it.	Solution
10	BNS1	251	29	1	2	0.12	1.418582271405106
50	BNS1	301	8	1	1	0.03	1.418576255595023

The total number of secular equation solves is greatly reduced. Moreover, we compute fewer SVDs. With $\nu = 50$, since we are monitoring the convergence of the TLS parameter to stop the Lanczos iteration, we are doing 50 more Lanczos iterations than necessary because for 250 iterations the criteria is not satisfied. This means that the stopping criterion at convergence is much less than 10^{-6}; in fact its value is $3.75867 \, 10^{-9}$.

Another possibility for stopping the Lanczos iterations is to monitor the convergence of the smallest singular value of B_k without solving the TLS secular equation. Figure 14.5 shows the convergence (relative difference between successive iterates) of the TLS parameter σ^2 (solid) and the convergence of the smallest singular value of A squared. We see that (at least for this example) they decay in the same way since the curves are almost superimposed except at the very beginning.

Thus, we "just" have to find a cheap way to compute the smallest singular value of B_k. We can solve the TLS secular equation and compute the full SVD of B_k (with singular values and singular vectors) only when we decide to stop.

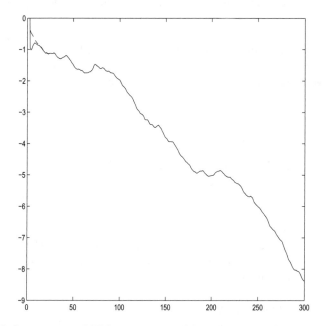

Figure 14.5 Convergence of TLS parameter (solid) and smallest singular value squared (dashed)

In the computations of table 14.8, we monitor the convergence of the smallest singular value (which is computed using the SVD of B_k). The Lanczos iterations are stopped when the relative difference is smaller than the TLS criteria. Then, the full SVD (with the singular vectors) is computed to solve the secular equation. Moreover, the convergence of the smallest singular value is checked only every ν iterations. We see that we have reduced the number of secular iterations to a minimum.

Table 14.8 Example TLS3, $m = 10000, n = 5000$, noise = 0.3, $\varepsilon = 10^{-6}$

ν	Method	Lanc. it.	Secul. it.	Min it.	Max it.	Solution
10	BNS1	250	1	1	1	1.418582932414374
50	BNS1	301	1	1	1	1.418576255595023

The Gauss–Radau Rule

We begin with the small Example TLS1. The Gauss–Radau approximations are computed with $\sqrt{z}$ chosen as the exact smallest and largest singular values of A. As we will see later this is not a problem if we compute the Gauss estimate first because we can then obtain good estimates of the extreme singular values. Results are given in tables 14.9 and 14.10 for which we solve the secular equation at each iteration. We use the Newton method (Newt) and a rational interpolation (Rat).

Table 14.9 Example TLS1, $m = 100, n = 50$, Gauss–Radau, noise = 0.3, $\varepsilon = 10^{-6}$

Method	Lanc. it.	$\sqrt{z}$	Secul. it.	Min it.	Max it.	Av. it.	Solution
Newt	28	σ_{min}	130	2	14	4.48	1.464891376927382
Newt	28	σ_{max}	79	2	4	2.72	1.464892626809155
Rat	28	σ_{min}	98	2	5	3.38	1.464891376927382
Rat	28	σ_{max}	74	2	3	2.55	1.464892626809155

Table 14.10 Example TLS1, $m = 100, n = 50$, Gauss–Radau, noise = 0.3, $\varepsilon = 10^{-10}$

Method	Lanc. it.	$\sqrt{z}$	Secul. it.	Min it.	Max it.	Av. it.	Solution
Newt	38	σ_{min}	172	2	15	4.41	1.464891451263721
Newt	38	σ_{max}	118	2	4	3.03	1.464891451304268
Rat	38	σ_{min}	133	2	6	3.41	1.464891451263721
Rat	38	σ_{max}	113	2	4	2.90	1.464891451304268

Table 14.11 Example TLS2, $m = 1000, n = 500$, Gauss–Radau, noise = 0.3, $\varepsilon = 10^{-6}$

Method	Lanc. it.	$\sqrt{z}$	Secul. it.	Min it.	Max it.	Av. it.	Solution
Newt	75	σ_{min}	370	3	11	4.87	5.200942068935540
Newt	75	σ_{max}	216	2	4	2.84	5.200960368809433
Rat	75	σ_{min}	320	3	5	4.21	5.200942068935540
Rat	75	σ_{max}	208	2	8	2.74	5.200960368809433

Table 14.12 Example TLS2, $m = 1000, n = 500$, Gauss–Radau, noise = 0.3, $\varepsilon = 10^{-10}$

Method	Lanc. it.	$\sqrt{z}$	Secul. it.	Min it.	Max it.	Av. it.	Solution
Newt	101	σ_{min}	488	2	12	4.87	5.200943687659306
Newt	101	σ_{max}	323	2	4	3.17	5.200943728774683
Rat	101	σ_{min}	422	2	6	4.14	5.200943687659305
Rat	101	σ_{max}	321	2	8	3.15	5.200943728774683

Results for large examples are in tables 14.11-14.14. The number of secular iterations is large. For these examples we can just compute the Gauss solution

Table 14.13 Example TLS3, $m = 10000, n = 5000$, Gauss–Radau, noise = 0.3, $\varepsilon = 10^{-6}$

Method	Lanc. it.	$\sqrt{z}$	Secul. it.	Min it.	Max it.	Av. it.	Solution
Newt	250	σ_{min}	2572	3	31	10.25	1.418576232676234
Newt	250	σ_{max}	1926	3	26	7.67	1.418583305908228
Rat	250	σ_{min}	837	2	5	3.33	1.418576232676233
Rat	250	σ_{max}	653	2	4	2.60	1.418583305908227

Table 14.14 Example TLS3, $m = 10000, n = 5000$, Gauss–Radau, noise = 0.3, $\varepsilon = 10^{-10}$

Method	Lanc. it.	$\sqrt{z}$	Secul. it.	Min it.	Max it.	Av. it.	Solution
Newt	328	σ_{min}	3000	2	32	9.12	1.418576233569240
Newt	328	σ_{max}	2335	2	26	7.10	1.418576233598322
Rat	328	σ_{min}	2093	2	51	6.36	1.418576233174081
Rat	328	σ_{max}	2142	2	51	6.51	1.418576233598324

and compute the Gauss–Radau solution when we decide that the Gauss estimate has converged. Moreover, we can compute the Gauss solution only every ν iterations. Results are given in table 14.15 for $\nu = 10$ where "Secul. it." is the number of secular iterations for Gauss–Radau at the end of the iterations. The Gauss estimate needs 29 secular equation solver (BNS1) iterations and the solution is 1.418582271405106. Table 14.16 gives the results for $\nu = 50$. The total number of secular equation solver iterations is 8 and the solution is 1.418576255595023.

Table 14.15 Example TLS3, $m = 10000, n = 5000$, Gauss–Radau, noise = 0.3, $\varepsilon = 10^{-6}$

Method	ν	Lanc. it.	$\sqrt{z}$	Secul. it.	Solution
Newt	10	251	σ_{min}	3	1.418576232734915
Newt	10	251	σ_{max}	3	1.418582638031528
Rat	10	251	σ_{min}	2	1.418576232734914
Rat	10	251	σ_{max}	2	1.418582638031525

Table 14.16 Example TLS3, $m = 10000, n = 5000$, Gauss–Radau, noise = 0.3, $\varepsilon = 10^{-6}$

Method	ν	Lanc. it.	$\sqrt{z}$	Secul. it.	Solution
Newt	50	301	σ_{min}	2	1.418576233173016
Newt	50	301	σ_{max}	2	1.418576256849264
Rat	50	301	σ_{min}	2	1.418576233173017
Rat	50	301	σ_{max}	2	1.418576256849263

Now, we monitor the convergence of the smallest singular value of A (computed using the SVD) every ν iterations to stop the Lanczos bidiagonalization iterations. Therefore, we compute only the Gauss and Gauss–Radau solutions at the end of the Lanczos iterations. The Gauss–Radau computations are done with the smallest and largest singular values of B_k at convergence. The results are given in table 14.17.

Table 14.17 Example TLS3, $m = 10000, n = 5000$, Gauss–Radau, noise $= 0.3$, $\varepsilon = 10^{-6}$

Method	ν	Lanc. it.	$\sqrt{z}$	Secul. it.	Solution
Gauss	10	250	-	1	1.418582932414374
Newt	10	250	$\sigma_{\min}(B_k)$	1	1.418582932414377
Newt	10	250	$\sigma_{\max}(B_k)$	3	1.418583305908215
Rat	10	250	$\sigma_{\min}(B_k)$	1	1.418582932414372
Rat	10	250	$\sigma_{\max}(B_k)$	2	1.418583305908213

Example TLS4 is the same problem as TLS3 but with $m = 100000$ and $n = 50000$. The version checking the convergence of the smallest singular value of A gives the results of table 14.18. We observe that we have to do a large number of Lanczos bidiagonalization iterations.

Table 14.18 Example TLS4, $m = 100000, n = 50000$, Gauss–Radau, noise $= 0.3$, $\varepsilon = 10^{-6}$

Method	ν	Lanc. it.	$\sqrt{z}$	Secul. it.	Solution
Gauss	50	801	-	1	0.8721024989570489
Newt	50	801	$\sigma_{\min}(B_k)$	1	0.8721024989575631
Newt	50	801	$\sigma_{\max}(B_k)$	2	0.8721025571544376
Rat	50	801	$\sigma_{\min}(B_k)$	1	0.8721024989574467
Rat	50	801	$\sigma_{\max}(B_k)$	2	0.8721025571545402

In Example TLS5 we keep the same value of $m = 100000$ and reduce the value of n to $n = 5000$. Results are given in table 14.19.

Table 14.19 Example TLS5, $m = 100000, n = 5000$, Gauss–Radau, noise $= 0.3$, $\varepsilon = 10^{-6}$

Method	ν	Lanc. it.	$\sqrt{z}$	Secul. it.	Solution
Gauss	50	351	-	1	1.168845952730205
Newt	50	351	$\sigma_{\min}(B_k)$	1	1.168845952730189
Newt	50	351	$\sigma_{\max}(B_k)$	2	1.168845953038015
Rat	50	351	$\sigma_{\min}(B_k)$	1	1.168845952729973
Rat	50	351	$\sigma_{\max}(B_k)$	2	1.168845953037930

In Example TLS6 we keep the same value of $m = 100000$ and change the value of n to $n = 500$. The problem is easier to solve. Results are given in table 14.20.

In the next computations, we monitor the convergence of the smallest singular value by solving a secular equation at every Lanczos iteration. The total number

Table 14.20 Example TLS6, $m = 100000, n = 500$, Gauss–Radau, noise = 0.3, $\varepsilon = 10^{-6}$

Method	ν	Lanc. it.	$\sqrt{z}$	Secul. it.	Solution
Gauss	50	150	-	1	1.291969588729442
Newt	50	150	$\sigma_{\min}(B_k)$	1	1.291969588729447
Newt	50	150	$\sigma_{\max}(B_k)$	1	1.291969588729463
Rat	50	150	$\sigma_{\min}(B_k)$	1	1.291969588729486
Rat	50	150	$\sigma_{\max}(B_k)$	2	1.291969588729502

of iterations to solve these secular equations is given under the name "trid". If the rational approximation method does not converge well, we use bisection. The Gauss and Gauss–Radau solutions are computed only at the end using the SVD of B_k and Newton iterations.

Table 14.21 Example TLS3, $m = 10000, n = 5000$, noise = 0.3, $\varepsilon = 10^{-6}$

Method	Lanc. it.	Trid	$\sqrt{z}$	Secul. it.	Solution
-	250	551			
Gauss			-	1	1.418582932414377
G-R			$\sigma_{\min}(B_k)$	1	1.418582932414377
G-R			$\sigma_{\max}(B_k)$	1	1.418583305908215

We see that for Example TLS4 in table 14.22, the iterations are stopped a little too early. Therefore we may want to use a smaller stopping criterion when computing the smallest singular value in this way. In tables 14.23 and 14.24 we decrease the value of n.

Table 14.22 Example TLS4, $m = 100000, n = 50000$, noise = 0.3, $\varepsilon = 10^{-6}$

Method	Lanc. it.	Trid	$\sqrt{z}$	Secul. it.	Solution
-	750	1603			
Gauss			-	1	0.8721144459858997
G-R			$\sigma_{\min}(B_k)$	1	0.8721144459856636
G-R			$\sigma_{\max}(B_k)$	3	0.8721146975760565

In the next computations (tables 14.25–14.28) we solve all the secular equations by using a third-order rational approximation and by solving tridiagonal systems to compute the function and its derivatives instead of using the SVD of B_k. Note that we have to solve the equation for the smallest singular value at every iteration since we have to know in which interval we have to look for the solution. This algorithm is more or less the best we can do without using SVDs.

Then, we change the singular value distribution. In descending order the singular values are $\sigma_i = \exp(-3\,10^{-8}i)$, $i = 1, \ldots, n$. First, in Example TLS7 we use $m =$

Table 14.23 Example TLS5, $m = 100000, n = 5000$, noise = 0.3, $\varepsilon = 10^{-6}$

Method	Lanc. it.	Trid	$\sqrt{z}$	Secul. it.	Solution
-	294	634			
Gauss			-	1	1.168850294618886
G-R			$\sigma_{\min}(B_k)$	1	1.168850294618983
G-R			$\sigma_{\max}(B_k)$	3	1.168850447388590

Table 14.24 Example TLS6, $m = 100000, n = 500$, noise = 0.3, $\varepsilon = 10^{-6}$

Method	Lanc. it.	Trid	$\sqrt{z}$	Secul. it.	Solution
-	92	211			
Gauss			-	1	1.291970910578612
G-R			$\sigma_{\min}(B_k)$	1	1.291970910578655
G-R			$\sigma_{\max}(B_k)$	3	1.291971161042421

Table 14.25 Example TLS3, $m = 10000, n = 5000$, noise = 0.3, $\varepsilon = 10^{-6}$

Method	Lanc. it.	Trid	$\sqrt{z}$	Secul. it.	Solution
-	250	551			
Gauss			-	2	1.418582932414440
G-R			$\sigma_{\min}(B_k)$	2	1.418582932414443
G-R			$\sigma_{\max}(B_k)$	3	1.418583305908306

Table 14.26 Example TLS4, $m = 100000, n = 50000$, noise = 0.3, $\varepsilon = 10^{-6}$

Method	Lanc. it.	Trid	$\sqrt{z}$	Secul. it.	Solution
-	755	1775			
Gauss			-	1	0.8721122166701496
G-R			$\sigma_{\min}(B_k)$	2	0.8721122166735605
G-R			$\sigma_{\max}(B_k)$	3	0.8721124331415380

Table 14.27 Example TLS5, $m = 100000, n = 5000$, noise = 0.3, $\varepsilon = 10^{-6}$

Method	Lanc. it.	Trid	$\sqrt{z}$	Secul. it.	Solution
-	293	634			
Gauss			-	1	1.168850294619013
G-R			$\sigma_{\min}(B_k)$	2	1.168850294618839
G-R			$\sigma_{\max}(B_k)$	3	1.168850513547429

Table 14.28 Example TLS6, $m = 100000, n = 500$, noise = 0.3, $\varepsilon = 10^{-6}$

Method	Lanc. it.	Trid	$\sqrt{z}$	Secul. it.	Solution
-	92	211			
Gauss			-	1	1.291999525873749
G-R			$\sigma_{\min}(B_k)$	2	1.291970910578657
G-R			$\sigma_{\max}(B_k)$	3	1.291971119694316

$100, n = 50$. The smallest singular value of A is 0.1611919831064825 and the TLS parameter, which is 0.1611917031832353, is very close to the smallest singular value. Monitoring the convergence of the smallest singular value, we obtain the results of table 14.29. Results in table 14.30 are obtained with $\varepsilon = 10^{-10}$. Note that most of the solutions are upper bounds.

Table 14.29 Example TLS7, $m = 100, n = 50$, noise = 0.3, $\varepsilon = 10^{-6}$

Method	Lanc. it.	Trid	$\sqrt{z}$	Secul. it.	Solution
-	28	70			
Gauss			-	2	0.1611917162057354
G-R			$\sigma_{\min}(B_k)$	2	0.1611917162057353
G-R			$\sigma_{\max}(B_k)$	2	0.1611917212163983

Table 14.30 Example TLS7, $m = 100, n = 50$, noise = 0.3, $\varepsilon = 10^{-10}$

Method	Lanc. it.	Trid	$\sqrt{z}$	Secul. it.	Solution
-	33	280			
Gauss			-	3	0.1611917031833536
G-R			$\sigma_{\min}(B_k)$	3	0.1611917031831545
G-R			$\sigma_{\max}(B_k)$	3	0.1611917031833930

Example TLS8 uses $m = 10000$ and $n = 5000$. Results are in table 14.31, monitoring the smallest singular value to stop the Lanczos iterations and using tridiagonal systems to compute functions and derivatives. However, in this computation we do not obtain a good approximation of the smallest singular value of A. The approximate value is $1.155072478369476 \ 10^{-3}$ whereas the "exact" result is $1.153611860471415 \ 10^{-3}$. Moreover, the computed TLS parameters are larger than the exact $\sigma_{\min}(A)$. Increasing the precision, we have the results of table 14.32. Then, the computed minimum singular value is $1.153611860609424 \ 10^{-3}$. However, as we can see, this computation is much more expensive and the secular equations are difficult to solve, some of the tridiagonal systems being almost singular.

If we use $m = 10000$ and $n = 500$ (Example TLS9) the problem is much easier to solve. The exact smallest singular value of A is 0.2379927701875446 and the computed one is 0.2379928280610862. The results are given in table 14.33.

Table 14.31 Example TLS8, $m = 10000, n = 5000$, noise = 0.3, $\varepsilon = 10^{-6}$

Method	Lanc. it.	Trid	$\sqrt{z}$	Secul. it.	Solution
-	379	884			
Gauss			-	2	$1.154993988896189 \ 10^{-3}$
G-R			$\sigma_{\min}(B_k)$	2	$1.154993976869066 \ 10^{-3}$
G-R			$\sigma_{\max}(B_k)$	3	$1.154994276316134 \ 10^{-3}$

Table 14.32 Example TLS8, $m = 10000, n = 5000$, noise = 0.3, $\varepsilon = 10^{-10}$

Method	Lanc. it.	Trid	$\sqrt{z}$	Secul. it.	Solution
-	503	10445			
Gauss			-	87	$1.153540337993690 \ 10^{-3}$
G-R			$\sigma_{\min}(B_k)$	87	$1.153540337993690 \ 10^{-3}$
G-R			$\sigma_{\max}(B_k)$	3	$1.153540341197251 \ 10^{-3}$

Table 14.33 Example TLS9, $m = 10000, n = 500$, noise = 0.3, $\varepsilon = 10^{-6}$

Method	Lanc. it.	Trid	$\sqrt{z}$	Secul. it.	Solution
-	92	260			
Gauss			-	3	0.2379928280610862
G-R			$\sigma_{\min}(B_k)$	3	0.2379902725954008
G-R			$\sigma_{\max}(B_k)$	3	0.2379902838087289

To summarize, we have seen in this chapter that we can compute the parameter σ_{n+1} of the TLS problem by using the approximation from the Golub–Kahan bidiagonalization algorithm and efficient solvers for the secular equations. We have optimized the algorithm. For Example TLS3 with $m = 10000$, $n = 5000$ and $\varepsilon = 10^{-6}$, solving the secular equations at each iteration, the computing time was 117 seconds. For the last optimized version using a rational approximation and solving tridiagonal systems the computing time is 12 seconds.

Chapter Fifteen

Discrete Ill-Posed Problems

15.1 Introduction to Ill-Posed Problems

Problems are generally defined as ill-posed (as defined by J. Hadamard) when the solution is not unique or does not depend continuously on the data. More practically, problems are said to be ill-posed when a small change in the data may cause a large change in the solution. Strictly speaking, problems cannot be ill-posed in Hadamard's sense in finite dimension, but problems whose solution is sensitive to perturbations of the data are called discrete ill-posed problems (DIP); see Hansen [178]. This is typically what may happen in the solution of least squares problems. Looking back at the solution of $Ax \approx c$ given in equation (13.2), we see that if the matrix A has small nonzero singular values (and if the corresponding projections of the perturbed right-hand side $(u^i)^T(c + \Delta c)$ are not small) then the perturbed solution can be much different from x_{LS}.

As a small example (close to one defined by P. C. Hansen), let us define

$$A = \begin{pmatrix} 0.15 & 0.1 \\ 0.16 & 0.1 \\ 2.02 & 1.3 \end{pmatrix}, \quad c + \Delta c = A \begin{pmatrix} 1 \\ 1 \end{pmatrix} + \begin{pmatrix} 0.01 \\ -0.032 \\ 0.01 \end{pmatrix}.$$

We added a small perturbation to a right-hand side corresponding to a solution with all components equal to 1. The solution of the perturbed least squares problem (rounded to four decimals) using the QR factorization of A is

$$x_{QR} = \begin{pmatrix} -2.9977 \\ 7.2179 \end{pmatrix},$$

which is quite different from the unperturbed solution. However, the norm of the residual is small being 0.0295. Rounded to four decimal digits the SVD of A is

$$U = \begin{pmatrix} -0.0746 & 0.7588 & -0.6470 \\ -0.0781 & -0.6513 & -0.7548 \\ -0.9942 & -0.0058 & 0.1078 \end{pmatrix}, \quad \Sigma = \begin{pmatrix} 2.4163 & 0 \\ 0 & 0.0038 \\ 0 & 0 \end{pmatrix},$$

$$V = \begin{pmatrix} -0.8409 & -0.5412 \\ -0.5412 & 0.8409 \end{pmatrix}.$$

The component $(u^2)^T \Delta c / \sigma_2$ (u^2 being the second column of U) corresponding to the smallest nonzero singular value is large, being 6.2161. This gives the large change in the solution. Of course, the computed solution is not really useful. We would have preferred computing something close to $(1 \quad 1)^T$ which is the solution without noise.

The solution of ill-posed linear systems arise in many areas of scientific computing; see, for instance, Engl [102], Hanke and Hansen [173], Tikhonov [324] and Varah [334]. Many problems in seismology, signal processing, medical imaging and image restoration (Adorf [1], Bardsley, Jefferies, Nagy and Plemmons [21], Hanke [170], Hanke and Nagy [174], Nagy, Palmer and Perrone [246], Thompson, Brown, Kay and Titterington [326], Vio, Nagy, Tenorio and Wamsteker [335]) as well as other scientific areas lead to integral equations of the first kind like

$$\int_0^1 k(s,t) f(t)\, dt = g(s) + e(s),$$

where $k(s,t)$ is the kernel, the right-hand side corresponds to measurements, e being an unknown noise, and f is the solution we are seeking. When discretized with quadrature and collocation, this often leads to discrete ill-posed problems. These problems are also called inverse problems since they amount to computing the cause from the results of some measurements.

We now consider the finite dimensional case with an overdetermined linear system

$$Ax \approx c = \bar{c} - e,$$

where A is a matrix of dimension $m \times n, m \geq n$ and the right-hand side $\bar{c}$ is contaminated by a (generally) unknown noise vector e. The standard solution of the least squares problem $\min \|c - Ax\|$ (even using backward stable methods like QR) may give a vector x severely contaminated by noise. This may seem hopeless. However, a way to compute something useful is to modify the problem into another problem whose solution can be computed more reliably. This process is called regularization. Of course this will not give the solution of the original unperturbed problem. Hence, we have to find a balance between obtaining a problem that we can solve reliably and obtaining a solution which is not too far from the solution without noise.

15.1.1 Truncated SVD

When the dimension of the problem is small enough and the SVD of A can be computed, one can obtain an approximate solution by neglecting the terms corresponding to the smallest singular values (which are labeled with the largest indices in our notation). The LS solution is

$$x_{LS} = \sum_{i=1}^{r} \frac{(u^i)^T c}{\sigma_i},$$

where r is the rank of A. The regularized solution by truncated SVD (TSVD) is

$$x_{TSVD} = \sum_{i=1}^{k} \frac{(u^i)^T c}{\sigma_i},$$

with $k < r$. This threshold integer k can be seen as the regularization parameter. The TSVD solution can also be written as a filtered solution

$$x_{TSVD} = \sum_{i=1}^{r} f_i \frac{(u^i)^T c}{\sigma_i}.$$

Here, the values of the filter coefficients f_i are particularly simple being 1 for $i \leq k$ and 0 otherwise.

15.1.2 Tikhonov Regularization

One of the most popular regularization methods to obtain a meaningful solution is Tikhonov regularization (see Tikhonov [324], Tikhonov and Arsenin [325]) in which the linear system or the LS problem is replaced by the minimization problem

$$\min_x \{\|c - Ax\|^2 + \mu\|x\|^2\}, \tag{15.1}$$

where $\mu \geq 0$ is a regularization parameter to be chosen. For some problems (particularly in image restoration) it is better to consider

$$\min_x \{\|c - Ax\|^2 + \mu\|Lx\|^2\}, \tag{15.2}$$

where L is typically the discretization of a derivative operator of first or second order. Numerically the problem (15.2) is solved by considering it as a least squares problem

$$\min \left\| \begin{pmatrix} A \\ \sqrt{\mu}L \end{pmatrix} x - \begin{pmatrix} c \\ 0 \end{pmatrix} \right\|,$$

whose solution is obtained using orthogonal transformations or iterative methods.

One can also consider other norms than the l_2 norm, for instance the l_1 norm. Another possible extension is to use more than one penalty term by writing the functional to be minimized as

$$k \left(\|c - Ax\|^2 + \sum_{i=1}^{k} \mu_i \|L_i x\|^2 \right).$$

For such an approach, see, for instance, Brezinski, Redivo–Zaglia, Rodriguez and Seatzu [41] and Belge, Kilmer and Miller [26]. Tikhonov regularization has also been used with total least squares by Golub, Hansen and O'Leary [161].

The solution x_μ of the problem (15.1) solves the linear system

$$(A^T A + \mu I)x = A^T c. \tag{15.3}$$

This is the so-called normal equations. The choice of the regularization parameter μ is crucial since if μ is too small the solution is contaminated by the noise in the right-hand side (as it is with $\mu = 0$), on the other hand if μ is too large the solution is a poor approximation of the original problem. Many methods have been devised for choosing μ; see Golub and von Matt [158] for an overview. As we will see, most of these methods lead to the evaluation of bilinear forms with different matrices. We will consider in more detail two of these techniques: generalized cross-validation and the L-curve criterion.

When the dimension is small enough the regularized solution x_μ can be computed using the SVD of A as

$$x_\mu = \sum_{i=1}^{n} \frac{\sigma_i (u^i)^T c}{\sigma_i^2 + \mu} v^i,$$

if A has full rank. This can be seen as a filtered solution, the filter coefficients being

$$f_i = \frac{\sigma_i^2}{\sigma_i^2 + \mu}.$$

For the large singular values σ_i the filter f_i is almost 1, when it is close to zero for the smallest singular values if they are near zero.

If $L = I$ the problem is said to be in standard form. Otherwise it is in general form. In this case the generalized singular value decomposition (GSVD; see, for instance, Golub and Van Loan [154]) can be used. The factorization of the pair (A, L) is

$$A = U \begin{pmatrix} \Gamma & 0 \\ 0 & I_{n-p} \\ 0 & 0 \end{pmatrix} W^{-1}, \quad L = V \begin{pmatrix} \Upsilon & 0 \end{pmatrix} W^{-1}, \tag{15.4}$$

where L is $p \times n$ and $m \geq n \geq p$. The matrices $U(m \times m)$ and $V(p \times p)$ are orthonormal and the nonsingular matrix W is $n \times n$. The matrices Γ and Υ with diagonal elements γ_i and τ_i are diagonal with $\gamma_i^2 + \tau_i^2 = 1$. The generalized singular values of the matrix pencil $(A \quad L)$ are defined as $\sigma_i = \gamma_i / \tau_i$. The last $n - p$ columns of W form a basis of the null space of L. If $L = I$ the GSVD reduces to the SVD of A.

For the generalized form, the regularized solution is written as

$$x_\mu = \sum_{i=1}^{p} \frac{\sigma_i^2}{\sigma_i^2 + \mu} \frac{(u^i)^T c}{\gamma_i} w^i + \sum_{i=p+1}^{n} (u^i)^T c w^i.$$

The second term in the right-hand side is the (unregularized) component of the solution in the null space of L.

All problems in general form can be reduced to problems in standard form; see, for instance, Hanke and Hansen [173]. In the next paragraphs we describe some ways for choosing the regularization parameter μ for problems in standard form. For other methods, see O'Leary [250].

15.1.3 Morozov's Discrepancy Principle

This method (see Morozov [244]) can be used only if the (norm of the) noise vector e is known. The value of the regularization parameter μ is chosen such that the norm of the residual equals the norm of the noise vector using the mathematical solution from equation (15.3),

$$\|c - A(A^T A + \mu I)^{-1} A^T c\| = \|e\|.$$

This equation can also be written as

$$\mu^2 c^T (AA^T + \mu I)^{-2} c = \|e\|^2,$$

and has a unique positive solution which can be obtained by using the SVD of A if the dimension of the problem is not too large. We have $A = U \Sigma V^T$ and by denoting $d = U^T c$, the previous relation is written as

$$\mu^2 d^T (\Sigma + \mu I)^{-2} d = \|e\|^2.$$

Considered as a function of μ the left-hand side is an increasing function whose value is 0 for $\mu = 0$. When $\mu \to \infty$ the left-hand side tends to $\|c\|^2$. Hence, if $\|e\| \leq \|c\|$ which is usually the case, there is a unique solution that can be computed with any scalar root finding algorithm or from the secular equation solvers of chapter 9.

15.1.4 The Gfrerer/Raus Method

This method (see Gfrerer [136], Hanke and Raus [175]) also needs knowledge of the norm of the noise vector. The parameter is chosen such that

$$\mu^3 c^T (AA^T + \mu I)^{-3} c = \|e\|^2.$$

As for the discrepancy principle, there is usually a unique solution to this equation. In fact, the paper [136] considers the problem posed in Hilbert spaces and uses the iterated Tikhonov regularization in which one computes a sequence of solutions x^j given by

$$(A^T A + \mu I)x^j = A^T c + \mu x^{j-1}, \quad j = 1, \dots, k.$$

The Gfrerer/Raus criterion is obtained by minimizing an upper bound of

$$\|(A^T A + \mu I)^{-1} c - (A^T A)^{-1} \bar{c}\|.$$

15.1.5 The quasi-Optimality Criterion

This criterion (Leonov [224]) chooses μ as a positive minimizer of

$$\mu^2 c^T A (A^T A + \mu I)^{-4} A^T c.$$

Note that with this criterion we do not have to know the norm of the noise vector. It involves computations of a quadratic form.

15.1.6 The L-curve Criterion

In [223], Lawson and Hanson observed that a "good" way to see how the regularized solution x_μ depends on the parameter μ is to plot the curve $(\|x_\mu\|, \|b - Ax_\mu\|)$ obtained by varying the value of $\mu \in [0, \infty)$. This curve is known as the L-curve since it is (in many circumstances) shaped as the letter "L". Actually it is even more illuminating to look at this curve in a log-log scale as suggested by Hansen and O'Leary [180]. Lawson and Hanson [223] proposed to choose the value μ_L corresponding to the "vertex" or the "corner" of the L-curve that is the point with maximal curvature; see also Hansen and O'Leary [180] and Hansen [176], [178].

A motivation for choosing the "vertex" is, as we said before, to have a balance between μ being too small and the solution contaminated by noise, and μ being too large giving a poor approximation of the solution. The "vertex" of the L-curve gives an average value between these two extremes.

The properties of the L-curve have been investigated in [180]. Let $\eta^2 = \|x_\mu\|^2$ and $\rho^2 = \|c - Ax_\mu\|^2$ be functions of μ. In [180] it is proved that

$$\frac{d(\eta^2)}{d(\rho^2)} = -\frac{1}{\mu},$$

and this derivative is negative since $\mu > 0$. Computation of the second derivative shows that the L-curve in linear scale is convex and steeper as μ approaches the smallest singular value. The L-curve has also been studied by Regińska [277] for different scaling functions, mainly the square root and the logarithm. In this paper it is proved that the L-curve remains decreasing in any differentiable strictly monotonic scale (which is the case for the logarithm). However, Regińska exhibited conditions under which the L-curve in logarithmic scale is (partly) strictly concave. It was suggested to choose a μ giving a local minimum of $\|x_\mu\| \|c - Ax_\mu\|^\lambda$. Conditions for having these minima and their relations to the L-curve are given in [277]. For the limitations of the L-curve criterion, see also Hanke [169] and Vogel [336], who showed that the L-curve approach may give regularized solutions which fail to converge for certain classes of problems.

A major problem in determining an approximate value of μ_L in large ill-posed problems for which the SVD of A is not available is that it is expensive to compute points on the L-curve. The computation of each point needs the solution of a minimization problem (15.1). Usually one computes only a few points on the curve and determines a value of the regularization parameter (not necessarily "optimal") by interpolation. This is discussed in Hansen and O'Leary [180].

In [51] Calvetti, Golub and Reichel proposed to use the techniques developed in the previous chapters to inexpensively compute approximations of points on the L-curve without having to solve large minimization problems. Let us now describe how we can efficiently compute these approximations.

From the solution (15.3) of the regularized problem we can compute the norm of the solution and the corresponding residual

$$\|x_\mu\|^2 = c^T A (A^T A + \mu I)^{-2} A^T c$$

and

$$\|c - Ax_\mu\|^2 = c^T c + c^T A (A^T A + \mu I)^{-1} A^T A (A^T A + \mu I)^{-1} A^T c$$
$$- 2c^T A (A^T A + \mu I)^{-1} A^T c.$$

The expression of the residual norm can be simplified by using the identity

$$I - A(A^T A + \mu I)^{-1} A^T = \mu (AA^T + \mu I)^{-1}.$$

Hanke [172] remarked that the norm of the residual can be written more simply as

$$\|c - Ax_\mu\|^2 = \mu^2 c^T (AA^T + \mu I)^{-2} c. \tag{15.5}$$

Nevertheless, in the paper [51] the more complex form of the norm of the residual was used. By denoting $K = A^T A$ and $d = A^T c$, we have

$$\|c - Ax_\mu\|^2 = c^T c + d^T K (K + \mu I)^{-2} d - 2d^T (K + \mu I)^{-1} d.$$

Then, if we define two functions

$$\phi_1(t) = (t + \mu)^{-2},$$
$$\phi_2(t) = t(t + \mu)^{-2} - 2(t + \mu)^{-1},$$

we are interested in $s_i = d^T \phi_i(K) d$, $i = 1, 2$ from which we can obtain points on the L-curve

$$\|x_\mu\| = s_1^{1/2},$$
$$\|c - Ax_\mu\| = (c^T c + s_2)^{1/2},$$

or we may be interested in the logarithms of these expressions. We can obtain bounds for s_1 and s_2 using our usual machinery by running the Lanczos algorithm with the symmetric matrix $A^T A$, which needs only multiplications by A and A^T. However, for this particular matrix, we can use the Lanczos bidiagonalization algorithm developed by Golub and Kahan [144]; see chapter 4. We note that this algorithm (as well as the Lanczos algorithm) is independent of μ. At iteration k, the algorithm computes a Jacobi matrix $J_k = B_k^T B_k$ and the approximation given by the Gauss rule for a function ϕ_i defined previously is

$$I_k^G(\phi_i) = \|d\|^2 (e^1)^T \phi_i(J_k) e^1.$$

The Gauss–Radau rule with one assigned node $a \leq \lambda_1$, where λ_1 is the smallest eigenvalue of $A^T A$, is obtained as

$$I_k^{GR}(\phi_i) = \|d\|^2 (e^1)^T \phi_i(\hat{J}_k) e^1,$$

where $\hat{J}_k$ is obtained by modifying the (k, k) element of J_k. In particular, if we prescribe a node at the origin $a = 0$, then we have $\hat{J}_k = \hat{B}_k^T \hat{B}_k$ where $\hat{B}_k$ is obtained from B_k by setting the last diagonal element $\delta_k = 0$; see Golub and von Matt [158].

To know if the approximations are lower or upper bounds we have the following results.

LEMMA 15.1 *Let* $\phi_1(t) = (t + \mu)^{-2}, \phi_2(t) = t(t + \mu)^{-2} - 2(t + \mu)^{-1}$. *Then, for* $t \geq 0$ *and* $k \geq 1$, *the derivatives are such that*

$$\phi_1^{(2k-1)}(t) < 0, \quad \phi_1^{(2k)}(t) > 0,$$

$$\phi_2^{(2k-1)}(t) > 0, \quad \phi_2^{(2k)}(t) < 0.$$

THEOREM 15.2 *The Gauss quadrature rule gives a lower bound for* ϕ_1 *defined in lemma 15.1. The Gauss–Radau rule with* $a \leq \lambda_1$ *gives an upper bound for* ϕ_1. *We have*

$$I_k^G(\phi_1) \leq s_1 \leq I_k^{GR}(\phi_1),$$

where

$$I_k^G(\phi_1) = \|d\|^2 (e^1)^T (B_k^T B_k + \mu I)^{-2} e^1,$$
$$I_k^{GR}(\phi_1) = \|d\|^2 (e^1)^T (\hat{B}_k^T \hat{B}_k + \mu I)^{-2} e^1.$$

The computation of the quadrature rules requires solving for vectors y satisfying

$$(B^T B + \mu I) y = e^1 \quad \text{with } B = B_k \text{ or } \hat{B}_k. \tag{15.6}$$

The solution y can be computed by solving a (small) least squares problem

$$\min_y \left\| \begin{pmatrix} B \\ \sqrt{\mu} I \end{pmatrix} y - \frac{1}{\sqrt{\mu}} z \right\|,$$

where z is a vector of dimension $2k$ with zero components except $z_{k+1} = 1$. One can check that the normal equation for this problem is identical to equation (15.6). The solution can be obtained incrementally and efficiently using Givens rotations to zero the lower part of the matrix since the upper part is already upper triangular. Let us denote by y_μ^k and $\hat{y}_\mu^k$ the solution of equations (15.6) for $B = B^k$ and $B = \hat{B}^k$. Then,

$$I_k^G(\phi_1) = \|d\|^2 \|y_\mu^k\|^2,$$
$$I_k^{GR}(\phi_1) = \|d\|^2 \|\hat{y}_\mu^k\|^2.$$

THEOREM 15.3 *The Gauss quadrature rule gives an upper bound for ϕ_2 defined in lemma 15.1. The Gauss–Radau rule with $a \leq \lambda_1$ gives a lower bound for ϕ_2. For the norm of the residual, we have*

$$I_k^{GR}(\phi_2) \leq s_2 \leq I_k^G(\phi_2),$$

where

$$I_k^G(\phi_2) = \|d\|^2[(e^1)^T B_k^T B_k (B_k^T B_k + \mu I)^{-2} e^1 - 2(e^1)^T (B_k^T B_k + \mu I)^{-1} e^1],$$
$$I_k^{GR}(\phi_2) = \|d\|^2[(e^1)^T \hat{B}_k^T \hat{B}_k (\hat{B}_k^T \hat{B}_k + \mu I)^{-2} e^1 - 2(e^1)^T (\hat{B}_k^T \hat{B}_k + \mu I)^{-1} e^1].$$

Proof. We use the fact that $B(B^T B + \mu I)^{-1} = (BB^T + \mu I)^{-1} B$ to simplify the expressions for ϕ_2. □

Using the solutions computed previously for ϕ_1, the bounds for ϕ_2 are written

$$I_k^G(\phi_2) = \|d\|^2[\|B_k y_\mu^k\|^2 - 2(e^1)^T y_\mu^k],$$
$$I_k^{GR}(\phi_2) = \|d\|^2[\|\hat{B}_k \hat{y}_\mu^k\|^2 - 2(e^1)^T \hat{y}_\mu^k].$$

Hanke [172] proved that the lower and upper bounds improve when the number of iterations is increased.

From these bounds we can define an approximation of the L-curve. Let

$$x^-(\mu) = \sqrt{I_k^G(\phi_1)},$$

$$x^+(\mu) = \sqrt{I_k^{GR}(\phi_1)},$$

$$y^-(\mu) = \sqrt{c^T c + I_k^{GR}(\phi_2)},$$

$$y^+(\mu) = \sqrt{c^T c + I_k^G(\phi_2)}.$$

For a given value of $\mu > 0$ the bounds are

$$x^-(\mu) \leq \|x_\mu\| \leq x^+(\mu),$$

$$y^-(\mu) \leq \|c - Ax_\mu\| \leq y^+(\mu).$$

This defines a rectangle. Calvetti, Golub and Reichel [51] defined the L-ribbon as the union of these rectangles for all $\mu > 0$,

$$\bigcup_{\mu > 0} \{ \{x(\mu), y(\mu)\} : x^-(\mu) \leq x(\mu) \leq x^+(\mu), \ y^-(\mu) \leq y(\mu) \leq y^+(\mu) \}.$$

The previous techniques compute bounds on the norms of the residual and the solution. After choosing a point inside the L-ribbon (that is, a value of μ), the solution must be computed. One can use the approximation

$$x_\mu^k = \|d\| Q_k y_\mu^k,$$

where Q_k is the matrix computed by the Lanczos bidiagonalization algorithm. Note that with this choice of point inside the L-ribbon $\|x_\mu^k\| = x^-(\mu)$.

This line of research has been pursued by Calvetti, Reichel and some collaborators; see [53], [48], [47], [50], [55], [56] and [57]. A possible improvement in the choice of μ is to look directly at the curvature of the L-curve and to select a point of (approximate) maximum curvature. In log-log scale the curvature is given (see Golub and von Matt [158]) by

$$C_\mu = 2\frac{\rho''\eta' - \rho'\eta''}{((\rho')^2 + (\eta')^2)^{3/2}},$$

where the prime denotes differentiation with respect to μ and

$$\rho(\mu) = \frac{1}{2}\log\|c - Ax_\mu\| = \log\mu^2 c^T \phi(AA^T)c,$$

$$\eta(\mu) = \frac{1}{2}\log\|x_\mu\| = \log c^T A\phi(A^T A)A^T c,$$

where $\phi(t) = (t + \mu)^{-2}$. The first derivatives can be computed as

$$\rho'(\mu) = \frac{c^T A(A^T A + \mu I)^{-3}A^T c}{\mu c^T(AA^T + \mu I)^{-2}c},$$

$$\eta'(\mu) = -\frac{c^T A(A^T A + \mu I)^{-3}A^T c}{c^T A(A^T A + \mu I)^{-2}A^T c}.$$

The numerator is more complicated

$$\rho'\eta'' - \rho''\eta' = \left(\frac{c^T A(A^T A + \mu I)^{-3}A^T c}{\mu c^T(AA^T + \mu I)^{-2}c \cdot c^T A(A^T A + \mu I)^{-2}A^T c}\right)^2$$
$$(c^T(AA^T + \mu I)^{-2}c \cdot c^T A(A^T A + \mu I)^{-2}A^T c$$
$$+2\mu c^T(AA^T + \mu I)^{-3}c \cdot c^T A(A^T A + \mu I)^{-2}A^T c$$
$$-2\mu c^T(AA^T + \mu I)^{-2}c \cdot c^T A(A^T A + \mu I)^{-3}A^T c).$$

It is important to note that this involves matrices of the form $(AA^T + \mu I)^{-p}$ and $A(A^T A + \mu I)^{-q}A^T$ with the powers p and q taking values 2 and 3 and corresponding bilinear forms for which lower and upper bounds can be computed.

In [53] Calvetti, Hansen and Reichel, using a slightly different formulation and Gauss and Gauss–Radau rules, obtained bounds for the curvature of the L-curve. This defines a curvature-ribbon around the curvature curve from which it is easier to find the value of μ giving the largest curvature. As with the L-curve, the bounds for the curvature are tighter when the number of Lanczos (or bidiagonalization) steps increases.

Numerical experiments using the L-ribbon and the curvature-ribbon will be described later in this chapter.

15.1.7 Minimization of the Error Norm

Brezinski, Rodriguez and Seatzu [39] proposed to use the estimates of the error devised by Brezinski [37] to select the parameter in the Tikhonov regularization method when the matrix is square. The estimate of the error is tailored to this special problem. One possibility is to use

$$e_3 = \frac{\|r_\mu\|^2}{\mu \|x_\mu\|},$$

where $r_\mu = c - Ax_\mu$ is the residual corresponding to the solution x_μ of the regularized problem. One samples e_3 for a set of values of μ and selects the value of μ which gives the minimum of the estimate. Brezinski, Rodriguez and Seatzu [40] extended their results to rectangular matrices.

Reichel, Rodriguez and Seatzu [280] used error estimates obtained from Gauss quadrature to obtain the regularization parameter.

15.1.8 Generalized Cross-Validation

Cross-validation and generalized cross-validation are techniques for model fitting for given data and model evaluation. These two tasks can be accomplished using independent data samples. The available data can be split into two sets, one for fitting and one for evaluation. This is not very efficient if the sample is not very large. The idea of cross-validation, introduced by Allen [5] for linear regression, is to recycle the data.

Suppose we have a measure given for the model evaluation; for instance, in a linear case the norm of the residual. Then, for cross-validation we split the data set D into N disjoint samples $D_1, \ldots, D_N$. For each $j = 1, \ldots, N$ the model is fit to the sample $\cup_{i \neq j} D_i$ and we compute the discrepancies $d_j(\mu)$ using D_j and the given measure. Then, the method finds the vector of the tuning parameters μ of the model as the minimizer of the total discrepancy over the N data samples,

$$d(\mu) = \sum_{j=1}^{N} d_j(\mu).$$

As an example, consider a regression model with m given data as $y_i = f(t_i) + \varepsilon_i$, where t represents the time, y are the observations and ε are random fluctuations. Then, remove only one observation each time to define the data samples D_j. Let $y^{(i)}$ be the vector of length $m - 1$ where the ith observation is removed from the response vector y. Let $f_\mu^{(i)}$ be the estimate of the response function based on the observations $y^{(i)}$. The optimal cross-validation parameter μ is the minimizer of

$$\frac{1}{m} \sum_{i=1}^{m} (y_i - f_\mu^{(i)}(t_i))^2.$$

Computing this minimizer can be computationally expensive. Let us now assume that a vector $\tilde{y}$ is obtained from y by replacing the ith coordinate by the estimate $f_\mu^{(i)}(t_i)$ and let $\tilde{f}_\mu^{(i)}$ be the estimate obtained from $\tilde{y}$. Then, in many cases one can show (Wahba [340]) that $\tilde{f}_\mu^{(i)} = f_\mu^{(i)}, \forall i$. For some models we have $f_\mu = C(\mu)y$ where $C(\mu)$ is a matrix with elements $c_{i,j}$. Then we have

$$(f_\mu)(t_i) = \sum_{j=1}^{m} c_{i,j} y_j$$

and

$$f_\mu^{(i)}(t_i) = \tilde{f}_\mu^{(i)}(t_i) = \sum_{j \neq i} c_{i,j} y_j + c_{i,i}(f_\mu^{(i)})(t_i).$$

These two relations lead to

$$f_\mu(t_i) - f_\mu^{(i)}(t_i) = c_{i,i}(y_i - f_\mu^{(i)}(t_i)).$$

Then

$$y_i - f_\mu^{(i)}(t_i) = \frac{y_i - f_\mu(t_i)}{1 - c_{i,i}}$$

and the prediction error is

$$\frac{1}{m} \sum_{i=1}^{m} \left(\frac{y_i - f_\mu(t_i)}{1 - c_{i,i}} \right)^2. \tag{15.7}$$

The generalized cross-validation (GCV) (see Craven and Wahba [70]) replaces $c_{i,i}$ in equation (15.7) by the average of the diagonal elements of C, so the denominator can be taken out of the sum. The criterion becomes

$$\frac{\sum_{i=1}^{m}(y_i - f_\mu(t_i))^2}{m(1 - \text{tr}(C)/m)^2},$$

where tr is the trace of the matrix. GCV is explained for that problem in Golub, Heath and Wahba [162], and some of its properties are studied using the singular value decomposition, which can also be used to compute an estimate of μ. This paper considered also GCV for subset selection and general linear model building and compared GCV with the maximum likelihood estimate. GCV has been used in many scientific areas. Applications in remote sensing, ridge regression, spline regression, likelihood estimation and log-hazard estimation are cited in Gu, Batres, Chen and Wahba [166]. Other applications include smoothing splines (Burrage, Williams, Ehrel and Pohl [46], Craven and Wahba [70], Hutchinson [194], Hutchinson and De Hoog [196], Lu and Mathis [229], Schumaker and Utreras [295], Utreras [329], [330], Wahba [339], Wahba and Wold [341], Williams and Burrage [350]), numerical weather forecasting (Wahba, Johnson, Gao and Gong [342]), image processing (Berman [29], Shahraray and Anderson [297], Thompson, Brown, Kay and Titterington [326]), statistics (Li [225]) and so on. There are of course many other examples of the use of GCV.

GCV for large-scale linear ill-posed problems is considered in the paper [159] by Golub and von Matt. This paper uses the techniques of estimation of quadratic forms to compute the parameter of the model. The regularized problem is written as

$$\min\{\|c - Ax\|^2 + m\mu\|x\|^2\},$$

where $\mu \geq 0$ is the regularization parameter and the matrix A is $m \times n$. The GCV estimate of the parameter μ is the minimizer of

$$G(\mu) = \frac{\frac{1}{m}\|(I - A(A^T A + m\mu I)^{-1}A^T)c\|^2}{(\frac{1}{m}\text{tr}(I - A(A^T A + m\mu I)^{-1}A^T))^2}. \tag{15.8}$$

The numerator is, up to a scaling factor, the square of the norm of the residual corresponding to the solution of the normal equations of the regularized problem. A weighted GCV method has been proposed by Chung, Nagy and O'Leary [66] in which the denominator in equation (15.8) is replaced by

$$\left(\frac{1}{m}\operatorname{tr}(I - \omega A(A^T A + m\mu I)^{-1} A^T)\right)^2,$$

where ω is a real parameter. In some cases and if ω is chosen properly, this gives better results than GCV.

We are concerned with the numerical evaluation and minimization of $G(\mu)$ in equation (15.8). The evaluation of the trace term in the denominator is based on Hutchinson's result (see [195] and proposition 11.9). In the following we will assume that A has full rank r, which is the minimum of m and n. Let $\nu = m\mu$; the function t we are interested in is

$$\operatorname{tr}(I - A(A^T A + \nu I)^{-1} A^T).$$

Let $A = U\Sigma V^T$ be the singular value decomposition of A where U and V are orthonormal. We can write

$$A(A^T A + \nu I)^{-1} A^T = U\Sigma(\Sigma^T\Sigma + \nu I)^{-1}\Sigma U^T.$$

Therefore,

$$-\operatorname{tr}(A(A^T A + \nu I)^{-1} A^T) = -\|\Sigma(\Sigma^T\Sigma + \nu I)^{-1/2}\|_F^2$$

$$= -\sum_{i=1}^{r}\frac{\sigma_i^2}{\sigma_i^2 + \nu} = -r + \sum_{i=1}^{r}\frac{\nu}{\sigma_i^2 + \nu}.$$

Hence, t can be written as

$$t(\mu) = m - n + \nu\operatorname{tr}(A^T A + \nu I)^{-1}$$

if $m \geq n$ and

$$t(\mu) = \nu\operatorname{tr}(AA^T + \nu I)^{-1}$$

if $m < n$. Let $\tilde{t}(\mu)$ be the stochastic estimator of t, $v(\mu)$ be the variance and $K = (AA^T + \nu I)^{-1}$. Then,

$$K = V(\Sigma^T\Sigma + \nu I)^{-1}V^T,$$

if $m \geq n$ and

$$K = U(\Sigma\Sigma^T + \nu I)^{-1}U^T$$

if $m < n$. The variance $v(\mu)$ is bounded by

$$v(\mu) \leq 2\nu^2\|A\|_F^2 = 2\sum_{i=1}^{r}\left(\frac{\nu}{\sigma_i^2 + \nu}\right)^2,$$

where the σ_i's are the singular values of A. The matrix A being of full rank,

$$v(\mu) \leq \frac{2r}{\sigma_r^4}\nu^2.$$

Therefore $v(\mu)$ goes to zero as μ goes to zero. One can also show that $v(\mu)$ goes to zero if $\mu \to \infty$. If $m > n$ the relative standard error satisfies

$$\frac{\sqrt{v(\mu)}}{t(\mu)} \leq \frac{\sqrt{2r}}{\sigma_r^2} \frac{\nu}{m-n},$$

and it goes to zero as $\mu \to 0$. In all cases the relative standard error is bounded. We now concentrate on the case $m \geq n$. Using our machinery for quadratic forms we can obtain an estimate of $s_z(\nu) = z^T (A^T A + \nu I)^{-1} z$, where z is a random vector as in proposition 11.9, using Lanczos bidiagonalization. In fact using the Gauss and Gauss–Radau rules we have lower and upper bounds

$$g_z(\nu) \leq s_z(\nu) \leq r_z(\nu).$$

We can also estimate $s_c^{(p)}(\nu) = c^T A (A^T A + \nu I)^{-p} A^T c$, $p = 1, 2$. Using the Gauss and Gauss–Radau rules we obtain bounds satisfying

$$g_c^{(p)}(\nu) \leq s_c^{(p)}(\nu) \leq r_c^{(p)}(\nu).$$

We want to compute approximations of

$$\tilde{G}(\mu) = m \frac{c^T c - s_c^{(-1)}(\nu) - \nu s_c^{(-2)}(\nu)}{(m - n + \nu s_z(\nu))^2}.$$

We remark that the numerator of $G(\mu)$ could have been simplified since

$$\|(I - A(A^T A + m\mu I)^{-1} A^T) c\|^2 = m^2 \mu^2 \|(AA^T + m\mu I)^{-1} c\|^2.$$

Nevertheless, the paper [159] defined

$$L_0(\nu) = m \frac{c^T c - r_c^{(-1)}(\nu) - \nu r_c^{(-2)}(\nu)}{(m - n + \nu r_z(\nu))^2},$$

$$U_0(\nu) = m \frac{c^T c - g_c^{(-1)}(\nu) - \nu g_c^{(-2)}(\nu)}{(m - n + \nu g_z(\nu))^2}.$$

These quantities, L_0 and U_0, are lower and upper bounds for the estimate of $G(\mu)$. The evaluation of the bounds was considered at length in [159]. We write

$$U_0(\mu) = m \frac{p(\nu)}{q^2(\nu)},$$

whose derivative is

$$U_0'(\nu) = m \left(\frac{p'(\nu)}{q^2(\nu)} - 2 \frac{p(\nu) q'(\nu)}{q^3(\nu)} \right),$$

where

$$p(\nu) = c^T c - g_c^{(-1)}(\nu) - \nu g_c^{(-2)}(\nu),$$

$$q(\nu) = m - n + \nu g_z(\nu).$$

The function $p(\nu)$ can be written as

$$p(\nu) = \|c\|^2 \|e^1 - B_k (B_k^T B_k + \nu I)^{-1} B_k^T e^1\|^2,$$

where B_k is the upper bidiagonal matrix obtained from the Lanczos bidiagonalization algorithm (defined in chapter 4) on A. The derivative is given by

$$p'(\nu) = 2m^2 \mu \|c\|^2 (e^1)^T B_k (B_k^T B_k + \nu I)^{-3} B_k^T e^1.$$

Defining

$$\xi = (B_k^T B_k + \nu I)^{-1} B_k^T e^1,$$

$$\eta = \sqrt{\nu} (B_k^T B_k + \nu I)^{-1} \xi,$$

we have

$$p(\nu) = \|c\|^2 \|e^1 - B_k \xi\|^2,$$

$$p'(\nu) = 2m\sqrt{\nu} \|c\|^2 \xi^T \eta.$$

The vectors ξ and η can be obtained as solutions of two (small) least squares problems

$$\begin{pmatrix} B_k \\ \sqrt{\nu} I \end{pmatrix} \xi \approx \begin{pmatrix} e^1 \\ 0 \end{pmatrix},$$

$$\begin{pmatrix} B_k \\ \sqrt{\nu} I \end{pmatrix} \eta \approx \begin{pmatrix} 0 \\ \xi \end{pmatrix}.$$

The denominator $q(\nu)$ is written as

$$q(\nu) = m - n + \nu \|z\|^2 (e^1)^T (B_k^T B_k + \nu I)^{-1} e^1.$$

Let us define

$$\xi = \sqrt{\nu} (B_k^T B_k + \nu I)^{-1} e^1,$$

which can be computed as the solution of the least squares problem

$$\begin{pmatrix} B_k \\ \sqrt{\nu} I \end{pmatrix} \xi \approx \begin{pmatrix} 0 \\ e^1 \end{pmatrix}.$$

Then

$$q(\nu) = m - n + \sqrt{\nu} \|z\|^2 \xi_1.$$

The derivative of q is

$$q'(\nu) = m\|z\|^2 (e^1)^T (B_k^T B_k + \nu I)^{-1} e^1 - m\nu \|z\|^2 (e^1)^T (B_k^T B_k + \nu I)^{-2} e^1.$$

Since $(B_k^T B_k + \nu I)^{-1} B_k = B_k (B_k B_k^T + \nu I)^{-1}$, it can also be written as

$$q'(\nu) = m\|z\|^2 (e^1)^T B_k^T (B_k B_k^T + \nu I)^{-2} B_k e^1.$$

Then, if $\eta = (B_k B_k^T + \nu I)^{-1} B_k e^1$, which can be obtained by solving

$$\begin{pmatrix} B_k^T \\ \sqrt{\nu} I \end{pmatrix} \eta \approx \begin{pmatrix} e^1 \\ 0 \end{pmatrix},$$

the derivative is given by

$$q'(\nu) = m\|z\|^2 \|\eta\|^2.$$

Similarly, the lower bound $L_0(\nu)$ and its derivative can be written as

$$L_0(\nu) = m \frac{r(\nu)}{s^2(\nu)},$$

$$L_0'(\nu) = m \left(\frac{r'(\nu)}{s^2(\nu)} - 2 \frac{r(\nu) s'(\nu)}{s^3(\nu)} \right),$$

where

$$r(\nu) = c^T c - r_c^{-1}(\nu) - \nu r_c^{-2}(\nu),$$

$$s(\nu) = m - n + \nu r_z^{-1}(\nu).$$

The numerator $r(\nu)$ can be written

$$r(\nu) = \|c\|^2 - \|A^T c\|^2 (e^1)^T (U_k^T U_k + \nu I)^{-1} e^1 - \nu \|A^T c\|^2 (e^1)^T (U_k^T U_k + \nu I)^{-2} e^1,$$

where U_k is defined by applying Givens rotations (whose product is W^T) to transform the lower bidiagonal matrix B_k into a $(k+1) \times k$ upper bidiagonal matrix,

$$W^T B_k = \tilde{U}_k,$$

and then the (k, k) element of $\tilde{U}_k$ is set to zero to obtain U_k for the Gauss–Radau rule. The derivative is

$$r'(\nu) = 2m\nu \|A^T b\|^2 (e^1)^T (U_k^T U_k + \nu I)^{-3} e^1.$$

Now we consider the QR factorization

$$\begin{pmatrix} U_k \\ \sqrt{\nu} I \end{pmatrix} = QR,$$

where R is a $k \times k$ upper bidiagonal matrix. Let $\xi = R^{-T} e^1$ and

$$\eta = \sqrt{\nu} (U_k^T U_k + \nu I)^{-1} e^1,$$

which can be computed by solving the least squares problem

$$\begin{pmatrix} U_k \\ \sqrt{\nu} I \end{pmatrix} \eta \approx \begin{pmatrix} 0 \\ e^1 \end{pmatrix}.$$

The function r can be written as

$$r(\nu) = \|c\|^2 - \|A^T c\|^2 (\|\xi\|^2 + \|\eta\|^2).$$

Defining $\zeta = R^{-T} \eta$, we have

$$r'(\nu) = 2m \|A^T c\|^2 \|\zeta\|^2.$$

The denominator $s(\nu)$ can be computed in the same way as $q(\nu)$.

The bounds $L_0(\nu)$ and $U_0(\nu)$ allow computation of an approximation of the global minimizer ν_* of the estimate of $G(\nu)$. Of course the tightness of these bounds depends on the number of the Lanczos (bidiagonalization) iterations. Moreover, $L_0(\nu)$ tends to $-\infty$ as $\nu \to 0$, but U_0 remains finite. Therefore, U_0 is a better approximation than L_0 for small ν's. The algorithm proposed in [159] to compute ν_* is to first do

$$k_{\min} = \lceil 3 \log \min(m, n) \rceil$$

Lanczos iterations. Then the global minimizer $\hat{\nu}$ of $U_0(\nu)$ is computed. If we can find a ν such that $0 < \nu < \hat{\nu}$ and $L_0(\nu) > L_0(\hat{\nu})$, the algorithm stops and returns $\hat{\nu}$ as the approximation to the parameter. Otherwise, the algorithm executes one more Lanczos iteration and repeats the convergence test. The range of ν is restricted to

$$\nu < \nu_{\max} = \frac{\|A\|_F^2}{\varepsilon},$$

where ε is the roundoff unit. It was noted in [159] that this algorithm does not guarantee that the global minimizer is found. After some numerical experiments, we will study how to improve the Golub and von Matt algorithm.

15.2 Iterative Methods for Ill-Posed Problems

As we said before, when solving large ill-posed problems it may not be practical to compute the SVD of the matrix A. Very early in the 1950s people turned to iterative methods since most of them only require the ability to compute the product of the matrix with a vector. Iterative methods can be used in two ways: either to solve the regularized system (for a given or several values of μ) or as regularizing procedures by themselves. When they are used as a regularization procedure, the iteration index takes the role of the regularization parameter.

The use of iterative methods can also be seen as a projection of the original problem on a subspace where the problem is less ill-posed or as a model reduction process. Of course, regularization and iterative methods can be combined since one can use an iterative method as a projection technique and then solve the reduced problem by regularization; this is considered, for instance, by O'Leary and Simmons [251] and Hanke [171].

Most popular methods are based on the Lanczos algorithm applied explicitly or implicitly to the normal equations. One important remark about the Lanczos algorithm is that it is shift invariant: if J_k is the Jacobi matrix produced at iteration k for a matrix B, then $J_k + \mu I$ is the Jacobi matrix produced for the matrix $B + \mu I$.

Suppose we want to solve a linear system $By = c$ for a symmetric matrix B; then the Lanczos algorithm we have studied in chapter 4 generates basis vectors v^i, $i = 1, \ldots, k$ which are the columns of an orthogonal matrix V_k and a Jacobi matrix J_k. For solving a linear system, the iterates are sought as

$$y^k = y^0 + V_k z^k,$$

and the residual $r^k = c - By^k = r^0 - BV_k z^k$ is asked to be orthogonal to the subspace spanned by the columns of V_k which gives that the coordinates z^k are solution of a tridiagonal linear system of order k,

$$J_k z^k = \|r^0\| e^1.$$

Therefore, if we want to solve a Tikhonov regularized system $(A^T A + \mu I)x_\mu = A^T c$, the iterates $x_\mu^k = x^0 + V_k z_\mu^k$ are determined by $(J_k + \mu I)z_\mu^k = \|r^0\| e^1$ if we run the Lanczos algorithm on $A^T A$.

However, we have to be careful for several reasons that we have already mentioned in chapter 4. First, it is well known (see, for instance, [239] or the review paper [242]) that the Lanczos algorithm in finite precision arithmetic does not fulfill its theoretical properties. In particular, after a first eigenvalue of J_k has converged to an eigenvalue of $A^T A$, the Lanczos vectors v^i are no longer mutually orthogonal. Moreover, there is a delay in the convergence of the other eigenvalues. Second, if the matrix has small (and close) eigenvalues, then it is difficult to obtain convergence towards these eigenvalues.

In fact, this last issue can be used to our advantage since, particularly for small values of μ, the noisy components of the right-hand side can be amplified by the smallest eigenvalues. If the smallest eigenvalues are not well approximated in the first iterations, then the approximate solution x^k is not polluted by the components over the corresponding eigenvectors. This is why the Lanczos algorithm can give a "regularized" solution. However, we have to stop the iterations before there is convergence toward the smallest eigenvalues.

Instead of using the Lanczos algorithm on $A^T A$, one may use instead the Golub–Kahan Lanczos bidiagonalization algorithm (see chapter 4) which is well suited for least squares problems. It produces an orthogonal matrix Q_k which is the equivalent of V_k and an upper bidiagonal matrix B_k, which is the Cholesky factor of $J_k = B_k^T B_k$. However, the matrix B_k cannot be used directly for computing the approximation of the solution of the regularized problem since B_k is not shift invariant. The Cholesky factor of $J_k + \mu I$ is not the shift of B_k. To solve this problem, one can use the QD algorithm with shifts; see chapter 3.

Usually, when the matrix is symmetric and positive definite (which is the case for $A^T A$ if A has full rank), one prefers to use the conjugate gradient (CG) method instead of the Lanczos algorithm. The reason is that in the Lanczos algorithm we have to store the matrix V_k of the Lanczos vectors to compute the solution, whereas CG uses only short recurrences. Unfortunately, CG is not shift invariant since if was obtained from the Lanczos algorithm by an LU factorization of the Lanczos matrix J_k. This can, somehow, be circumvented as we will see later. When applied to the normal equations $A^T A x = A^T c$, CG is usually denoted as CGNR. In this case, the CG method minimizes the $A^T A$-norm of the error $x - x^k$ but we have

$$\|x - x^k\|_{A^T A}^2 = (A(x - x^k), A(x - x^k)) = \|c - Ax^k\|^2.$$

Hence, CGNR minimizes the l_2 norm of the residual $c - Ax^k$.

Of course, one does not compute the product $A^T A$. The multiplication by a vector is done by successive multiplications with A and A^T. Nevertheless, for ill-posed problems, $A^T A$ is ill conditioned and it is usually preferred to use a variant denoted as CGLS [30], which is the following for $A^T A + \mu I$:
$z^0 = c - Ax^0$, $r^0 = A^T z^0 - \mu x^0$, $p^0 = r^0$, and for $k = 0, 1, \ldots$

$$q^k = Ap^k,$$

$$\gamma_k = \frac{\|r^k\|^2}{\|q^k\|^2 + \mu \|p^k\|^2},$$

$$x^{k+1} = x^k + \gamma_k p^k,$$

$$z^{k+1} = z^k - \gamma_k q^k,$$

$$r^{k+1} = A^T z^{k+1} - \mu x^{k+1},$$

$$\omega_k = \frac{\|r^{k+1}\|^2}{\|r^k\|^2},$$

$$p^{k+1} = r^{k+1} + \omega_k p^k.$$

Note that this algorithm does the recursive update on z^k which is the residual $c - Ax^k$ (the quantity minimized by the algorithm) and not on the residuals of the normal equations.

As we said before, a difficult problem is to know when to stop the Lanczos or CG iterations. For instance, if the norm of the noise e is approximately known, one can choose to stop when the discrepancy principle is satisfied

$$\|c - Ax^k_\mu\| \approx \tau \|e\|,$$

where e is the known noise vector.

When solving a least squares problem $\min \|c - Ax\|$, it is usually preferred to use the LSQR algorithm of Paige and Saunders [256] since this method uses orthogonal transformations to solve the projected problems. The algorithm uses the Lanczos bidiagonalization (denoted as "Lanczos bidiagonalization II" in [159] and "Bidiag 1" in [256]) we have described in chapter 4. The relations describing the LSQR algorithm are the following as defined in [256].

Let $\beta_1 = \|c\|$, $u^1 = c/\beta_1$, $\alpha_1 = \|A^T u^1\|$, $v^1 = A^T u^1/\alpha_1$, $w^1 = v^1$, $x^0 = 0$, $\bar{\phi}_1 = \beta_1$, $\bar{\rho}_1 = \alpha_1$, then for $k = 1, \ldots$

$$\bar{u} = Av^k - \alpha_k u^k, \quad \beta_{k+1} = \|\bar{u}\|, \quad u^{k+1} = \bar{u}/\beta_{k+1},$$

$$\bar{v} = A^T u^{k+1} - \beta_{k+1} v^k, \quad \alpha_{k+1} = \|\bar{v}\|, \quad v^{k+1} = \bar{v}/\alpha_{k+1},$$

$$\rho_k = \sqrt{\bar{\rho}_k^2 + \beta_{k+1}^2},$$

$$c_k = \bar{\rho}_k/\rho_k, \quad s_k = \beta_{k+1}/\rho_k,$$

$$\theta_{k+1} = s_k \alpha_{k+1}, \quad \bar{\rho}_{k+1} = -c_k \alpha_{k+1},$$

$$\phi_k = c_k \bar{\phi}_k, \quad \bar{\phi}_{k+1} = s_k \bar{\phi}_k,$$

$$x^k = x^{k-1} + (\phi_k/\rho_k)w^k,$$

$$w^{k+1} = v^{k+1} - (\theta_{k+1}/\rho_k)w^k.$$

If we want to solve a regularized problem for a given value of μ, we apply this algorithm to the least squares problem

$$\min \left\| \begin{pmatrix} A \\ \mu I \end{pmatrix} x - \begin{pmatrix} c \\ 0 \end{pmatrix} \right\|.$$

Otherwise, when we want to solve the regularized system for several values of μ, we can use the correspondance between the coefficients of the Lanczos and CG algorithms (see chapter 4) and the fact that the Lanczos algorithm is shift invariant. If the CG vectors are obtained as

$$p^k = r^k + \beta_k p^{k-1}, \quad x^{k+1} = x^k + \gamma_k p^k.$$

Then, the coefficients α_i and η_i of the Lanczos algorithm which are the nonzero entries of J_k are given by

$$\alpha_k = \frac{1}{\gamma_{k-1}} + \frac{\beta_{k-1}}{\gamma_{k-2}}, \quad \beta_0 = 0, \quad \gamma_{-1} = 1, \quad \eta_k = \frac{\sqrt{\beta_k}}{\gamma_{k-1}}.$$

Moreover, γ_k is related to the diagonal elements of the Cholesky (LDL^T) factorization of the Jacobi matrices J_k,

$$\gamma_k = \frac{1}{\delta_{k+1}}.$$

Then, when running CG (or CGLS) with the matrix $A^T A$ we can compute J_k from the CG coefficients. Then, we can shift J_k by μI for a given μ and compute the LDL^T factorization to obtain the CG coefficients for the shifted system.

These ways of proceeding have the advantage of requiring only two matrix vector products (with A and A^T) per iteration for computing the solutions for as many values of μ as we wish. This kind of algorithms was advocated in a paper by Frommer and Maas [116].

Other aspects of iterative methods for ill-posed problems are developed in Kilmer and Stewart [204] and Kilmer and O'Leary [203].

15.3 Test Problems

We first use several examples from the Regutools toolbox (version 3.1) by Hansen [177]. A newer version of this toolbox is now available [179].

Example IP1 = Baart
   This problem arises from the discretization of a first-kind Fredholm integral equation with kernel K and right-hand side g given by

$$K(s, t) = \exp(s \cos(t)), \quad g(s) = 2 \sinh(s)/s,$$

and with integration intervals $s \in [0, \pi/2], t \in [0, \pi]$; see Baart [19]. The solution is given by $f(t) = \sin(t)$. The square matrix A of order 100 is dense and its smallest and largest singular values are $1.7170 \ 10^{-18}$ and 3.2286. The singular values are displayed in figure 15.1. We see that there are many small singular values (smaller than 10^{-12}).

Example IP2 = ILaplace
   This problem comes from the discretization of the inverse Laplace transform by means of a Gauss–Laguerre quadrature rule. The kernel K is given by

$$K(s, t) = \exp(-st),$$

and the integration interval is $[0, \infty[$. The solution f and the right-hand side g are

$$f(t) = \exp(-t/2), \quad g(s) = 1/(s + 0.5).$$

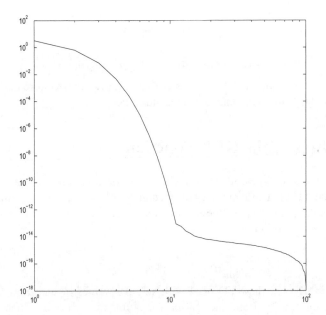

Figure 15.1 Singular values for the Baart problem, $m = n = 100$

The square matrix A of order 100 is partly dense and its smallest and largest singular values are $8.3948\ 10^{-33}$ and 2.3749. The distribution of the singular values is even worse than for Example IP1.

Example IP3 = Phillips

This problem arises from the discretization of a first-kind Fredholm integral equation devised by D. L. Phillips. Define the function

$$\phi(x) = 1 + \cos(x\pi/3) \text{ for } |x| < 3,\ 0 \text{ for } |x| >= 3.$$

The kernel K, the solution f and the right-hand side g are given by

$$K(s, t) = \phi(s - t), \quad f(t) = \phi(t),$$

$$g(s) = (6 - |s|)(1 + 0.5\cos(s\pi/3)) + 9/(2\pi)\sin(|s|\pi/3).$$

The integration interval is $[-6, 6]$. The square matrix A of order 200 is banded and its smallest and largest singular values are $1.3725\ 10^{-7}$ and 5.8029.

All the preceding examples are small and can be solved using the SVD of A.

Example IP4 = von Matt's problem

To obtain a large matrix A with similar properties as the matrices from ill-posed problems without having to store a dense matrix, Urs von Matt [158], [159] proposed the following example that we have already used in chapter 14. The matrix A is defined as $A = U\Sigma V^T$ where

$$U = I - 2\frac{uu^T}{\|u\|^2}, \quad U = I - 2\frac{vv^T}{\|v\|^2},$$

u and v being given vectors of length m and n. The diagonal matrix Σ (containing the singular values) is chosen such that we have a linear distribution of the singular values or as

$$\sigma_i = \exp(-ci), \ i = 1, \ldots, n \leq m$$

where c is suitably chosen. This choice of matrix allows one to store only u, v and the diagonal of Σ to do the matrix-vector multiplications.

15.4 Study of the GCV Function

GCV for large scale linear ill-posed problems is considered in the paper by Golub and von Matt [158]. This paper uses the techniques of estimation of quadratic forms to compute the parameters of the model. Remember that the GCV estimate of the parameter μ is the minimizer of

$$G(\mu) = \frac{\frac{1}{m}\|(I - A(A^T A + m\mu I)^{-1}A^T)c\|^2}{(\frac{1}{m}\mathrm{tr}(I - A(A^T A + m\mu I)^{-1}A^T))^2}. \tag{15.9}$$

Let $A = U\Sigma V^T$ be the SVD of A with singular values σ_i and $d = U^T c$. Let us assume that $m \geq n$ and the matrix A has full rank $r = n$. Then the GCV function (15.9) can be written as

$$G(\nu) = \frac{m\left\{\sum_{i=1}^r d_i^2 \left(\frac{\nu}{\sigma_i^2 + \nu}\right)^2 + \sum_{i=r+1}^m d_i^2\right\}}{[m - n + \sum_{i=1}^r \frac{\nu}{\sigma_i^2 + \nu}]^2}, \tag{15.10}$$

where $\nu = m\mu$. An example of GCV function (as a function of μ) is displayed in figure 15.2 for the IP1 (Baart) example with $m = n = 100$. The right-hand side c is generated from the exact solution $\tilde{x}$ by $c = A\tilde{x} + e$ where e is a random vector with a normal distribution of norm noise $= 10^{-3}$. The cross on the curve shows the minimum of the function. The GCV functions for different noise levels are displayed in figure 15.3. We observe that these functions are rather flat near the minimum and this can be a problem.

From the figures, we see that G is almost constant when ν is very small or large, at least in log-log scale. When $\nu \to \infty$, the ratios involving ν in the numerator and denominator tend to 1. The numerator tends to

$$m \sum_{i=1}^m d_i^2$$

and the denominator tends to m^2. Therefore $G(\nu) \to \|c\|^2/m$ since

$$\sum_{i=1}^m d_i^2 = \|U^T c\|^2 = \|c\|^2.$$

When $\nu \to 0$ the situation is different wether $m = n$ or not. The numerator of G tends to

$$m \sum_{i=r+1}^m d_i^2$$

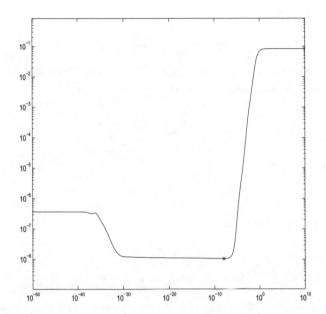

Figure 15.2 GCV function for the IP1 (Baart) problem, $m = n = 100$, noise= 10^{-3}

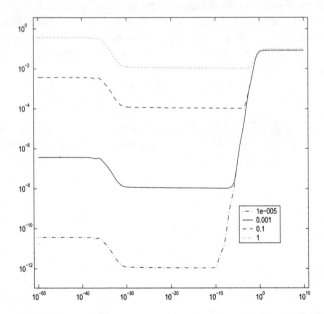

Figure 15.3 GCV functions for the IP1 (Baart) problem, $m = n = 100$ for different noise
          levels

if $m \neq n$ or 0 when $m = n$. When $m \neq n$ the denominator tends to m and to 0 otherwise. Therefore, the limit value of G is

$$\frac{m \sum_{i=r+1}^{m} d_i^2}{m - n}$$

for $m \neq n$. When $m = n$ we obtain $0/0$ but we can simplify the GCV function by ν^2 in the numerator and denominator. The function is written as

$$G(\nu) = \frac{m \sum_{i=1}^{r} d_i^2 \left(\frac{1}{\sigma_i^2 + \nu}\right)^2}{[\sum_{i=1}^{r} \frac{1}{\sigma_i^2 + \nu}]^2}. \tag{15.11}$$

This can be written as

$$G(\nu) = \frac{m \|(AA^T + \nu I)^{-1} c\|}{[\mathrm{tr}((AA^T + \nu I)^{-1})]^2}. \tag{15.12}$$

Therefore, the limit value when $\nu \to 0$ is

$$\frac{m \sum_{i=1}^{r} d_i^2 \left(\frac{1}{\sigma_i^2}\right)^2}{\left(\sum_{i=1}^{r} \frac{1}{\sigma_i^2}\right)^2}. \tag{15.13}$$

When the smallest singular value σ_n is much smaller than most of the other ones (as in many ill-posed problems), this limit value is $m d_n^2$. Moreover, it turns out that the GCV function is almost constant for $\nu < \sigma_n^2$ and for $\nu > \sigma_1^2$. This is illustrated in figure 15.4, where the horizontal dashed lines give the limit values and the vertical lines show the squares of the smallest and largest singular values. Therefore when looking for the minimum value of G (or its approximations) it is enough to consider the interval $[\sigma_n^2, \sigma_1^2]$. We note that the largest singular value can be estimated with the bidiagonalization algorithm. In von Matt's software, the interval which is considered for ν is

$$\frac{(\varepsilon \|A\|)^2}{m} \leq \nu \leq \frac{1}{m} \left(\frac{\|A\|}{\varepsilon}\right)^2,$$

where ε is the machine epsilon. This is shown in figure 15.5, where these values are the vertical dot-dashed lines. We see that the upper bound is greatly overestimated. Therefore, we propose to use the interval

$$[\max(\sigma_n^2, \varepsilon \|A\|)^2, \sigma_1^2]$$

for ν if σ_n is known or

$$[(\varepsilon \|A\|)^2, \sigma_1^2]$$

otherwise.

The GCV functions for the other examples are shown in figures 15.6 and 15.7. The GCV function G is approximated by $\tilde{G}$ where the trace in the denominator is computed using Hutchinson's estimator. The function $\tilde{G}$ computed using the singular values of A is plotted with dotted lines for the IP1 example in figure 15.8. The circle gives the minimum of $\tilde{G}$. We see that G and $\tilde{G}$ do not have the same limit value when $\nu \to 0$. In most cases $\tilde{G}$ is a poor approximation of G for small

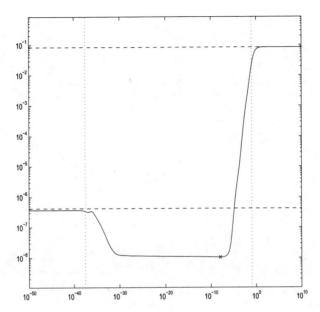

Figure 15.4 GCV function for the IP1 (Baart) problem, $m = n = 100$, noise$= 10^{-3}$

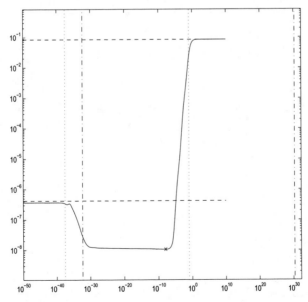

Figure 15.5 GCV function for the IP1 (Baart) problem, $m = n = 100$, noise$= 10^{-3}$

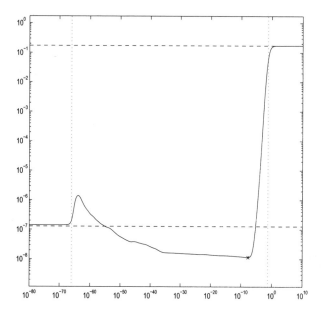

Figure 15.6 GCV function for the IP2 (inverse Laplace equation) problem, $m = n = 100$, noise$= 10^{-3}$

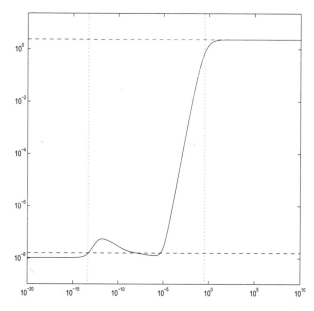

Figure 15.7 GCV function for the IP3 (Phillips) problem, $m = n = 100$, noise$= 10^{-3}$

values of ν. This comes, of course, from the computation of the denominator since the numerator is computed in the same way for both functions. However, near the location of the minimum $\tilde{G}$ is quite accurate for this problem. Unfortunately, this is not always the case as we can see with figure 15.9 for the IP1 example using a large perturbation with noise= 0.1. The function $\tilde{G}$ is then a poor approximation of G and its minimizer is quite different from the one for G. The parameter ν that is found using the minimum of $\tilde{G}$ is not correct.

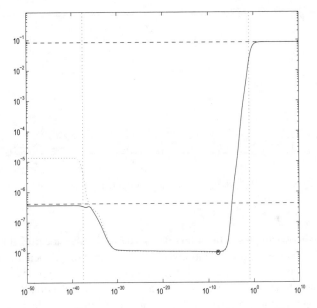

Figure 15.8 G (solid) and $\tilde{G}$ (dotted) functions for the IP1 (Baart) problem, $m = n = 100$, noise= 10^{-3}

15.5   Optimization of Finding the GCV Minimum

The problem we have is to locate the minimum of the approximation $\tilde{G}$ of the GCV function G as cheaply as possible. Von Matt [159] computed the minimum of the upper bound obtained by Lanczos bidiagonalization by sampling this function on 100 points with an exponential distribution which gives a regular distribution using a log scale. After locating the minimum of these samples, if the neighbors of the point giving the minimum do not have the same values, he looked at the derivative and sought for a local minimum in either the left or right interval depending on the sign of the derivative. The local minimum is found by using bisection.

To try to find a minimum in a more efficient way, we propose first to work with the logarithms of ν and $\tilde{G}$ instead of the function itself. Then we compute 50 samples on a regular mesh. We locate the minimum, say the point k, we then

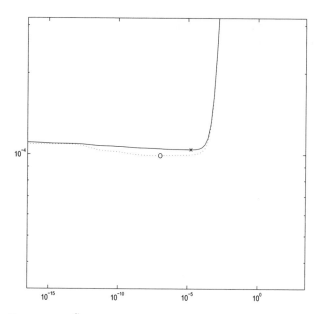

Figure 15.9 G (solid) and $\tilde{G}$ (dotted) functions for the IP1 (Baart) problem, $m = n = 100$, noise$= 10^{-1}$

compute again 50 samples in the interval $[k-1, \ k+1]$. After locating the minimum in this interval we use the von Matt algorithm for computing a local minimum.

To compare both approaches, we use the number of function evaluations as a criteria. Each function evaluation leads to solving several (small) least squares problems. The results are given for some test problems in table 15.1 where vm refers to the von Matt implementation and gm to ours. Note that for $m = n = 100$, 14 is the minimum number of iterations that can be done with this algorithm. In most cases, the values of μ from the two minimization methods are close. Although there are some exceptions, in many cases, the number of function evaluations is about the same.

Looking for improving these results, one has to look more closely at these algorithms. In the Golub and von Matt algorithm the lower bound is used only in the stopping criterion. The algorithm is stopped when the maximum of the lower bound is larger than the minimum of the upper bound. This is done to have the upper bound close enough to the function $\tilde{G}$ around the minimum. The value of the minimizer μ that is returned corresponds to the minimum value of the upper bound.

The difficulty can be understood by looking at figure 15.10 for the Baart problem for a level of noise of 10^{-3} which displays upper bounds at different Lanczos iterations. We see that the upper bound does not have the right asymptotic behavior when ν is small since the upper bound tends to infinity, whence the function is bounded. The convergence of the minimum is rather slow and it is difficult to know when the minimum has converged. The solution chosen by Golub and von Matt is to look also at the lower bound. This problem arises because we are in the

Table 15.1 Minimizer values and number of function evaluations

Example	Noise	No. it.	μ	f min	f max	f total
Baart vm	10^{-7}	17	$9.6482\ 10^{-15}$	494	613	1107
	10^{-5}	14	$9.7587\ 10^{-12}$	125	132	257
	10^{-3}	14	$1.2018\ 10^{-8}$	130	123	253
	10^{-1}	14	$1.0336\ 10^{-7}$	128	126	254
	10	14	$8.8817\ 10^{-8}$	127	119	246
Baart gm	10^{-7}	14	$1.1496\ 10^{-14}$	146	145	291
	10^{-5}	14	$1.1470\ 10^{-11}$	146	118	264
	10^{-3}	14	$1.3702\ 10^{-8}$	147	115	262
	10^{-1}	14	$1.1208\ 10^{-7}$	148	114	262
	10	14	$9.9400\ 10^{-8}$	147	114	261
ILaplace vm	10^{-7}	112	$2.1520\ 10^{-15}$	12438	10216	22654
	10^{-5}	47	$5.2329\ 10^{-12}$	4242	3428	7670
	10^{-3}	18	$2.2111\ 10^{-8}$	620	541	1161
	10^{-1}	14	$1.9484\ 10^{-5}$	120	125	245
	10	14	$6.5983\ 10^{-3}$	124	126	250
ILaplace gm	10^{-7}	82	$7.9939\ 10^{-15}$	10101	9788	19889
	10^{-5}	47	$5.9072\ 10^{-12}$	4977	4175	9152
	10^{-3}	17	$2.4905\ 10^{-8}$	589	416	1005
	10^{-1}	14	$2.2036\ 10^{-5}$	148	120	268
	10	14	$7.0253\ 10^{-3}$	149	123	272
Phillips vm	10^{-7}	221	$8.7929\ 10^{-11}$	26299	21616	47915
	10^{-5}	122	$4.5432\ 10^{-9}$	13435	10759	24194
	10^{-3}	32	$4.3674\ 10^{-7}$	2111	1736	3847
	10^{-1}	16	$3.8320\ 10^{-5}$	130	128	258
	10	16	$8.4751\ 10^{-3}$	118	121	239
Phillips gm	10^{-7}	136	$2.5684\ 10^{-10}$	17761	17306	35067
	10^{-5}	116	$5.4911\ 10^{-9}$	14857	12265	27122
	10^{-3}	31	$4.6715\ 10^{-7}$	2364	1640	4004
	10^{-1}	16	$4.2203\ 10^{-5}$	147	119	266
	10	14	$9.3842\ 10^{-3}$	149	116	265

case $m = n$ and from the way the function is computed with the general formula (as a norm of a residual) the numerator is bounded away from zero when the denominator tends to zero with ν^2. This is different when $m \neq n$ as we can see in figure 15.11 for IP4 with $m = 200, n = 100$ because in this case the denominator does not tend to zero when $\nu \to 0$ and the approximation has the right behavior.

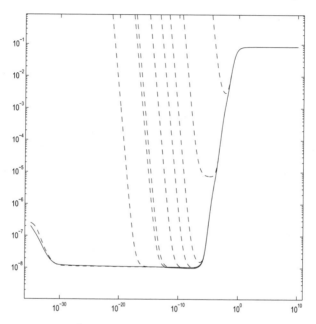

Figure 15.10 G (solid) and $\tilde{G}$ (dashed) functions and upper bounds for the IP1 (Baart) problem, $m = n = 100$, noise$= 10^{-3}$

To try to better understand the issues, let us use the SVD of the bidiagonal matrix B_k (arising from the Lanczos bidiagonalization algorithm) which is a $(k + 1) \times k$ matrix. The numerator of the approximation is

$$p(\nu) = \|c\|^2 \, \|e^1 - B_k(B_k^T B_k + \nu I)^{-1} B_k^T e^1\|^2. \qquad (15.14)$$

Using the SVD of $B_k = U_k S_k V_k^T$ (S_k being a $(k + 1) \times k$ matrix) and denoting the singular values as θ_i (with no reference to k for the sake of simplicity), from equation (15.14) we have

$$p(\nu) = \|c\|^2 \, \|(I - S_k(S_k^T S_k + \nu I)^{-1} S_k^T) U_k^T e^1\|^2.$$

This can be written as

$$p(\nu) = \|c\|^2 \left\{ \sum_{i=1}^{k} \frac{\nu^2 f_i^2}{(\theta_i^2 + \nu)^2} + f_{k+1}^2 \right\}, \qquad (15.15)$$

where $f = U_k^T e^1$. Therefore, we see that when $\nu \to \infty$, we have $p(\nu) \to \|c\|^2$ and when $\nu \to 0$, then $p(\nu) \to \|c\|^2 f_{k+1}^2$. The limit value is not the same as for the numerator of the function $\tilde{G}$ and it is bounded away from zero (at least when k is small enough).

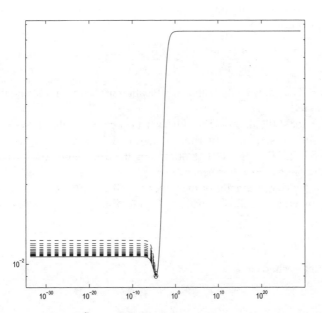

Figure 15.11 G (solid) and $\tilde{G}$ (dashed) functions and upper bounds for the IP4 example, $m = 200, n = 100$, noise= 1

The denominator is

$$q(\nu) = m - n + \nu \|u\|^2 \, (e^1)^T (\bar{B}_k^T \bar{B}_k + \nu I)^{-1} e^1, \qquad (15.16)$$

where $\bar{B}_k$ is a bidiagonal square matrix of order k and u is a random vector. Using the SVD of $\bar{B}_k = Z_k T_k W_k^T$ with singular values t_i, we can write

$$q(\nu) = m - n + \|u\|^2 \sum_{i=1}^{k} \frac{\bar{f}_i^2 \nu}{t_i^2 + \nu}, \qquad (15.17)$$

where $\bar{f} = W_k^T e^1$. When $\nu \to \infty$ we have $q(\nu) \to m - n + k$ and when $\nu \to 0$, then $q(\nu) \to m - n$. Hence, when $m = n$ as in the IP1 example, we have a numerator that is bounded away from zero and a denominator that goes to zero. Thus, the limit value for $\nu \to 0$ is infinity. We see that the problem comes from the term f_{k+1}^2 that we have in the numerator since it prevents us from being able to divide the numerator and the denominator by ν^2.

To improve upon the implementation of von Matt, we propose the following algorithm:

1) Working in log-log scale (that is, $\log_{10}(\nu)$ and $\log_{10}(U_k^0)$), we compute an upper bound of the function (saving the computation of the lower bound) for which we seek a minimizer.

2) We evaluate the numerator of the approximation (that is, equation (15.14)) by computing the SVD of B_k and using formula (15.15). The computation of the SVD is done only once per iteration and there is no need to solve a least squares problem for each value of ν.

3) After locating a minimum ν_k with a value of the upper bound U_k^0 at iteration k, the stopping criterion is

$$\left| \frac{\nu_k - \nu_{k-1}}{\nu_{k-1}} \right| + \left| \frac{U_k^0 - U_{k-1}^0}{U_{k-1}^0} \right| \leq \epsilon,$$

for a given ϵ. That is, we look both at the location of the minimum and the value of the upper bound.

Of course, the difficult problem with such an approach is the choice of a suitable value of ϵ for the stopping criterion in step 3. If ϵ is large we will save some computation time but there is a risk of missing the true minimum. If ϵ is too small, the computation may be too expensive.

To obtain the right behavior of the upper bound close to zero when $m = n$, we modify the function as follows: instead of using $mp(\nu)/q(\nu)^2$ we can use

$$m \frac{p(\nu)}{q(\nu)^2 + \alpha}.$$

We may want to choose

$$\alpha = \bar{\alpha} = \frac{\|c\|^2 f_{k+1}^2}{d_n^2}.$$

Such a choice is not really feasible for large problems since we do not know d_n which can only be obtained from the SVD of A. It turns out that the value of $\bar{\alpha}$ is too large during the first iterations; this value must be limited and we use $\alpha = \min(\bar{\alpha}, 500)$. Of course, such a modification may be problem dependent. For Example IP1 the computation using mp/q^2 was given in figure 15.10. When using $mp/(q^2 + \alpha)$ we obtain figure 15.12, where we can see the truncation given by introducing α in the denominator.

As we said before, when solving real problems, it is likely that d_n is not available. Hence, we use $\alpha = \|c\|^2$ instead. There is not much difference in the results, except that the asymptotic value when $\nu \to 0$ is not the correct one, but the smallest values of ν do not have any influence on the minimizer. However, we will see that we can use them to reduce the number of function evaluations that is needed to find the minimum.

It would be nice if we could know when this costly process of function minimization is really needed. The Golub and von Matt algorithm [158] imposes to perform $kmin$ Lanczos iterations before starting to compute the minimum of the approximation of the GCV function. The value of $kmin$ was selected empirically by Golub and von Matt as $kmin = \lceil 3 \log \min(m, n) \rceil$. Imposing to perform $kmin - 1$ iterations before looking for the minimum may give us a penalty. But, we can take advantage that we now have a better asymptotic behavior of the upper bound when $\nu \to 0$. We choose a (small) value of ν (denoted as ν_0) for which we monitor the convergence of the upper bound at each iteration from the beginning. When it satisfies the condition

$$\left| \frac{U_k^0(\nu_0) - U_{k-1}^0(\nu_0)}{U_{k-1}^0(\nu_0)} \right| \leq \epsilon_0,$$

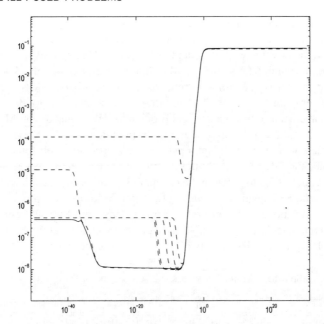

Figure 15.12 G (solid) and $\tilde{G}$ (dashed) functions and upper bounds for the IP1 (Baart) problem, $m = n = 100$, noise$= 10^{-3}$

we start computing the minimum of the upper bound. We denote this algorithm by gm-opt. The results are given in table 15.2 with $\epsilon_0 = 10^{-5}$. It has to be compared with table 15.1. We see that when the convergence is not fast we can save some function evaluations.

Table 15.2 Minimizer values and number of function evaluations, gm-opt, $\epsilon = 10^{-6}$, $kmin = 2$

Example	Noise	No. it.	μ	f min	f max	f total
Baart	10^{-7}	12	$1.0706\ 10^{-14}$	436	0	436
	10^{-5}	12	$1.0581\ 10^{-11}$	437	0	437
	10^{-3}	8	$1.3077\ 10^{-8}$	293	0	293
	10^{-1}	7	$1.1104\ 10^{-7}$	294	0	294
	10	7	$9.1683\ 10^{-8}$	294	0	294
ILaplace	10^{-7}	58	$4.2396\ 10^{-14}$	5239	0	5239
	10^{-5}	28	$5.4552\ 10^{-11}$	1453	0	1453
	10^{-3}	17	$2.3046\ 10^{-8}$	440	0	440
	10^{-1}	15	$2.0896\ 10^{-5}$	293	0	293
	10	10	$6.8436\ 10^{-3}$	296	0	296
Phillips	10^{-7}	157	$1.6343\ 10^{-10}$	17179	0	17179
	10^{-5}	103	$5.3835\ 10^{-9}$	11086	0	11086
	10^{-3}	39	$4.1814\ 10^{-7}$	1759	0	1759
	10^{-1}	17	$4.1875\ 10^{-5}$	438	0	438
	10	13	$8.7084\ 10^{-3}$	294	0	294

When a regularization parameter has been computed, an important issue for comparing the different methods is to assess the quality of the solution we can compute with the parameter value we have found. This can be done by looking at the norms of the solution x and the residual $c - Ax$. In our examples, we also know the solution x_0 of the noise-free problem from which the right-hand side was computed. So we can compute $\|x - x_0\|$. The solution x is computed using the SVD of A (this, of course, cannot be done for real problems). We compare von Matt's original implementation and the gm-opt algorithm in table 15.3. The computing times (using Matlab on a Dell D600 laptop) should be considered with some care since, with Matlab, they depend very much on the implementation. For instance, von Matt used C mex-files to solve the least squares problem for each value of ν and gm-opt uses the SVD, which is a built-in function. Moreover, the problem sizes are quite small. Results on a larger problem (on a Sony Vaio RC102, Pentium D 2.8 Ghz) are given in table 15.4. The algorithm gm-opt seems generally faster than von Matt's implementation. The quality of the solutions is the same.

Table 15.3 Minimizer values and comparison of the solutions

Example	Noise	μ	$\|c - Ax\|$	$\|x - x_0\|$	Time (s)
Baart vm	10^{-7}	$9.6482\ 10^{-15}$	$9.8049\ 10^{-8}$	$5.9424\ 10^{-2}$	0.38
	10^{-5}	$9.7587\ 10^{-12}$	$9.8566\ 10^{-6}$	$6.5951\ 10^{-2}$	0.18
	10^{-3}	$1.2018\ 10^{-8}$	$9.8573\ 10^{-4}$	$1.5239\ 10^{-1}$	0.16
	10^{-1}	$1.0336\ 10^{-7}$	$9.8730\ 10^{-2}$	1.6614	–
	10	$8.8817\ 10^{-8}$	9.8728	16.722	–
Baart gm-opt	10^{-7}	$1.0706\ 10^{-14}$	$9.8058\ 10^{-8}$	$5.9519\ 10^{-2}$	0.18
	10^{-5}	$1.0581\ 10^{-11}$	$9.8588\ 10^{-6}$	$6.5957\ 10^{-2}$	0.27
	10^{-3}	$1.3077\ 10^{-8}$	$9.8582\ 10^{-4}$	$1.5205\ 10^{-1}$	0.14
	10^{-1}	$1.1104\ 10^{-7}$	$9.8736\ 10^{-2}$	1.6227	–
	10	$9.1683\ 10^{-8}$	9.8730	16.569	–
ILaplace vm	10^{-7}	$2.1520\ 10^{-15}$	$9.5132\ 10^{-8}$	$1.4909\ 10^{-2}$	10.06
	10^{-5}	$5.2329\ 10^{-12}$	$9.6965\ 10^{-6}$	$6.8646\ 10^{-2}$	2.37
	10^{-3}	$2.2111\ 10^{-8}$	$9.7215\ 10^{-4}$	$1.9890\ 10^{-1}$	0.35
	10^{-1}	$1.9484\ 10^{-5}$	$9.8196\ 10^{-2}$	$3.4627\ 10^{-1}$	0.22
	10	$6.5983\ 10^{-3}$	9.9095	$8.8165\ 10^{-1}$	0.12
ILaplace gm-opt	10^{-7}	$4.2396\ 10^{-14}$	$1.1004\ 10^{-7}$	$2.7130\ 10^{-2}$	2.03
	10^{-5}	$5.4552\ 10^{-11}$	$1.0560\ 10^{-5}$	$9.6771\ 10^{-2}$	0.53
	10^{-3}	$2.3046\ 10^{-8}$	$9.7243\ 10^{-4}$	$1.9937\ 10^{-1}$	0.29
	10^{-1}	$2.0896\ 10^{-5}$	$9.8235\ 10^{-2}$	$3.4634\ 10^{-1}$	0.09
	10	$6.8436\ 10^{-3}$	9.9115	$8.8791\ 10^{-1}$	0.14
Phillips vm	10^{-7}	$8.7929\ 10^{-11}$	$9.0162\ 10^{-8}$	$2.2391\ 10^{-4}$	29.50
	10^{-5}	$4.5432\ 10^{-9}$	$9.0825\ 10^{-6}$	$2.2620\ 10^{-3}$	6.09
	10^{-3}	$4.3674\ 10^{-7}$	$9.7826\ 10^{-4}$	$1.0057\ 10^{-2}$	1.14
	10^{-1}	$3.8320\ 10^{-5}$	$9.8962\ 10^{-2}$	$9.3139\ 10^{-2}$	0.16
	10	$8.4751\ 10^{-3}$	10.012	$5.2677\ 10^{-1}$	0.10
Phillips gm-opt	10^{-7}	$1.6343\ 10^{-10}$	$1.1260\ 10^{-7}$	$2.2163\ 10^{-4}$	15.30
	10^{-5}	$5.3835\ 10^{-9}$	$9.1722\ 10^{-6}$	$2.1174\ 10^{-3}$	6.10
	10^{-3}	$4.1814\ 10^{-7}$	$9.7737\ 10^{-4}$	$1.0375\ 10^{-2}$	0.66
	10^{-1}	$4.1875\ 10^{-5}$	$9.9016\ 10^{-2}$	$9.0659\ 10^{-2}$	0.22
	10	$8.7084\ 10^{-3}$	10.015	$5.2683\ 10^{-1}$	0.15

On these examples we see that the norm of the residual is approximately equal

Table 15.4 Minimizer values and comparison of the solutions

Example	Noise	No. it.	μ	$\|c - Ax\|$	$\|x - x_0\|$	Time (s)
Baart n=500, vm	10^{-3}	19	$7.5768\ 10^{-10}$	$9.9810\ 10^{-4}$	$1.4113\ 10^{-1}$	0.26
Baart n=500, gm-opt	10^{-3}	11	$8.1350\ 10^{-10}$	$9.8811\ 10^{-4}$	$1.4132\ 10^{-1}$	0.16

to the level of noise. The larger is the noise, the larger is the norm of the difference with the unperturbed solution.

15.6 Study of the L-Curve

15.6.1 Properties of the L-Curve

The L-curve is the plot of $\|x\|$ versus $\|c - Ax\|$ where x is obtained as a function of ν by solving a regularized problem. In general, one uses the log-log plot of these curves. An example of an L-curve is given in figure 15.13 in log-log scale for the IP1 example. The circles give the values for the sample of $\nu = m\mu$ we used.

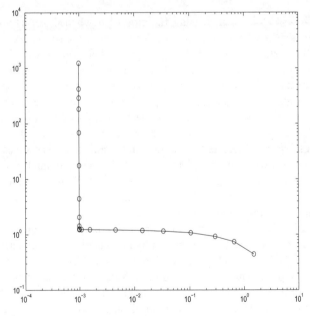

Figure 15.13 The L-curve for the IP1 (Baart) problem, $m = n = 100$, noise $= 10^{-3}$

Of course, the plot of the L-curve depends on the range which has been chosen for the regularization parameter $\nu = m\mu$. Properties of the L-curve have been studied in many papers. However, as we will see, it is sometimes more informative to look separately at $\|x(\nu)\|$ and $\|c - Ax(\nu)\|$ as functions of ν. If $A = U\Sigma V^T$ is the SVD of A, a full rank $m \times n$ matrix with $m \geq n$ and ν is the Tikhonov regularization parameter, we have

$$\|c - Ax\|^2 = \sum_{i=1}^{n} \left(\frac{\nu d_i}{\sigma_i^2 + \nu} \right)^2, \tag{15.18}$$

and

$$\|x\|^2 = \sum_{i=1}^{n} \left(\frac{\sigma_i d_i}{\sigma_i^2 + \nu} \right)^2. \tag{15.19}$$

Computing the derivatives with respect to ν we have

$$[\|c - Ax\|^2]' = 2 \sum_{i=1}^{n} \frac{\nu \sigma_i^2 d_i^2}{(\sigma_i^2 + \nu)^3} \tag{15.20}$$

and

$$[\|x\|^2]' = -2 \sum_{i=1}^{n} \frac{\sigma_i^2 d_i^2}{(\sigma_i^2 + \nu)^3}. \tag{15.21}$$

This implies that there is a simple relation between the derivatives, since

$$[\|x\|^2]' = -\frac{1}{\nu}[\|c - Ax\|^2]'.$$

Moreover, $\|c - Ax\|$ is an increasing function of ν whence $\|x\|$ is a decreasing function of ν. This is what is seen in figure 15.14 where the solid curve is the $\log_{10}$ of $\|x\|$ and the dashed curve is the $\log_{10}$ of $\|c - Ax\|$. For this example there is a large range of values of ν for which both the logarithms of $\|x\|$ and $\|c - Ax\|$ are almost constant. These values correspond to the accumulation of points close to the corner of the L-curve in figure 15.13. We note that it can be more interesting to locate this range of values rather than just the corner of the L-curve. We will come back to this point later. To know where the curves of figure 15.14 are more or less contant, we are interested in the (absolute values of) derivatives relative to $\log_{10} \nu$. This is shown in figure 15.15. Both derivatives are small for ν between 10^{-8} and 10^{-6} which corresponds to the plateaus of the curves of the logarithms of $\|x\|$ and $\|c - Ax\|$. It is likely that one can choose any value of ν in the interval corresponding to the intersection of both plateaus.

Other quantities that can be used to characterize the L-curve are distances and angles. The distances between two consecutive points of the L-curve are displayed in figure 15.16. There is a well-located minimum of the distances. We will see that it corresponds approximately to the minimum of the norm of the error. The angles are not much different whatever the value of ν is. If we look more closely around the corner, as in figure 15.17, we see that all the angles are far from $\pi/2$.

Of course, what is also interesting is the distance of the regularized solution x to the unperturbed solution x_0. Since for the IP1 example we know the unperturbed solution, this distance is displayed in figure 15.18, where the minimum is shown by a circle. Of course, this is not feasible for real problems.

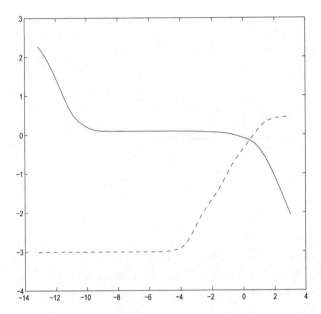

Figure 15.14 $\log_{10}$ of $\|x\|$ (solid) and $\|c - Ax\|$ (dashed) for the IP1 (Baart) problem, $m = n = 100$, noise= 10^{-3}

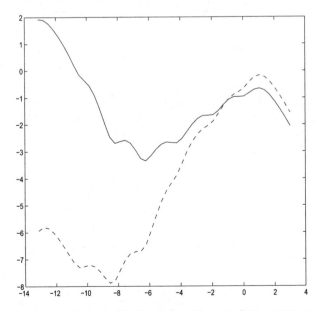

Figure 15.15 $\log_{10}$ of derivatives of $\|x\|$ (solid) and $\|c - Ax\|$ (dashed) for the IP1 (Baart) problem, $m = n = 100$, noise= 10^{-3}

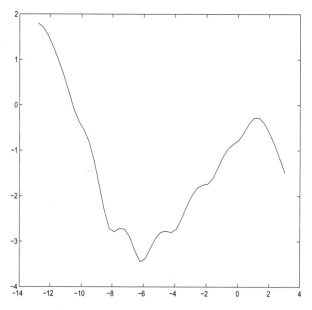

Figure 15.16 $\log_{10}$ of distances for the IP1 (Baart) problem, $m = n = 100$, noise= 10^{-3}

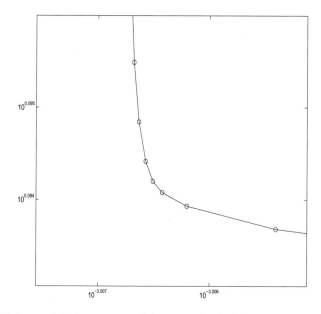

Figure 15.17 Zoom of the L-curve around the corner for the IP1 (Baart) problem, $m = n = 100$, noise= 10^{-3}

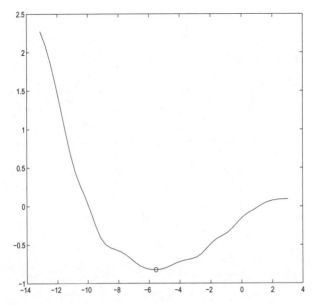

Figure 15.18 $\log_{10}$ of $\|x - x_0\|$ for the IP1 (Baart) problem, $m = n = 100$, noise$= 10^{-3}$

15.6.2 Locating the Corner of the L-Curve

When using the L-curve, the regularization parameter is the value ν corresponding to the "corner" of the L-curve. If it is feasible to compute the SVD of the matrix A, the corner of the L-curve can be determined as the maximum of the convexity of the curve. However, if the matrix is too large for computing the SVD, this method cannot be used. Here, we assume that we can only compute values of points on the L-curve. We will compare our results with algorithms using the SVD.

The idea we want to exploit here is that if we rotate the L-curve by, say, $-\pi/4$, (see figure 15.19, using 25 sampling points) then finding the corner is almost equivalent to finding the minimum of the curve, at least when the L-curve is really L-shaped. Since the curve is composed of a sequence of discrete values and parameterized by ν, we can obtain the value of ν corresponding to the corner by finding the index corresponding to the minimum. Therefore, we propose doing several passes of the following algorithm: we select the index k of the minimum value and apply the same algorithm iteratively computing new (say 25) sampling points on the curve in the interval $[k - 1, k + 1]$. Portions of the L-curves and rotated L-curves for the first two passes are displayed in figures 15.20 and 15.21. If in one of the passes the minimum is located at one of the ends of the interval, then we select the index with minimum angle. The center of rotation is found by the intersection of two linear fits respectively for the smallest and largest values of μ. We denote this algorithm as lc1.

Another possibility is to consider the differences of consecutive points of the

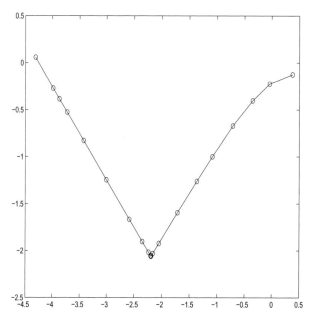

Figure 15.19 The rotated L-curve for the IP1 (Baart) problem, $m = n = 100$, noise= 10^{-3}

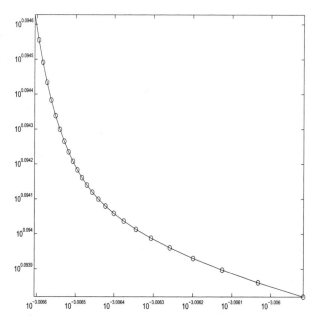

Figure 15.20 The L-curve for the IP1 (Baart) problem, $m = n = 100$, noise= 10^{-3}, second pass

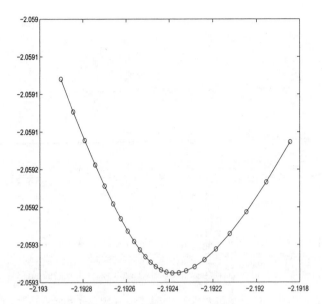

Figure 15.21 The rotated L-curve for the IP1 (Baart) problem, $m = n = 100$, noise$= 10^{-3}$, second pass

L-curve. We compute the differences

$$\Delta_i^r = |\log_{10}(\|c - Ax(\nu_i)\|) - \log_{10}(\|c - Ax(\nu_{i-1})\|)|,$$

$$\Delta_i^x = |\log_{10}(\|x(\nu_i)\|) - \log_{10}(\|x(\nu_{i-1})\|)|.$$

The logarithms of these quantities are shown in figure 15.22. Let ϵ_Δ be a given threshold. The algorithm lc2 returns the ends of intervals for which both Δ_i^x and Δ_i^r are smaller than ϵ_Δ. If there is no such interval, the algorithm returns the minimum of the angles.

15.6.3 Comparison of L-Curve Algorithms

In tables 15.5-15.7 we compare the algorithm (lc) for finding the corner of the L-curve in Regutools (version 3.1) [177], the pruning algorithm (lp) of Hansen, Jensen and Rodriguez [181] with lc1 and lc2 ($\epsilon_\Delta = 10^{-2}$) for which there are two lines because we give the ends of the interval found by the algorithm. The value opt gives the point on the L-curve (discretized with 200 points) with the smallest error. Note that to be able to compare with the GCV results we give the value of $\mu = \nu/m$. The different methods give results of comparable accuracy.

15.6.4 Approximations of the L-Curve

In practical problems we are not always able to compute the SVD of A and therefore we cannot compute points on the real L-curve. However, the techniques for

Table 15.5 L-curve algorithms, IP1 (Baart) problem, $n = 100$

Noise	Method	μ	$\|c - Ax\|$	$\|x - x_0\|$
10^{-7}	opt	$8.9107\ 10^{-17}$	$9.7917\ 10^{-8}$	$4.9820\ 10^{-2}$
	lc	$1.0908\ 10^{-16}$	$9.7919\ 10^{-8}$	$4.9940\ 10^{-2}$
	lp	$1.6360\ 10^{-16}$	$9.7923\ 10^{-8}$	$5.0453\ 10^{-2}$
	lc1	$6.0889\ 10^{-17}$	$9.7912\ 10^{-8}$	$5.0041\ 10^{-2}$
	lc2	$7.3065\ 10^{-17}$	$9.7915\ 10^{-8}$	$4.9842\ 10^{-2}$
		$2.0622\ 10^{-14}$	$9.8182\ 10^{-8}$	$5.9956\ 10^{-2}$
10^{-5}	opt	$7.3023\ 10^{-12}$	$9.8509\ 10^{-6}$	$6.5944\ 10^{-2}$
	lc	$6.5087\ 10^{-13}$	$9.8433\ 10^{-6}$	$6.8269\ 10^{-2}$
	lp	failed		
	lc1	$6.1717\ 10^{-13}$	$9.8433\ 10^{-6}$	$6.8445\ 10^{-2}$
	lc2	$9.2095\ 10^{-15}$	$9.8344\ 10^{-6}$	$3.5326\ 10^{-1}$
		$1.3033\ 10^{-11}$	$9.8666\ 10^{-6}$	$6.5985\ 10^{-2}$
10^{-3}	opt	$2.4990\ 10^{-8}$	$9.8720\ 10^{-4}$	$1.5080\ 10^{-1}$
	lc	$4.5414\ 10^{-9}$	$9.8524\ 10^{-4}$	$1.6030\ 10^{-1}$
	lp	$8.2364\ 10^{-9}$	$9.8545\ 10^{-4}$	$1.5454\ 10^{-1}$
	lc1	$6.3232\ 10^{-9}$	$9.8534\ 10^{-4}$	$1.5669\ 10^{-1}$
	lc2	$5.8203\ 10^{-12}$	$9.8463\ 10^{-4}$	$4.1492\ 10^{-1}$
		$4.1297\ 10^{-8}$	$9.8996\ 10^{-4}$	$1.5153\ 10^{-1}$
10^{-1}	opt	$1.1750\ 10^{-5}$	$9.9264\ 10^{-2}$	$2.9455\ 10^{-1}$
	lc	$4.1328\ 10^{-5}$	$9.9681\ 10^{-2}$	$3.2147\ 10^{-1}$
	lp	failed		
	lc1	$7.2928\ 10^{-5}$	$1.0021\ 10^{-1}$	$3.4666\ 10^{-1}$
	lc2	$2.3246\ 10^{-6}$	$9.9127\ 10^{-2}$	$3.9737\ 10^{-1}$
		$5.8440\ 10^{-5}$	$9.9958\ 10^{-2}$	$3.3650\ 10^{-1}$
10	opt	$5.5250\ 10^{-3}$	9.9772	$6.1260\ 10^{-1}$
	lc	$1.7086\ 10^{-2}$	9.9990	$7.1387\ 10^{-1}$
	lp	$1.8518\ 10^{-1}$	10.257	$9.9211\ 10^{-1}$
	lc1	$2.7442\ 10^{-2}$	10.018	$7.6407\ 10^{-1}$
	lc2	$8.2700\ 10^{-2}$	10.120	$8.8160\ 10^{-1}$
		$8.2700\ 10^{-2}$	10.120	$8.8160\ 10^{-1}$

Table 15.6 L-curve algorithms, IP2 (ILaplace) problem, $n = 100$

Noise	Method	μ	$\|c - Ax\|$	$\|x - x_0\|$
10^{-7}	opt	$3.7606\ 10^{-15}$	$9.5730\ 10^{-8}$	$1.3306\ 10^{-2}$
	lc	$2.1645\ 10^{-17}$	$9.3028\ 10^{-8}$	$1.8717\ 10^{-1}$
	lp	failed		
	lc1	$1.5871\ 10^{-17}$	$9.2893\ 10^{-8}$	$2.2020\ 10^{-1}$
	lc2	$6.0929\ 10^{-17}$	$9.3464\ 10^{-8}$	$1.0878\ 10^{-1}$
		$3.2817\ 10^{-15}$	$9.5557\ 10^{-8}$	$1.3391\ 10^{-2}$
10^{-5}	opt	$7.5376\ 10^{-13}$	$9.6065\ 10^{-6}$	$6.1880\ 10^{-2}$
	lc	$2.5207\ 10^{-13}$	$9.5870\ 10^{-6}$	$8.0606\ 10^{-2}$
	lp	$1.7675\ 10^{-13}$	$9.5797\ 10^{-6}$	$1.0110\ 10^{-1}$
	lc1	$1.6951\ 10^{-13}$	$9.5788\ 10^{-6}$	$1.0407\ 10^{-1}$
	lc2	$1.6166\ 10^{-14}$	$9.5211\ 10^{-6}$	$4.9916\ 10^{-1}$
		$9.5200\ 10^{-12}$	$9.7844\ 10^{-6}$	$7.3388\ 10^{-2}$
10^{-3}	opt	$4.9065\ 10^{-10}$	$9.6720\ 10^{-4}$	$1.4798\ 10^{-1}$
	lc	$1.7737\ 10^{-9}$	$9.6762\ 10^{-4}$	$1.6240\ 10^{-1}$
	lp	failed		
	lc1	$1.7535\ 10^{-9}$	$9.6762\ 10^{-4}$	$1.6221\ 10^{-1}$
	lc2	$4.6897\ 10^{-11}$	$9.6561\ 10^{-4}$	$5.3631\ 10^{-1}$
		$2.7617\ 10^{-8}$	$9.7386\ 10^{-4}$	$2.0140\ 10^{-1}$
10^{-1}	opt	$1.9712\ 10^{-5}$	$9.8202\ 10^{-2}$	$3.4628\ 10^{-1}$
	lc	$1.2893\ 10^{-5}$	$9.8020\ 10^{-2}$	$3.4741\ 10^{-1}$
	lp	$1.6264\ 10^{-5}$	$9.8109\ 10^{-2}$	$3.4644\ 10^{-1}$
	lc1	$1.8139\ 10^{-5}$	$9.8159\ 10^{-2}$	$3.4628\ 10^{-1}$
	lc2	$1.4875\ 10^{-6}$	$9.7534\ 10^{-2}$	$4.9989\ 10^{-1}$
		$3.6097\ 10^{-5}$	$9.8705\ 10^{-2}$	$3.4921\ 10^{-1}$
10	opt	$3.2468\ 10^{-3}$	9.8835	$8.2361\ 10^{-1}$
	lc	$5.5909\ 10^{-3}$	9.9014	$8.5642\ 10^{-1}$
	lp	$2.1257\ 10^{-2}$	10.041	1.1757
	lc1	$1.1683\ 10^{-2}$	9.9532	1.0038
	lc2	$2.1257\ 10^{-2}$	10.041	1.1757
		$2.1257\ 10^{-2}$	10.041	1.1757

Table 15.7 L-curve algorithms, IP3 (Phillips) problem, $n = 200$

Noise	Method	μ	$\|c - Ax\|$	$\|x - x_0\|$
10^{-7}	opt	$1.1147\ 10^{-10}$	$9.6750\ 10^{-8}$	$2.1751\ 10^{-4}$
	lc	$7.8289\ 10^{-16}$	$6.2009\ 10^{-9}$	$2.2252\ 10^{-2}$
	lp	$5.0212\ 10^{-16}$	$5.2779\ 10^{-9}$	$2.4073\ 10^{-2}$
	lc1	$6.3760\ 10^{-16}$	$5.7625\ 10^{-9}$	$2.3071\ 10^{-2}$
	lc2	$9.1009\ 10^{-11}$	$9.1001\ 10^{-8}$	$2.2254\ 10^{-4}$
		$9.1009\ 10^{-14}$	$9.1001\ 10^{-8}$	$2.2254\ 10^{-4}$
10^{-5}	opt	$1.6035\ 10^{-8}$	$1.0595\ 10^{-5}$	$1.6452\ 10^{-3}$
	lc	$3.6730\ 10^{-14}$	$2.4301\ 10^{-6}$	$7.9811\ 10^{-1}$
	lp	$1.2677\ 10^{-14}$	$1.7596\ 10^{-6}$	1.1377
	lc1	$2.8635\ 10^{-14}$	$2.2673\ 10^{-6}$	$8.6888\ 10^{-1}$
	lc2	$1.1545\ 10^{-8}$	$9.9214\ 10^{-6}$	$1.7001\ 10^{-3}$
		$1.1545\ 10^{-8}$	$9.9214\ 10^{-6}$	$1.7001\ 10^{-3}$
10^{-3}	opt	$8.5392\ 10^{-7}$	$9.9864\ 10^{-4}$	$7.3711\ 10^{-3}$
	lc	$7.1966\ 10^{-10}$	$8.5111\ 10^{-4}$	$5.3762\ 10^{-1}$
	lp	$4.5729\ 10^{-10}$	$8.3869\ 10^{-4}$	$6.8849\ 10^{-1}$
	lc1	$3.6084\ 10^{-10}$	$8.3172\ 10^{-4}$	$7.8603\ 10^{-1}$
	lc2	$1.0250\ 10^{-9}$	$8.6013\ 10^{-4}$	$4.4563\ 10^{-1}$
		$2.9147\ 10^{-7}$	$9.7098\ 10^{-4}$	$1.3595\ 10^{-2}$
10^{-1}	opt	$1.4985\ 10^{-4}$	$1.0164\ 10^{-1}$	$7.5250\ 10^{-2}$
	lc	$7.9348\ 10^{-6}$	$9.8383\ 10^{-2}$	$1.8558\ 10^{-1}$
	lp	failed		
	lc1	$5.3451\ 10^{-6}$	$9.8260\ 10^{-2}$	$2.2869\ 10^{-1}$
	lc2	$1.4645\ 10^{-6}$	$9.7757\ 10^{-2}$	$4.9688\ 10^{-1}$
		$3.6975\ 10^{-5}$	$9.8941\ 10^{-2}$	$9.4203\ 10^{-2}$
10	opt	$7.9797\ 10^{-3}$	10.060	$5.2728\ 10^{-1}$
	lc	$6.1453\ 10^{-3}$	9.9867	$5.3855\ 10^{-1}$
	lp	failed		
	lc1	$2.1489\ 10^{-2}$	10.222	$6.5404\ 10^{-1}$
	lc2	$1.0514\ 10^{-2}$	10.038	$5.3249\ 10^{-1}$
		$1.0514\ 10^{-2}$	10.038	$5.3249\ 10^{-1}$

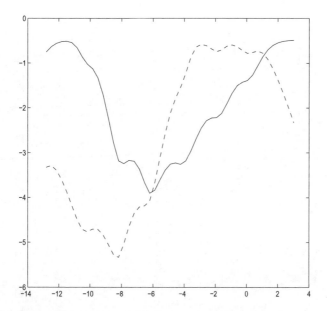

Figure 15.22 $\log_{10}$ of Δ^x (solid) and Δ^r (dashed) for the IP1 (Baart) problem, $m = n = 100$, noise$= 10^{-3}$

computing lower and upper bounds of quadratic forms allow the computation of the L-ribbon approximation introduced in Calvetti, Golub and Reichel [51]. An example of an L-ribbon for the IP1 example with $m = n = 100$ and six iterations is shown in figure 15.23. We display one of the boxes of the L-ribbon. The cross is the point on the exact L-curve for the given value of μ.

In table 15.8 we give the number of iterations and the values of μ obtained by the following algorithms. At each iteration of the Lanczos bidiagonalization algorithm we compute the corners of the curves given by the lower left and upper right corners of the rectangles using algorithm lc1. When these values of μ have both converged (up to 10^{-2}) and are close enough to each other, we stop the iterations. It turns out that there is a large difference in the number of iterations with and without full reorthogonalization for the Lanczos bidiagonalization algorithm. Therefore we give the number of iterations for both algorithms (in the third and fifth columns). Note that the Phillips problem requires a large number of iterations. The smaller is the noise level, the larger is the number of iterations.

Anyway, these results show that with only a few iterations of the Lanczos bidiagonalization algorithm (which requires only one multiplication by A and one by A^T per iteration), we can obtain values of the regularization parameter close to the ones of the "exact" curve.

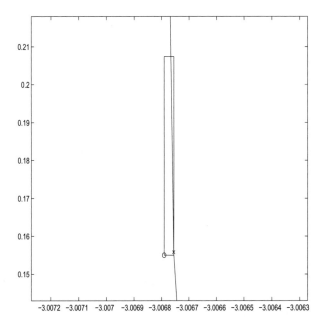

Figure 15.23 One rectangle of the L-ribbon for the IP1 (Baart) problem, $m = n = 100$, noise= 10^{-3}

Table 15.8 L-ribbon

Example	Noise	No. it.	μ	No. it. without reorth.
Baart	10^{-7}	11	$6.0889 \; 10^{-17}$	40
	10^{-5}	9	$6.1717 \; 10^{-13}$	19
	10^{-3}	8	$6.3232 \; 10^{-9}$	10
	10^{-1}	6	$7.2928 \; 10^{-5}$	6
	10	5	$3.260 \; 10^{-2}$	5
ILaplace	10^{-7}	23	$1.5871 \; 10^{-17}$	> 200
	10^{-5}	20	$1.6951 \; 10^{-13}$	93
	10^{-3}	15	$1.7535 \; 10^{-9}$	33
	10^{-1}	10	$1.8139 \; 10^{-5}$	11
	10	6	$1.3850 \; 10^{-2}$	6
Phillips	10^{-7}	188	$7.0255 \; 10^{-16}$	> 200
	10^{-5}	197	$2.6269 \; 10^{-14}$	> 200
	10^{-3}	48	$3.5697 \; 10^{-10}$	> 200
	10^{-1}	17	$5.3451 \; 10^{-6}$	26
	10	7	$2.4194 \; 10^{-2}$	7

15.6.5 Approximation of the Curvature

In table 15.9 we give the number of iterations and the values of μ obtained by computing lower and upper bounds for the curvature of the L-curve. Convergence is measured is the same way as for the L-ribbon. We see that in some cases convergence is more difficult. The advantage of the L-curvature is that we do not have to rely on algorithms like lc1 to find the corner. We just have to compute the maximum of vectors. However, the results (as well as the number of iterations) seem to be quite dependent on the interval that is chosen for the values of ν.

Table 15.9 L-curvature

Example	Noise	No. it.	μ	No. it. without reorth.
Baart	10^{-7}	10	$1.5220\ 10^{-16}$	27
	10^{-5}	10	$6.4404\ 10^{-13}$	20
	10^{-3}	10	$5.8220\ 10^{-9}$	20
	10^{-1}	10	$5.2630\ 10^{-5}$	20
	10	10	$2.2839\ 10^{-2}$	20
ILaplace	10^{-7}	23	$2.7322\ 10^{-17}$	237
	10^{-5}	22	$2.2216\ 10^{-13}$	160
	10^{-3}	22	$1.8064\ 10^{-9}$	160
	10^{-1}	22	$1.4689\ 10^{-5}$	159
	10	23	$5.9395\ 10^{-3}$	205
Phillips	10^{-7}	200	$5.0113\ 10^{-16}$	> 300
	10^{-5}	200	$4.7952\ 10^{-14}$	-
	10^{-3}	197	$9.3901\ 10^{-10}$	-
	10^{-1}	196	$8.5979\ 10^{-6}$	-
	10	200	$8.0479\ 10^{-3}$	-

15.7 Comparison of Methods for Computing the Regularization Parameter

15.7.1 Results on Moderate Size Problems

In this section we compare the different methods for computing the regularization parameter. The chosen methods are the von Matt implementation of the Golub and von Matt method (vm) [158] and the gm-opt algorithm for GCV. The other methods are computed using the SVD of A (which is not feasible for large problems), using the discrepancy principle (disc), the Gfrerer/Raus method (gr), finding the minimum of the GCV function G (gcv), locating the corner of the L-curve (lc) and looking at the minimum of the quasi-optimality function (qo). The results for lc and qo come from the regularization toolbox Regutools from P. C. Hansen [177] version 3.1. The parameters given by the L-ribbon and L-curvature algorithms are denoted respectively as L-rib and L-cur. The selection of the parameter from Brezinski et al. by minimizing an estimate of the error is denoted as err. We give the results for μ. Note that the regularization parameter is $\nu = m\mu$. μ opt is the parameter which approximately minimizes the error with the unperturbed solution.

Example IP1

The results for the Baart problem are given in table 15.10. Plots of the unperturbed solution (solid line) and some of its approximations are displayed in figure 15.24.

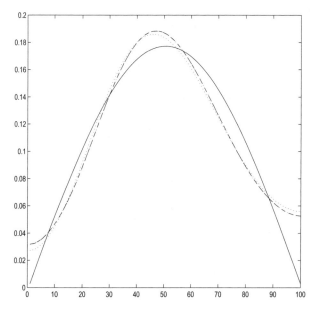

Figure 15.24 Solutions for the IP1 (Baart) problem, $m = n = 100$, noise$= 10^{-3}$, unperturbed solution (solid), vm (dashed) and gm-opt (dotted)

On this problem, which is not too difficult to solve, all the methods give more or less the same results for the norms of the residual and of the error relative to the unperturbed solution. However, as we have seen before, vm and gm-opt give poor results (as we see when comparing to gcv) for noise$= 10^{-1}$ and 10 (the distance to the unperturbed solution being two or three times larger) because in these cases $\tilde{G}$ is a poor approximation of the GCV function G around the minimizer.

Table 15.10 IP1 (Baart) problem, $n = 100$

Noise	Method	μ	$\|c - Ax\|$	$\|x - x_0\|$
10^{-7}	μ opt	$8.6975\ 10^{-17}$	$2.0335\ 10^{-7}$	$4.9814\ 10^{-2}$
	vm	$9.6482\ 10^{-15}$	$9.8049\ 10^{-8}$	$5.9424\ 10^{-2}$
	gm-opt	$1.0706\ 10^{-14}$	$9.8058\ 10^{-8}$	$5.9519\ 10^{-2}$
	gcv	$8.0440\ 10^{-15}$	$9.8035\ 10^{-8}$	$5.9238\ 10^{-2}$
	disc	$7.6875\ 10^{-14}$	$1.0000\ 10^{-7}$	$6.0364\ 10^{-2}$
	gr	$2.0434\ 10^{-13}$	$1.0853\ 10^{-7}$	$6.0621\ 10^{-2}$
	lc	$1.0908\ 10^{-16}$	$9.7919\ 10^{-8}$	$4.9856\ 10^{-2}$
	qo	$2.0937\ 10^{-14}$	$9.8187\ 10^{-8}$	$5.9964\ 10^{-2}$
	L-rib	$6.0889\ 10^{-17}$	$9.7912\ 10^{-8}$	$5.0041\ 10^{-2}$
	L-cur	$1.5220\ 10^{-16}$	$9.7923\ 10^{-8}$	$5.0341\ 10^{-2}$
	err	$2.9151\ 10^{-12}$	$2.6025\ 10^{-7}$	$6.3637\ 10^{-2}$
10^{-5}	μ opt	$8.9022\ 10^{-12}$	$2.9008\ 10^{-5}$	$6.5946\ 10^{-2}$
	vm	$9.7587\ 10^{-12}$	$9.8566\ 10^{-6}$	$6.5951\ 10^{-2}$
	gm-opt	$1.0581\ 10^{-11}$	$9.8588\ 10^{-6}$	$6.5957\ 10^{-2}$
	gcv	$8.7357\ 10^{-12}$	$9.8540\ 10^{-6}$	$6.5945\ 10^{-2}$
	disc	$3.5344\ 10^{-11}$	$1.0000\ 10^{-5}$	$6.6498\ 10^{-2}$
	gr	$1.0058\ 10^{-10}$	$1.0832\ 10^{-5}$	$6.9237\ 10^{-2}$
	lc	$6.5087\ 10^{-13}$	$9.8433\ 10^{-6}$	$6.8263\ 10^{-2}$
	qo	$3.3484\ 10^{-12}$	$9.8450\ 10^{-6}$	$6.6072\ 10^{-2}$
	L-rib	$6.1717\ 10^{-13}$	$9.8433\ 10^{-6}$	$6.8445\ 10^{-2}$
	L-cur	$6.4404\ 10^{-13}$	$9.8433\ 10^{-6}$	$6.8303\ 10^{-2}$
	err	$8.3022\ 10^{-9}$	$3.9008\ 10^{-5}$	$1.3679\ 10^{-1}$
10^{-3}	μ opt	$2.7826\ 10^{-8}$	$2.3501\ 10^{-3}$	$1.5084\ 10^{-1}$
	vm	$1.2018\ 10^{-8}$	$9.8573\ 10^{-4}$	$1.5239\ 10^{-1}$
	gm-opt	$1.3077\ 10^{-8}$	$9.8582\ 10^{-4}$	$1.5205\ 10^{-1}$
	gcv	$9.4870\ 10^{-9}$	$9.8554\ 10^{-4}$	$1.5362\ 10^{-1}$
	disc	$8.4260\ 10^{-8}$	$1.0000\ 10^{-3}$	$1.5556\ 10^{-1}$
	gr	$1.7047\ 10^{-7}$	$1.0235\ 10^{-3}$	$1.6373\ 10^{-1}$
	lc	$4.5414\ 10^{-9}$	$9.8524\ 10^{-4}$	$1.6028\ 10^{-1}$
	qo	$1.2586\ 10^{-8}$	$9.8450\ 10^{-4}$	$6.6072\ 10^{-1}$
	L-rib	$6.3232\ 10^{-9}$	$9.8534\ 10^{-4}$	$1.5669\ 10^{-1}$
	L-cur	$5.8220\ 10^{-9}$	$9.8531\ 10^{-4}$	$1.5749\ 10^{-1}$
	err	$2.3101\ 10^{-6}$	$1.6505\ 10^{-3}$	$2.0094\ 10^{-1}$
10^{-1}	μ opt	$1.1768\ 10^{-5}$	$2.2583\ 10^{-1}$	$2.9455\ 10^{-1}$
	vm	$1.0336\ 10^{-7}$	$9.8730\ 10^{-2}$	1.6614
	gm-opt	$1.1104\ 10^{-7}$	$9.8736\ 10^{-2}$	1.6267
	gcv	$3.0727\ 10^{-5}$	$9.9378\ 10^{-2}$	$2.9955\ 10^{-1}$
	disc	$6.0927\ 10^{-5}$	$1.0000\ 10^{-1}$	$3.3839\ 10^{-1}$
	gr	$1.5620\ 10^{-4}$	$1.0197\ 10^{-1}$	$3.8022\ 10^{-1}$
	lc	$4.1338\ 10^{-5}$	$9.9682\ 10^{-2}$	$3.2142\ 10^{-1}$
	qo	$1.9141\ 10^{-4}$	$1.0291\ 10^{-1}$	$3.8810\ 10^{-1}$
	L-rib	$7.2928\ 10^{-5}$	$1.0021\ 10^{-1}$	$3.4666\ 10^{-1}$
	L-cur	$5.2630\ 10^{-5}$	$9.9862\ 10^{-2}$	$3.3180\ 10^{-1}$
	err	$1.2328\ 10^{-3}$	$1.5756\ 10^{-1}$	$4.6015\ 10^{-1}$
10	μ opt	$4.9770\ 10^{-3}$	26.307	$6.1180\ 10^{-1}$
	vm	$8.8817\ 10^{-8}$	9.8728	1.6722
	gm-opt	$9.1683\ 10^{-8}$	9.8730	1.6267
	gcv	$1.1189\ 10^{-2}$	9.9885	$6.6657\ 10^{-1}$
	disc	$1.7654\ 10^{-2}$	10.000	$7.1748\ 10^{-1}$
	gr	$3.6604\ 10^{-2}$	10.035	$7.9288\ 10^{-1}$
	lc	$1.7087\ 10^{-2}$	9.9990	$7.1386\ 10^{-1}$
	qo	$2.3769\ 10^{-2}$	10.011	$7.4933\ 10^{-1}$
	L-rib	$3.260\ 10^{-2}$	10.027	$7.8135\ 10^{-1}$
	L-cur	$2.2839\ 10^{-2}$	10.009	$7.4516\ 10^{-1}$

Example IP2

For this problem there are more differences between the methods, especially for the error norm. However, there is no clear overall winner when we vary the noise level. The results are given in table 15.11. We see that most methods give a residual norm which is close to the noise level whereas the "optimal" value of the parameter gives a larger residual norm but a smaller norm of the error.

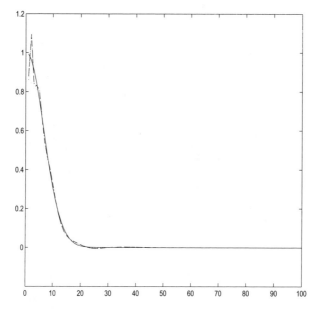

Figure 15.25 Solutions for the IP2 (ILaplace) problem, $m = n = 100$, noise= 10^{-3}, unperturbed solution (solid), vm (dashed), gm-opt (dot-dashed) and disc(dotted)

Table 15.11 IP2 (ILaplace) problem, $n = 100$

Noise	Method	μ	$\|c - Ax\|$	$\|x - x_0\|$
10^{-7}	μ opt	$4.3288\ 10^{-15}$	$3.2079\ 10^{-5}$	$1.3402\ 10^{-2}$
	vm	$2.1520\ 10^{-15}$	$9.5132\ 10^{-8}$	$1.4909\ 10^{-2}$
	gm-opt	$4.2396\ 10^{-14}$	$1.1004\ 10^{-7}$	$2.7130\ 10^{-2}$
	gcv	$1.0754\ 10^{-15}$	$9.4656\ 10^{-8}$	$2.1639\ 10^{-2}$
	disc	$1.5720\ 10^{-14}$	$1.0000\ 10^{-7}$	$1.9742\ 10^{-2}$
	gr	$3.3438\ 10^{-14}$	$1.0659\ 10^{-7}$	$2.5316\ 10^{-2}$
	lc	$2.1645\ 10^{-17}$	$9.3028\ 10^{-8}$	$1.8717\ 10^{-1}$
	qo	$2.1912\ 10^{-14}$	$1.0226\ 10^{-7}$	$2.2147\ 10^{-2}$
	L-rib	$1.5871\ 10^{-17}$	$9.2893\ 10^{-8}$	$2.2020\ 10^{-1}$
	L-cur	$2.7322\ 10^{-17}$	$9.3130\ 10^{-8}$	$1.6540\ 10^{-1}$
	err	$3.3516\ 10^{-13}$	$2.1248\ 10^{-7}$	$4.4507\ 10^{-2}$
10^{-5}	μ opt	$6.5793\ 10^{-13}$	$3.2418\ 10^{-3}$	$6.1866\ 10^{-2}$
	vm	$5.2329\ 10^{-12}$	$9.6965\ 10^{-6}$	$6.8646\ 10^{-2}$
	gm-opt	$5.4552\ 10^{-11}$	$1.0560\ 10^{-5}$	$9.6771\ 10^{-2}$
	gcv	$2.3495\ 10^{-12}$	$9.6371\ 10^{-6}$	$6.5557\ 10^{-2}$
	disc	$2.1272\ 10^{-11}$	$1.0000\ 10^{-5}$	$8.3321\ 10^{-2}$
	gr	$4.2645\ 10^{-11}$	$1.0362\ 10^{-5}$	$9.3222\ 10^{-2}$
	lc	$2.5207\ 10^{-13}$	$9.5870\ 10^{-6}$	$8.0606\ 10^{-2}$
	qo	$1.9841\ 10^{-12}$	$9.6301\ 10^{-6}$	$6.5035\ 10^{-2}$
	L-rib	$1.6951\ 10^{-13}$	$9.5788\ 10^{-6}$	$1.0407\ 10^{-1}$
	L-cur	$1.6951\ 10^{-13}$	$9.5788\ 10^{-6}$	$1.0407\ 10^{-1}$
	err	$1.2328\ 10^{-9}$	$2.5975\ 10^{-5}$	$1.4084\ 10^{-1}$
10^{-3}	μ opt	$5.3367\ 10^{-10}$	$3.2173\ 10^{-1}$	$1.4791\ 10^{-1}$
	vm	$2.2111\ 10^{-8}$	$9.7215\ 10^{-4}$	$1.9890\ 10^{-1}$
	gm-opt	$2.3046\ 10^{-8}$	$9.7243\ 10^{-4}$	$1.9937\ 10^{-1}$
	gcv	$2.0776\ 10^{-8}$	$9.7177\ 10^{-4}$	$1.9819\ 10^{-1}$
	disc	$8.6814\ 10^{-8}$	$1.0000\ 10^{-3}$	$2.1391\ 10^{-1}$
	gr	$1.8660\ 10^{-7}$	$1.0555\ 10^{-3}$	$2.2402\ 10^{-1}$
	lc	$1.7737\ 10^{-9}$	$9.6762\ 10^{-4}$	$1.6240\ 10^{-1}$
	qo	$1.2759\ 10^{-8}$	$9.6973\ 10^{-4}$	$1.9227\ 10^{-1}$
	L-rib	$1.7535\ 10^{-9}$	$9.6762\ 10^{-4}$	$1.6221\ 10^{-1}$
	L-cur	$1.8064\ 10^{-9}$	$9.6763\ 10^{-4}$	$1.6271\ 10^{-1}$
	err	$2.9151\ 10^{-6}$	$2.3831\ 10^{-3}$	$2.7795\ 10^{-1}$
10^{-1}	μ opt	$1.8738\ 10^{-5}$	$3.1515\ 10^{1}$	$3.4627\ 10^{-1}$
	vm	$1.9484\ 10^{-5}$	$9.8196\ 10^{-2}$	$3.4627\ 10^{-1}$
	gm-opt	$2.0896\ 10^{-5}$	$9.8235\ 10^{-2}$	$3.4634\ 10^{-1}$
	gcv	$2.2562\ 10^{-5}$	$9.8282\ 10^{-2}$	$3.4649\ 10^{-1}$
	disc	$6.9079\ 10^{-5}$	$1.0000\ 10^{-1}$	$3.5847\ 10^{-1}$
	gr	$1.7227\ 10^{-4}$	$1.0568\ 10^{-1}$	$3.8490\ 10^{-1}$
	lc	$1.2893\ 10^{-5}$	$9.8020\ 10^{-2}$	$3.4741\ 10^{-1}$
	qo	$2.7439\ 10^{-5}$	$9.8426\ 10^{-2}$	$3.4724\ 10^{-1}$
	L-rib	$1.8139\ 10^{-5}$	$9.8159\ 10^{-2}$	$3.4628\ 10^{-1}$
	L-cur	$1.4689\ 10^{-5}$	$9.8067\ 10^{-2}$	$3.4675\ 10^{-1}$
	err	$6.1359\ 10^{-4}$	$1.4475\ 10^{-1}$	$4.4754\ 10^{-1}$
10	μ opt	$3.9442\ 10^{-3}$	$3.1352\ 10^{3}$	$8.2453\ 10^{-1}$
	vm	$6.5983\ 10^{-3}$	9.9095	$8.8165\ 10^{-1}$
	gm-opt	$6.8436\ 10^{-3}$	9.9115	$8.8791\ 10^{-1}$
	gcv	$6.0540\ 10^{-3}$	9.9051	$8.6785\ 10^{-1}$
	disc	$1.6819\ 10^{-2}$	10	1.1039
	gr	$3.2174\ 10^{-2}$	10.137	1.3146
	lc	$5.5909\ 10^{-3}$	9.9014	$8.5642\ 10^{-1}$
	qo	$6.0077\ 10^{-3}$	9.9047	$8.6669\ 10^{-1}$
	L-rib	$1.3850\ 10^{-2}$	9.8835	$8.2361\ 10^{-1}$
	L-cur	$5.9395\ 10^{-3}$	9.9728	1.0487

Example IP3

For this problem there are some differences in the error norm when the noise is between 10^{-5} and 10^{-1}. Results are given in table 15.12. The algorithms based on the L-curve fail on this problem. The norms of the residuals are smaller but the norms of the error are larger than with the other methods, except for large noise levels. Note that for this example, the "optimal" parameter gives residual norms which are much smaller than with the other choices of the parameter.

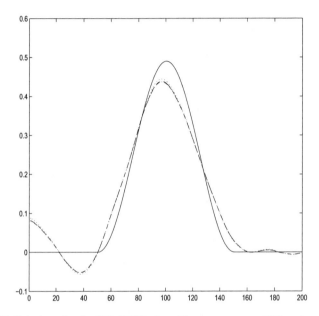

Figure 15.26 Solutions for the IP3 (Phillips) problem, $m = n = 200$, noise= 10, unperturbed solution (solid), vm (dashed), gm-opt (dot-dashed) and disc (dotted)

15.7.2 Results on Large Problems

In this section we solve problems with Example IP4 (von Matt example). We start with a medium size problem $m = 2000$ and $n = 1000$ with a linear distribution of the singular values

$$\sigma = 1 - 10^{-3} [0 : n - 1]'$$

The vectors u and v of lengths $= m$ or n are chosen with components $\sin(2\pi i/(l + 1))$, $i = 1, \ldots, l$. The solution is

$$x = 2 \exp(-6(-\pi/2 + [0.5 : m - 0.5]'\pi/m - 0.8)^2)$$
$$+ \exp(-2(-\pi/2 + [0.5 : m - 0.5]'\pi/n - 0.5)^2).$$

The results are obtained without reorthogonalization since this is impractical for large problems because of the storage issue. The results for lc were obtained by the L-curve code of von Matt because for this problem we cannot use directly the

Table 15.12 IP3 (Phillips) problem, $n = 200$

Noise	Method	μ	$\|c - Ax\|$	$\|x - x_0\|$
10^{-7}	μ opt	$1.3280 \ 10^{-10}$	$2.8193 \ 10^{-14}$	$2.1729 \ 10^{-4}$
	vm	$8.7929 \ 10^{-11}$	$9.0162 \ 10^{-8}$	$2.2391 \ 10^{-4}$
	gm-opt	$1.6343 \ 10^{-10}$	$1.1260 \ 10^{-7}$	$2.2163 \ 10^{-4}$
	gcv	$2.3940 \ 10^{-11}$	$7.3021 \ 10^{-8}$	$3.9089 \ 10^{-4}$
	disc	$1.2259 \ 10^{-10}$	$1.0000 \ 10^{-7}$	$2.1693 \ 10^{-4}$
	gr	$4.3205 \ 10^{-10}$	$2.0760 \ 10^{-7}$	$2.9550 \ 10^{-4}$
	lc	$7.8290 \ 10^{-16}$	$6.2011 \ 10^{-9}$	$2.2259 \ 10^{-2}$
	qo	$1.8148 \ 10^{-10}$	$1.1844 \ 10^{-7}$	$2.2559 \ 10^{-4}$
	L-rib	$7.0255 \ 10^{-16}$	$5.9666 \ 10^{-9}$	$2.2680 \ 10^{-2}$
	L-cur	$5.0113 \ 10^{-16}$	$5.2740 \ 10^{-9}$	$2.4082 \ 10^{-2}$
	err	$1.6758 \ 10^{-10}$	$1.1393 \ 10^{-7}$	$2.2248 \ 10^{-4}$
10^{-5}	μ opt	$1.3725 \ 10^{-7}$	$2.9505 \ 10^{-14}$	$1.6641 \ 10^{-3}$
	vm	$4.5432 \ 10^{-9}$	$9.0825 \ 10^{-6}$	$2.2620 \ 10^{-3}$
	gm-opt	$5.3835 \ 10^{-9}$	$9.1722 \ 10^{-6}$	$2.1174 \ 10^{-3}$
	gcv	$3.1203 \ 10^{-9}$	$8.9283 \ 10^{-6}$	$2.6499 \ 10^{-3}$
	disc	$1.2107 \ 10^{-8}$	$1.0000 \ 10^{-5}$	$1.6873 \ 10^{-3}$
	gr	$4.1876 \ 10^{-8}$	$1.5784 \ 10^{-5}$	$1.9344 \ 10^{-3}$
	lc	$3.6731 \ 10^{-14}$	$2.4301 \ 10^{-6}$	$7.9811 \ 10^{-1}$
	qo	$1.5710 \ 10^{-8}$	$1.0542 \ 10^{-5}$	$1.6463 \ 10^{-3}$
	L-rib	$2.6269 \ 10^{-14}$	$2.2118 \ 10^{-6}$	$8.9457 \ 10^{-1}$
	L-cur	$4.7952 \ 10^{-14}$	$2.6093 \ 10^{-6}$	$7.2750 \ 10^{-1}$
	err	$3.5274 \ 10^{-8}$	$1.4322 \ 10^{-5}$	$1.8362 \ 10^{-3}$
10^{-3}	μ opt	$9.1537 \ 10^{-7}$	$2.4133 \ 10^{-13}$	$7.3429 \ 10^{-3}$
	vm	$4.3674 \ 10^{-7}$	$9.7826 \ 10^{-4}$	$1.0057 \ 10^{-2}$
	gm-opt	$4.1814 \ 10^{-7}$	$9.7737 \ 10^{-4}$	$1.0375 \ 10^{-2}$
	gcv	$4.0669 \ 10^{-7}$	$9.7682 \ 10^{-4}$	$1.0585 \ 10^{-2}$
	disc	$8.7965 \ 10^{-7}$	$1.0000 \ 10^{-3}$	$7.3535 \ 10^{-3}$
	gr	$2.9376 \ 10^{-6}$	$1.1656 \ 10^{-3}$	$1.1260 \ 10^{-2}$
	lc	$7.1965 \ 10^{-10}$	$8.5111 \ 10^{-4}$	$5.3762 \ 10^{-1}$
	qo	$1.9308 \ 10^{-6}$	$1.0711 \ 10^{-3}$	$9.2198 \ 10^{-3}$
	L-rib	$3.5697 \ 10^{-10}$	$8.3172 \ 10^{-4}$	$7.8603 \ 10^{-1}$
	L-cur	$9.3901 \ 10^{-10}$	$8.5794 \ 10^{-4}$	$4.6668 \ 10^{-1}$
	err	$5.8841 \ 10^{-6}$	$1.5336 \ 10^{-3}$	$1.5434 \ 10^{-2}$
10^{-1}	μ opt	$1.5269 \ 10^{-4}$	$2.1773 \ 10^{-11}$	$7.5243 \ 10^{-2}$
	vm	$3.8320 \ 10^{-5}$	$9.8962 \ 10^{-2}$	$9.3139 \ 10^{-2}$
	gm-opt	$4.1875 \ 10^{-5}$	$9.9016 \ 10^{-2}$	$9.0659 \ 10^{-2}$
	gcv	$5.3006 \ 10^{-5}$	$9.9195 \ 10^{-2}$	$8.5119 \ 10^{-2}$
	disc	$9.3222 \ 10^{-5}$	$1.0000 \ 10^{-1}$	$7.7226 \ 10^{-2}$
	gr	$4.9453 \ 10^{-4}$	$1.2322 \ 10^{-1}$	$8.5662 \ 10^{-2}$
	lc	$7.9330 \ 10^{-6}$	$9.8383 \ 10^{-2}$	$1.8555 \ 10^{-1}$
	qo	$1.8894 \ 10^{-4}$	$1.0314 \ 10^{-1}$	$7.5516 \ 10^{-2}$
	L-rib	$5.3451 \ 10^{-6}$	$9.8260 \ 10^{-2}$	$2.2869 \ 10^{-1}$
	L-cur	$8.5979 \ 10^{-6}$	$9.8407 \ 10^{-2}$	$1.7808 \ 10^{-1}$
	err	$7.7784 \ 10^{-4}$	$1.5046 \ 10^{-1}$	$9.8279 \ 10^{-2}$
10	μ opt	$8.3405 \ 10^{-3}$	$2.1624 \ 10^{-9}$	$5.2682 \ 10^{-1}$
	vm	$8.4751 \ 10^{-3}$	10.012	$5.2677 \ 10^{-1}$
	gm-opt	$8.7084 \ 10^{-3}$	10.015	$5.2683 \ 10^{-1}$
	gcv	$4.0214 \ 10^{-3}$	9.9679	$5.8251 \ 10^{-1}$
	disc	$7.4140 \ 10^{-3}$	10	$5.2902 \ 10^{-1}$
	gr	$1.8947 \ 10^{-2}$	10.175	$6.2024 \ 10^{-1}$
	lc	$6.1445 \ 10^{-3}$	9.9867	$5.3837 \ 10^{-1}$
	qo	$5.0910 \ 10^{-3}$	9.9768	$5.5454 \ 10^{-1}$
	L-rib	$2.4194 \ 10^{-2}$	10.275	$6.9069 \ 10^{-1}$
	L-cur	$8.0479 \ 10^{-3}$	10.007	$5.2715 \ 10^{-1}$

code from Regutools since we do not store the matrix. Note that we are able to use the algorithms which need the singular value distribution because of the way the problem is constructed. Otherwise we would have had to compute the SVD of large matrices. Results are not given for vm when the noise is small because it needs too many iterations.

Table 15.13 IP4 (von Matt) problem, $m = 2000, n = 1000$, linear distribution

Noise	Method	No. it.	μ	$\|c - Ax\|$	$\|x - x_0\|$
10^{-3}	gm-opt	1102	$9.9271 \ 10^{-11}$	$6.9950 \ 10^{-4}$	$5.7874 \ 10^{-2}$
	disc	-	$6.2891 \ 10^{-9}$	$1.2199 \ 10^{-3}$	$8.7843 \ 10^{-2}$
	gr	-	$9.3420 \ 10^{-8}$	$1.4567 \ 10^{-2}$	0.1386
	gcv	-	failed	-	-
	lc	-	$8.6889 \ 10^{-15}$	$6.9919 \ 10^{-4}$	$6.1444 \ 10^{-2}$
	qo	-	$7.3848 \ 10^{-15}$	$6.9919 \ 10^{-4}$	$6.1445 \ 10^{-2}$
10^{-2}	gm-opt	366	$2.1224 \ 10^{-9}$	$7.0054 \ 10^{-3}$	0.1522
	disc	-	$6.3912 \ 10^{-8}$	$1.2199 \ 10^{-2}$	0.1398
	gr	-	$4.9023 \ 10^{-7}$	$7.5531 \ 10^{-2}$	0.3517
	gcv	-	$1.2408 \ 10^{-9}$	$6.9986 \ 10^{-3}$	0.1652
	lc	-	$8.6989 \ 10^{-15}$	$6.9919 \ 10^{-3}$	0.3137
	qo	-	$7.3848 \ 10^{-15}$	$6.9919 \ 10^{-3}$	0.3154
0.1	vm	219	$3.0202 \ 10^{-8}$	$7.0278 \ 10^{-2}$	0.7675
	gm-opt	197	$3.3106 \ 10^{-8}$	$7.9321 \ 10^{-2}$	0.7522
	disc	-	$6.5011 \ 10^{-7}$	$1.2199 \ 10^{-1}$	0.5651
	gr	-	$2.5391 \ 10^{-6}$	$3.7179 \ 10^{-1}$	1.5162
	gcv	-	$3.3217 \ 10^{-8}$	$7.0323 \ 10^{-2}$	0.7516
	lc	-	$8.6889 \ 10^{-15}$	$6.9919 \ 10^{-2}$	3.1093
	qo	-	$7.3848 \ 10^{-15}$	$6.9919 \ 10^{-2}$	3.1093

Then in table 15.14 we display the results with $m = 2000, n = 1000$ and an exponential distribution of the singular values. We see that with this problem we have to do a large number of iterations and it is likely that some form of preconditioning would have to be introduced in the method using the Lanczos bidiagonalization algorithm. Finally we solve a problem with $m = 20000$, $n = 2000$ and $c = 0.003$ (Table 15.15).

To summarize, we have shown in this chapter that the techniques for computing approximations of quadratic forms can be used to determine a good regularization parameter for solving discrete ill-posed problems.

Table 15.14 IP4 (von Matt) problem, $m = 2000, n = 1000$, exponential distribution, $c = 0.025$

Noise	Method	No. it.	μ	$\|c - Ax\|$	$\|x - x_0\|$
10^{-2}	gm-opt	1373	$5.5723\ 10^{-11}$	$9.4447\ 10^{-3}$	22.56
	disc	-	$5.7744\ 10^{-10}$	$1.2199\ 10^{-2}$	25.03
	gr	-	$1.1164\ 10^{-9}$	$1.4258\ 10^{-2}$	25.74
	gcv	-	$7.7201\ 10^{-12}$	$9.0380\ 10^{-3}$	21.60
	lc	-	failed	-	-
	qo	-	$4.7561\ 10^{-4}$	0.9821	31.53
10^{-1}	gm-opt	410	$2.2399\ 10^{-9}$	$9.4690\ 10^{-2}$	26.63
	disc	-	$9.8649\ 10^{-8}$	0.1220	29.46
	gr	-	$1.8484\ 10^{-7}$	0.1371	29.81
	gcv	-	$5.7442\ 10^{-10}$	$9.2855\ 10^{-2}$	26.37
	lc	-	failed	-	-
	qo	-	$4.7561\ 10^{-4}$	0.9880	31.53
1	gm-opt	80	$2.6542\ 10^{-7}$	0.9644	30.31
	disc	-	$7.2771\ 10^{-5}$	1.2199	31.43
	gr	-	$1.2238\ 10^{-4}$	1.2725	31.47
	gcv	-	$2.7099\ 10^{-7}$	0.9647	30.31
	lc	-	failed	-	-
	qo	-	$4.7561\ 10^{-4}$	1.4050	31.53

Table 15.15 IP4 (von Matt) problem, $m = 20000, n = 2000$, exponential distribution, $c = 0.003$

Noise	Method	No. it.	μ	$\|c - Ax\|$	$\|x - x_0\|$
10^{-2}	gm-opt	933	$4.8353\ 10^{-13}$	$9.4753\ 10^{-3}$	0.3578
	disc	-	$1.3263\ 10^{-10}$	$1.3776\ 10^{-2}$	1.4219
	gr	-	failed	-	-
	gcv	-	$3.0721\ 10^{-10}$	$2.3836\ 10^{-2}$	2.9420
	lc	-	$3.0721\ 10^{-10}$	$2.3836\ 10^{-2}$	2.9420
	qo	-	$4.9701\ 10^{-5}$	4.2021	1.4856
10^{-1}	gm-opt	585	$4.3687\ 10^{-11}$	$9.4816\ 10^{-2}$	3.394
	disc	-	$1.9861\ 10^{-9}$	0.1378	11.531
	gr	-	failed	-	-
	gcv	-	$3.0721\ 10^{-10}$	$9.7315\ 10^{-2}$	3.905
	lc	-	$3.0721\ 10^{-10}$	$9.7315\ 10^{-2}$	3.905
	qo	-	$4.9701\ 10^{-5}$	4.203	44.216
1	gm-opt	289	$1.6036\ 10^{-9}$	0.9554	17.915
	disc	-	$3.5982\ 10^{-7}$	1.3776	38.187
	gr	-	$6.8886\ 10^{-7}$	1.6325	39.343
	gcv	-	$1.0991\ 10^{-9}$	0.9527	18.868
	lc	-	$3.0721\ 10^{-10}$	0.9485	25.720
	qo	-	$4.9701\ 10^{-5}$	4.3195	44.216

Bibliography

[1] H-M. ADORF, *Hubble Space Telescope image restoration in its fourth year*, Inverse Problems, v 11 (1995), pp 639–653.

[2] M. AFANASJEW, M. EIERMANN, O.G. ERNST AND S. GÜTTEL, *On the steepest descent method for matrix functions*, Elec. Trans. Numer. Anal., v 28 (2007–2008), pp 206–222.

[3] M. AFANASJEW, M. EIERMANN, O.G. ERNST AND S. GÜTTEL, *Implementation of a restarted Krylov subspace method for the evaluation of matrix functions*, Linear Alg. Appl., v 429 n 10 (2008), pp 2293–2314.

[4] N.I. AKHIEZER, *The classical moment problem*, Oliver and Boyd, London (1965). Russian edition 1961.

[5] D.M. ALLEN, *The relationship between variable selection and data augmentation and a method for prediction*, Technometrics, v 16 (1974), pp 125-127.

[6] G.S. AMMAR, D. CALVETTI AND L. REICHEL, *Computation of Gauss–Kronrod quadrature rules with non-positive weights*, Elec. Trans. Numer. Anal., v 9 (1999), pp 26–38.

[7] A.I. APTEKAREV AND E.M. NIKISHIN, *The scattering problem for a discrete Sturm-Liouville operator*, Mat. Sb., v 121 n 163 (1983), pp 327–358; Math. USSR Sb., v 49 (1984), pp 325–355.

[8] P. ARBENZ AND G.H. GOLUB, *On the spectral decomposition of Hermitian matrices modified by low rank perturbations with applications*, SIAM J. Matrix Anal. Appl., v 9 (1988), pp 40–58.

[9] M. ARIOLI, *Stopping criterion for the conjugate gradient algorithm in a finite element method framework*, Numer. Math., v 97 (2004), pp 1–24.

[10] M. ARIOLI, D. LOGHIN AND A. WATHEN, *Stopping criteria for iterations in finite element methods*, Numer. Math., v 99 (2005), pp 381–410.

[11] W.E. ARNOLDI, *The principle of minimized iterations in the solution of the matrix eigenvalue problem*, Quarterly of Appl. Math., v 9 (1951), pp 17–29.

[12] K.S. ARUN, *A unitarily constrained total least squares problem in signal processing*, SIAM J. Matrix Anal. Appl., v 13 n 3 (1992), pp 729–745.

[13] F.V. ATKINSON, *Discrete and continuous boundary problems*, Academic Press, New York (1964).

[14] G. AUCHMUTY, *A posteriori error estimates for linear equations*, Numer. Math., v 61 n 1 (1992), pp 1–6.

[15] Z. BAI AND G.H. GOLUB, *Bounds for the trace of the inverse and the determinant of symmetric positive definite matrices*, Annals Numer. Math., v 4 (1997), pp 29–38.

[16] Z. BAI AND G.H. GOLUB, *Some unusual matrix eigenvalue problems*, in *Proceedings of Vecpar'98 – Third International Conference for Vector and Parallel Processing*, J. Palma, J. Dongarra and V. Hernandez Eds., Springer, Berlin (1999), pp 4–19.

[17] Z. BAI, M. FAHEY AND G.H. GOLUB, *Some large scale matrix computation problems*, J. Comput. Appl. Math., v 74 (1996), pp 71–89.

[18] Z. BAI, M. FAHEY, G.H. GOLUB, M. MENON AND E. RICHTER, *Computing partial eigenvalue sum in electronic structure calculation*, Report SCCM 98-03, Stanford University (1998).

[19] M.M. BAART, *The use of auto-correlation for pseudo-rank determination in noisy ill-conditioned least-squares problems*, IMA J. Numer. Anal., v 2 (1982), pp 241–247.

[20] J. BARANGER AND M. DUC-JACQUET, *Matrices tridiagonales symétriques et matrices factorisables*, RIRO, v 3 (1971), pp 61–66.

[21] J. BARDSLEY, S. JEFFERIES, J. NAGY, AND R. PLEMMONS, *Blind iterative restoration of images with spatially-varying blur*, Optics Express, v 14 (2006), pp 1767–1782.

[22] S. BASU AND N.K. BOSE, *Matrix Stieltjes series and network models*, SIAM J. Math. Anal., v 14 n 2 (1983), pp 209–222.

[23] B. BECKERMANN, *The stable computation of formal orthogonal polynomials*, Numer. Algo., v 11 (1996), pp 1–23.

[24] B. BECKERMANN, *The condition number of real Vandermonde, Krylov and positive definite Hankel matrices*, Numer. Math., v 85 (2000), pp 553–577.

[25] B. BECKERMANN AND B. BOURREAU, *How to choose modified moments?*, J. Comput. Appl. Math., v 98 (1998), pp 81–98.

[26] M. BELGE, M.E. KILMER AND E.L. MILLER, *Efficient determination of multiple regularization parameters in a generalized L-curve framework*, Inverse Problems, v 18 (2002), pp 1161–1183.

[27] M. BENZI AND G.H. GOLUB. *Bounds for the entries of matrix functions with applications to preconditioning*, BIT Numer. Math., v 39 n 3 (1999), pp 417–438.

[28] M. BENZI AND N. RAZOUK, *Decay bounds and O(N) algorithms for approximating functions of sparse matrices*, Elec. Trans. Numer. Anal., v 28 (2007), pp 16–39.

[29] M. BERMAN, *Automated smoothing of image and other regularly spaced data*, IEEE Trans. Pattern Anal. Machine Intell., v 16 (1994), pp 460–468.

[30] A. BJÖRCK, *Numerical methods for least squares problems*, SIAM, Philadelphia (1996).

[31] D. BOLEY AND G.H. GOLUB, *A survey of matrix inverse eigenvalue problems*, Inverse Problems, v 3 (1987), pp 595–622.

[32] G. BOUTRY, *Contributions à l'approximation et à l'algèbre linéaire numérique*, Thèse, Université des Sciences et Technologies de Lille, (2003).

[33] G. BOUTRY, *Construction of orthogonal polynomial wavelets as inverse eigenvalue problem and a new construction of Gauss–Kronrod matrices*, submitted (2007).

[34] C. BREZINSKI, *Padé-type approximation and general orthogonal polynomials*, ISNM v 50, Birkhäuser-Verlag, Basel (1980).

[35] C. BREZINSKI, *A direct proof of the Christoffel-Darboux identity and its equivalence to the recurrence relationship*, J. Comput. Appl. Math., v 32 (1990), pp 17–25.

[36] C. BREZINSKI, *Biorthogonality and its applications to numerical analysis*, Marcel Dekker, New York (1992).

[37] C. BREZINSKI, *Error estimates in the solution of linear systems*, SIAM J. Sci. Comput., v 21 n 2 (1999), pp 764–781.

[38] C. BREZINSKI, M. REDIVO-ZAGLIA AND H. SADOK, *Avoiding breakdown and near-breakdown in Lanczos type algorithms*, Numer. Algo., v 1 (1991), pp 261–284.

[39] C. BREZINSKI, G. RODRIGUEZ AND S. SEATZU, *Error estimates for linear systems with applications to regularization*, Numer. Algo., v 49 (2008), pp 85–104.

[40] C. BREZINSKI, G. RODRIGUEZ AND S. SEATZU, *Error estimates for the regularization of least squares problems*, Numer. Algo., v 51 n 1 (2009), pp 61–76.

[41] C. Brezinski, M. Redivo-Zaglia, G. Rodriguez and S. Seatzu, *Multi-parameter regularization techniques for ill-conditioned linear systems*, Numer. Math., v 94, (2003) pp 203–228.

[42] M.I. Bueno and F.M. Dopico, *A more accurate algorithm for computing the Christoffel transformation*, J. Comput. Appl. Math., v 205 n 1 (2007), pp 567–582.

[43] M.I. Bueno and F. Marcellán, *Darboux transformation and perturbation of linear functionals*, Linear Alg. Appl., v 384 (2004), pp 215–242.

[44] J.R. Bunch and C.P. Nielsen, *Updating the singular value decomposition*, Numer. Math., v 31 (1978), pp 11–129.

[45] J.R. Bunch, C.P. Nielsen and D.C. Sorensen, *Rank-one modification of the symmetric eigenproblem*, Numer. Math., v 31 (1978), pp 31–48.

[46] K. Burrage, A. Williams, J. Ehrel and B. Pohl, *The implementation of a generalized cross-validation algorithm using deflation techniques for linear systems*, J. Appl. Numer. Math., v 19 (1996), pp 17–31.

[47] D. Calvetti and L. Reichel, *Gauss quadrature applied to trust region computations*, Numer. Algo., v 34 (2003), pp 85–102.

[48] D. Calvetti and L. Reichel, *Tikhonov regularization of large linear problems*, BIT Numer. Math., v 43 n 2, (2003) pp 263–283.

[49] D. Calvetti and L. Reichel, *Symmetric Gauss–Lobatto and modified anti-Gauss rules*, BIT Numer. Math., v 43 (2003), pp 541–554.

[50] D. Calvetti and L. Reichel, *Tikhonov regularization with a solution constraint*, SIAM J. Sci. Comput., v 26 n 1 (2004), pp 224–239.

[51] D. Calvetti, G.H. Golub and L. Reichel, *Estimation of the L-curve via Lanczos bidiagonalization*, BIT, v 39 n 4 (1999), pp 603–619.

[52] D. Calvetti, G.H. Golub and L. Reichel, *A computable error bound for matrix functionals*, J. Comput. Appl. Math., v 103 n 2 (1999), pp 301–306.

[53] D. Calvetti, P.C. Hansen and L. Reichel, *L-curve curvature bounds via Lanczos bidiagonalization*, Elec. Trans. Numer. Anal., v 14 (2002), pp 20–35.

[54] D. Calvetti, L. Reichel and F. Sgallari, *Application of anti-Gauss rules in linear algebra*, in *Applications and computation of orthogonal polynomials*, W. Gautschi, G.H. Golub and G. Opfer Eds., Birkhauser, Boston (1999), pp 41–56.

[55] D. CALVETTI, L. REICHEL AND F. SGALLARI, *Tikhonov regularization with nonnegativity constraint*, Elec. Trans. Numer. Anal., v 18 (2004), pp 153–173.

[56] D. CALVETTI, L. REICHEL AND A. SHUIBI, *Tikhonov regularization of large symmetric problems*, Numer. Lin. Alg. Appl., v 12 n 2-3 (2004), pp 127–139.

[57] D. CALVETTI, L. REICHEL AND A. SHUIBI, *L-curve and curvature bounds for Tikhonov regularization*, Numer. Algo., v 35 n 2-4 (2004), pp 301–314.

[58] D. CALVETTI, SUN–MI KIM AND L. REICHEL, *Quadrature rules based on the Arnoldi process*, SIAM J. Matrix Anal. Appl., v 26 n 3 (2005), pp 765–781.

[59] D. CALVETTI, G.H. GOLUB, W.B. GRAGG AND L. REICHEL, *Computation of Gauss–Kronrod quadrature rules*, Math. Comput., v 69 n 231 (2000), pp 1035–1052.

[60] D. CALVETTI, S. MORIGI, L. REICHEL AND F. SGALLARI, *Computable error bounds and estimates for the conjugate gradient method*, Numer. Algo., v 25 (2000), pp 79–88.

[61] R.H. CHAN, C. GREIF AND D.P. O' LEARY EDS., *Milestones in matrix computation: the selected works of Gene H. Golub with commentaries*, Oxford University Press, Oxford (2007).

[62] X.W. CHANG, G.H. GOLUB AND C.C. PAIGE, *Towards a backward perturbation analysis for data least squares problems*, SIAM J. Matrix Anal. Appl., v 30 n 4 (2008), pp 1281–1301.

[63] P.L. CHEBYSHEV, *Sur l'interpolation par la méthode des moindres carrés*, Mém. Acad. Impér. Sci. St. Pétersbourg, v 7 n 15 (1859), pp 1–24. Also in Oeuvres I, pp 473–498.

[64] T.S. CHIHARA, *An introduction to orhogonal polynomials*, Gordon and Breach, New York (1978).

[65] T.S. CHIHARA, *45 years of orthogonal polynomials: a view from the wings*, J. Comput. Appl. Math., v 133 (2001), pp 13–21.

[66] J. CHUNG, J.G. NAGY AND D.P. O'LEARY, *A weighted-GCV method for Lanczos-hybrid regularization*, Elec. Trans. Numer. Anal., v 28 (2008), pp 149–167.

[67] C.W. CLENSHAW AND A.R. CURTIS, *A method for numerical integration on an automatic computer*, Numer. Math., v 2 (1960), pp 197–205.

[68] P. CONCUS, G.H. GOLUB AND D.P. O'LEARY, *A generalized conjugate gradient method for the numerical solution of elliptic partial differential equations*, in *Sparse matrix computations*, J.R. Bunch and D.J. Rose Eds., Academic Press, New York (1976), pp 309–332.

[69] P. CONCUS, D. CASSAT, G. JAEHNIG AND E. MELBY, *Tables for the evaluation of $\int_0^\infty x^\beta e^{-x} f(x) dx$ by Gauss–Laguerre quadrature*, Math. Comput., v 17 (1963), pp 245–256.

[70] P. CRAVEN AND G. WAHBA, *Smoothing noisy data with spline functions*, Numer. Math., v 31 (1979), pp 377–403.

[71] J.J.M. CUPPEN, *A divide and conquer method for the symmetric tridiagonal eigenvalue problem*, Numer. Math., v 36 (1981), pp 177–195.

[72] A. CUYT, *Floating-point versus symbolic computations in the QD-algorithm*, J. Symbolic Comput., v 24 (1997), pp 695–703.

[73] G. DAHLQUIST AND A. BJÖRCK, *Numerical methods*, Prentice-Hall, Englewood Cliffs, NJ (1974).

[74] G. DAHLQUIST AND A. BJÖRCK, *Numerical methods in scientific computing, volume I*, SIAM, Philadelphia (2008).

[75] G. DAHLQUIST, S.C. EISENSTAT AND G.H. GOLUB, *Bounds for the error of linear systems of equations using the theory of moments*, J. Math. Anal. Appl., v 37 (1972), pp 151–166.

[76] G. DAHLQUIST, G.H. GOLUB AND S.G. NASH, *Bounds for the error in linear systems*, in *Proceedings of the workshop on semi–infinite programming*, R. Hettich Ed., Springer, Berlin (1978), pp 154–172.

[77] P.J. DAVIS, *The thread: A mathematical yarn*, Harcourt Brace Jovanovich, Boston (1989).

[78] P.J. DAVIS AND P. RABINOWITZ, *Methods of numerical integration*, Second Edition, Academic Press, New York (1984).

[79] C. DE BOOR AND G.H. GOLUB, *The numerically stable reconstruction of a Jacobi matrix from spectral data*, Linear Alg. Appl., v 21 (1978), pp 245–260.

[80] H. DETTE AND W.J. STUDDEN, *Matrix measures, moment spaces and Favard's theorem for the interval $[0, 1]$ and $[0, \infty)$*, Linear Alg. Appl., v 345 (2002), pp 169–193.

[81] H. DETTE AND W.J. STUDDEN, *A note on the matrix q-d algorithm and matrix orthogonal polynomials on $[0, 1]$ and $[0, \infty)$*, J. Comput. Appl. Math., v 148 (2002), pp 349–361.

[82] H. DETTE AND W.J. STUDDEN, *Quadrature formulas for matrix measure – a geometric approach*, Linear Alg. Appl., v 364 (2003), pp 33–64.

[83] I.S. DHILLON, *A new $O(n^2)$ algorithm for the symmetric tridiagonal eigenvalue problem*, Ph.D. thesis, University of California, Berkeley (1997).

[84] I.S. DHILLON AND B.N. PARLETT, *Relatively robust representations of symmetric tridiagonals*, Linear Alg. Appl., v 309 (2000), pp 121–151.

[85] I.S. DHILLON AND B.N. PARLETT, *Multiple representations to compute orthogonal eigenvectors of symmetric tridiagonal matrices*, Linear Alg. Appl., v 387 (2004), pp 1–28.

[86] I.S. DHILLON, B.N. PARLETT AND C. VÖMEL, *The design and implementation of the MRRR algorithm*, LAPACK Working Note 162, Tech. Report UCB/CSD-04-1346, UC Berkeley (2004).

[87] I.S. DHILLON, B.N. PARLETT AND C. VÖMEL, *Glued matrices and the MRRR algorithm*, SIAM J. Sci. Comput., v 27 n 2 (2005), pp 496–510.

[88] S. DONG AND K. LIU, *Stochastic estimation with z_2 noise*, Phys. Lett. B, v 328 (1994), pp 130–136.

[89] J.J. DONGARRA AND D.C. SORENSEN, *A fully parallel algorithm for the symmetric eigenvalue problem*, SIAM J. Sci. Statist. Comput., v 2 n 2 (1987), pp 139–154.

[90] A. DRAUX, *Polynômes orthogonaux formels. Applications*, LNM 974, Springer-Verlag, Berlin (1983).

[91] V. DRUSKIN AND L. KNIZHNERMAN, *Two polynomial methods of calculating functions of symmetric matrices*, USSR Comput. Maths. Math. Phys., v 29 n 6 (1989), pp 112–121.

[92] V. DRUSKIN AND L. KNIZHNERMAN, *Error bounds in the simple Lanczos procedure for computing functions of symmetric matrices and eigenvalues*, Comput. Maths. Math. Phys., v 31 n 7 (1991), pp 20–30.

[93] V. DRUSKIN AND L. KNIZHNERMAN, *Krylov subspace approximation of eigenpairs and matrix functions in exact and computer arithmetic*, Numer. Linear Alg. Appl., v 2 n 3 (1995), pp 205–217.

[94] V. DRUSKIN AND L. KNIZHNERMAN, *Extended Krylov subspaces: approximation of the matrix square root and related functions*, SIAM J. Matrix Anal. Appl., v 19 n 3 (1998), pp 755–771.

[95] V. DRUSKIN, A. GREENBAUM AND L. KNIZHNERMAN, *Using nonorthogonal Lanczos vectors in the computation of matrix functions*, SIAM J. Sci. Comput., v 19 n 1 (1998), pp 38–54.

[96] S. EHRICH, *On stratified extensions of Gauss–Laguerre and Gauss–Hermite quadrature formulas*, J. Comput. Appl. Math., v 140 (2002), pp 291–299.

[97] M. EIERMANN AND O.G. ERNST, *A restarted Krylov subspace method for the evaluation of matrix functions*, SIAM J. Numer. Anal., v 44 (2006), pp 2481–2504.

[98] S. ELHAY AND J. KAUTSKY, *Generalized Kronrod Patterson type embedded quadratures*, Aplikace Matematiky, v 37 (1992), pp 81–103.

[99] S. ELHAY AND J. KAUTSKY, *Jacobi matrices for measures modified by a rational factor*, Numer. Algo., v 6 (1994), pp 205–227.

[100] S. ELHAY, G.H. GOLUB AND J. KAUTSKY, *Updating and downdating of orthogonal polynomials with data fitting applications*, SIAM J. Matrix. Anal. Appl., v 12 n 2 (1991), pp 327–353.

[101] S. ELHAY, G.H. GOLUB AND J. KAUTSKY, *Jacobi matrices for sums of weight functions*, BIT, v 32 (1992), pp 143–166.

[102] H.W. ENGL, *Regularization methods for the stable solution of inverse problems*, Surveys Math. Industry, v 3 (1993), pp 71–143.

[103] D. FASINO, *Spectral properties of Hankel matrices and numerical solutions of finite moment problems*, J. Comput. Appl. Math., v 65 (1995), pp 145–155.

[104] J. FAVARD, *Sur les polynômes de Tchebicheff*, C.R. Acad. Sci. Paris, v 200 (1935), pp 2052–2053.

[105] K.V. FERNANDO, *On computing an eigenvector of a tridiagonal matrix. Part I: basic results*, SIAM J. Matrix Anal., v 18 n 4 (1997), pp 1013–1034.

[106] K.V. FERNANDO AND B.N. PARLETT, *Accurate singular values and differential qd algorithms*, Numer. Math., v 67 (1994), pp 191–229.

[107] K.V. FERNANDO, B.N. PARLETT AND I.S. DHILLON, *A way to find the most redundant equation in a tridiagonal system*, Report PAM-635, Center for Pure and Applied Mathematics, University of California, Berkeley (1995).

[108] R.D. FIERRO, G.H. GOLUB, P.C. HANSEN AND D.P. O'LEARY, *Regularization by truncated total least squares*, SIAM J. Sci. Comput., v 18 n 4 (1997), pp 1223–1241.

[109] B. FISCHER, *Polynomial based iteration methods for symmetric linear systems*, Wiley-Tubner, Leipzig (1996).

[110] B. FISCHER AND G.H. GOLUB, *On generating polynomials which are orthogonal over several intervals*, Math. Comput., v 56 n 194 (1991), pp 711–730.

[111] B. FISCHER AND G.H. GOLUB, *On the error computation for polynomial based iteration methods*, in *Recent advances in iterative methods*, A. Greenbaum and M. Luskin Eds., Springer, Berlin (1993).

[112] R. FLETCHER, *Conjugate gradient methods for indefinite systems*, in *Proceedings of the Dundee conference on numerical analysis*, G.A. Watson Ed., Springer (1975), pp 73–89.

[113] G.E. FORSYTHE, *Generation and use of orthogonal polynomials for data fitting with a digital computer*, J. Soc. Indust. Appl. Math., v 5 (1957), pp 74–88.

[114] R.W. FREUND AND M. HOCHBRUCK, *Gauss quadratures associated with the Arnoldi process and the Lanczos algorithm*, in *Linear algebra for large scale and real-time applications*, M.S. Moonen, G.H. Golub and B. De Moor Eds., Kluwer, Dordrecht (1993), pp 377–380.

[115] R.W. FREUND, M.H. GUTKNECHT AND N.M. NACHTIGAL, *An implementation of the look-ahead Lanczos algorithm for non Hermitian matrices*, SIAM J. Sci. Comput., v 14 (1993), pp 137–158.

[116] A. FROMMER AND P. MAAS, *Fast CG-based methods for Tikhonov-Phillips regularization*, SIAM J. Sci. Comput., v 20 (1999), pp 1831–1850.

[117] A. FROMMER AND V. SIMONCINI, *Matrix functions*, in *Model order reduction: theory, research aspects and applications*, Mathematics in industry, WH. Schilders and H.A. van der Vorst Eds., Springer, Berlin (2006), pp 1–24.

[118] D. GALANT, *An implementation of Christoffel's theorem in the theory of orthogonal polynomials*, Math. Comput., v 25 (1971), pp 111–113.

[119] D. GALANT, *Algebraic methods for modified orthogonal polynomials*, Math. Comput., v 59 n 200 (1992), pp 541–546.

[120] W. GANDER, *Least squares with a quadratic constraint*, Numer. Math., v 36 (1981), pp 291–307.

[121] W. GANDER, G.H. GOLUB AND U. VON MATT, *A constrained eigenvalue problem*, Linear Alg. Appl., v 114/115 (1989), pp 815–839.

[122] F.R. GANTMACHER, *The theory of matrices*, vol. 1, Chelsea, New York (1959).

[123] K. GATES AND W.B. GRAGG, *Notes on TQR algorithms*, J. Comput. Appl. Math., v 86 (1997), pp 195–203.

[124] C.F. GAUSS, *Methodus nova integralium valores per approximationem inveniendi*, (1814), in Werke, v 3 , K. Gesellschaft Wissenschaft. Göttingen (1886), pp 163-196.

[125] W. GAUTSCHI, *Construction of Gauss-Christoffel quadrature formulas*, Math. Comput., v 22 (1968), pp 251–270.

[126] W. GAUTSCHI, *On the construction of Gaussian quadrature rules from modified moments*, Math. Comput., v 24 (1970), pp 245–260.

[127] W. GAUTSCHI, *On generating orthogonal polynomials*, SIAM J. Sci. Comput., v 3 n 3 (1982), pp 289–317.

[128] W. GAUTSCHI, *Orthogonal polynomials– constructive theory and applications*, J. Comput. Appl. Math., v 12–13 (1985), pp 61–76.

[129] W. GAUTSCHI, *Moments in quadrature problems*, Comput. Math. Appl., v 33 n 1/2 (1997), pp 105–118.

[130] W. GAUTSCHI, *Orthogonal polynomials and quadrature*, Elec. Trans. Numer. Anal., v 9 (1999), pp 65–76.

[131] W. GAUTSCHI, *Orthogonal polynomials: computation and approximation*, Oxford University Press, Oxford (2004).

[132] W. GAUTSCHI, *Orthogonal polynomials (in Matlab)*, J. Comput. Appl. Math., v 178 n 1-2 (2005), pp 215–234.

[133] W. GAUTSCHI, *A historical note on Gauss–Kronrod quadrature*, Numer. Math., v 100 (2005), pp 483–484.

[134] W. GAUTSCHI AND R.S. VARGA, *Error bounds for Gaussian quadrature of analytic functions*, SIAM J. Numer. Anal., v 20 n 6 (1983), pp 1170–1186.

[135] J.S. GERONIMO, *Scattering theory and matrix orthogonal polynomials on the real line*, Circuits Systems Signal Process., v 1 (1982), pp 471–495.

[136] H. GFRERER, *An a posteriori parameter choice for ordinary and iterated Tikhonov regularization of ill-posed problems leading to optimal convergence rates*, Math. Comput., v 49 (1987), pp 507–522.

[137] G. GOERTZEL, H.V. WALDINGER AND J. AGRESTA, *The method of spherical harmonics as applied to the one-velocity Boltzmann equation in infinite cylinders*, in *ACM'59: Preprints of papers presented at the 14th national meeting of the Association for Computing Machinery*, (1959), pp 1–5.

[138] G.H. GOLUB, *Bounds for eigenvalues of tridiagonal symmetric matrices computed by the LR method*, Math. Comput., v 16 (1962), pp 438–445.

[139] G.H. GOLUB, *Some modified matrix eigenvalue problems*, SIAM Rev., v 15 n 2 (1973), pp 318–334.

[140] G.H. GOLUB, *Bounds for matrix moments*, Rocky Mnt. J. Math., v 4 n 2 (1974), pp 207–211.

[141] G.H. GOLUB, *Matrix computation and the theory of moments*, in *Proceedings of the international congress of mathematicians*, Zürich, Switzerland 1994, Birkhäuser, Basel (1995), pp 1440–1448.

[142] G. H. GOLUB AND B. FISCHER, *How to generate unknown orthogonal polynomials out of known orthogonal polynomials*, J. Comput. Appl. Math., v 43 (1992), pp 99–115.

[143] G. H. GOLUB AND M.H. GUTKNECHT, *Modified moments for indefinite weight functions*, Numer. Math., v 57 (1990), pp 607–624.

[144] G. H. GOLUB AND W. KAHAN, *Calculating the singular values and pseudo-inverse of a matrix*, SIAM J. Numer. Anal., v 2 (1965), pp 205–224.

[145] G.H. GOLUB AND J. KAUTSKY, *Calculation of Gauss quadratures with multiple free and fixed knots*, Numer. Math., v 41 (1983), pp 147–163.

[146] G.H. GOLUB AND M.D. KENT, *Estimates of eigenvalues for iterative methods*, Math. Comput., v 53 n 188 (1989), pp 619–626.

[147] G.H. GOLUB AND H. MELBØ, *A stochastic approach to error estimates for iterative linear solvers, part 1*, BIT, v 41 n 5 (2001), pp 977–985.

[148] G.H. GOLUB AND H. MELBØ, *A stochastic approach to error estimates for iterative linear solvers, part 2*, Report SCCM-01-05, Stanford University, Computer Science Department (2001).

[149] G.H. GOLUB AND G. MEURANT, *Matrices, moments and quadrature*, in *Numerical analysis 1993*, D.F. Griffiths and G.A. Watson Eds., Pitman Research Notes in Mathematics, v 303 (1994), pp 105–156. Reprinted in [61].

[150] G.H. GOLUB AND G. MEURANT, *Matrices, moments and quadrature II or how to compute the norm of the error in iterative methods*, BIT, v 37 n 3 (1997), pp 687–705.

[151] G.H. GOLUB AND Z. STRAKŎS, *Estimates in quadratic formulas*, Numer. Algo., v 8 n II–IV, (1994).

[152] G.H. GOLUB AND R. UNDERWOOD, *The block Lanczos method for computing eigenvalues*, in *Mathematical software III*, J. Rice Ed., Academic Press, New York (1977), pp 361–377.

[153] G.H. GOLUB AND C. VAN LOAN, *An analysis of the total least squares problem*, SIAM J. Numer. Anal., v 17 n 6 (1980), pp 883–893.

[154] G.H. GOLUB AND C. VAN LOAN, *Matrix Computations*, Third Edition, Johns Hopkins University Press, Baltimore (1996).

[155] G.H. GOLUB AND R.S. VARGA, *Chebyshev semi-iterative methods, successive overrelaxation iterative methods and second order Richardson iterative methods, Part I*, Numer. Math., v 3 (1961), pp 147–156.

[156] G.H. GOLUB AND R.S. VARGA, *Chebyshev semi-iterative methods, successive overrelaxation iterative methods and second order Richardson iterative methods, Part II*, Numer. Math., v 3 (1961), pp 157–168.

[157] G.H. GOLUB AND U. VON MATT, *Quadratically constrained least squares and quadratic problems*, Numer. Math., v 59 (1991), pp 561–580.

[158] G.H. GOLUB AND U. VON MATT, *Tikhonov regularization for large scale problems*, in *Scientific computing*, G.H. Golub, S.H. Lui, F. Luk and R. Plemmons Eds., Springer, Berlin (1997), pp 3–26.

[159] G.H. GOLUB AND U. VON MATT, *Generalized cross-validation for large scale problems*, in *Recent advances in total least squares techniques and errors-in-variable modelling*, S. van Huffel Ed., SIAM, Philadelphia (1997), pp 139–148.

[160] G.H. GOLUB AND J.H. WELSCH, *Calculation of Gauss quadrature rules*, Math. Comput., v 23 (1969), pp 221–230. Reprinted in [61].

[161] G.H. GOLUB, P.C. HANSEN AND D.P. O'LEARY, *Tikhonov regularization and total least squares*, SIAM J. Matrix Anal. Appl., v 21 n 1 (1999), pp 185–194.

[162] G.H. GOLUB, M. HEATH AND G. WAHBA, *Generalized cross-validation as a method to choosing a good ridge parameter*, Technometrics, v 21 n 2 (1979), pp 215–223.

[163] G.H. GOLUB, M. STOLL AND A. WATHEN, *Approximation of the scattering amplitude*, Elec. Trans. Numer. Anal., v 31 (2008), pp 178–203.

[164] W.B. GRAGG AND W.J. HARROD, *The numerically stable reconstruction of Jacobi matrices from spectral data*, Numer. Math., v 44 (1984), pp 317–335.

[165] J. GRCAR, *Optimal sensitivity analysis of linear least squares problems*, Report LBNL–52434, Lawrence Berkeley National Laboratory (2003).

[166] C. GU, D.M. BATRES, Z. CHEN AND G. WAHBA, *The computation of generalized cross-validation functions through Householder tridiagonalization with applications to the fitting of interaction spline models*, SIAM J. Matrix Anal., v 10 (1989), pp 457–480.

[167] M. GU AND S.C. EISENSTAT, *A divide-and-conquer algorithm for the symmetric tridiagonal eigenvalue problem*, SIAM J. Matrix Anal., v 16 (1995), pp 172–191.

[168] J. HADAMARD, *Essai sur l'étude des fonctions données par leur développement de Taylor*, J. Math. Pures et Appl., v 4 (1892), pp 101–186.

[169] M. HANKE, *Limitations of the L-curve method in ill-posed problems*, BIT, v 36 (1996), pp 287–301.

[170] M. HANKE, *Iterative regularization techniques in image restoration*, in *Surveys on solution methods for inverse problems*, D. Colton et al. Eds., Springer-Verlag, Berlin (2000), pp 35–52.

[171] M. HANKE, *On Lanczos based methods for the regularization of discrete ill-posed problems*, BIT, v 41 (2001), pp 1008–1018.

[172] M. HANKE, *A note on Tikhonov regularization of large linear problems*, BIT, v 43 (2003), pp 449–451.

[173] M. HANKE AND P.C. HANSEN, *Regularization methods for large scale problems*, Surveys Math. Industry, v 3 (1993), pp 253–315.

[174] M. HANKE AND J.G. NAGY, *Restoration of atmospherically blurred images by symmetric indefinite conjugate gradient techniques*, Inverse Problems, v 12 (1996), pp 157–173.

[175] M. HANKE AND T. RAUS, *A general heuristic for choosing the regularization parameter in ill-posed problems*, SIAM J. Sci. Comput., v 17 (1996), pp 956–972.

[176] P.C. HANSEN, *Analysis of discrete ill-posed problems by means of the L-curve*, SIAM Rev., v 34 (1992), pp 561–580.

[177] P.C. HANSEN, *Regularization tools: a Matlab package for analysis and solution of discrete ill-posed problems*, Numer. Algo., v 6 (1994), pp 1–35.

[178] P.C. HANSEN, *Rank-deficient and discrete ill-posed problems. Numerical aspects of linear inversion*, SIAM, Philadelphia (1997).

[179] P.C. HANSEN, *Regularization Tools version 4.0 for Matlab 7.3*, Numer. Algo., v 46 (2007), pp 189–194.

[180] P.C. HANSEN AND D.P. O'LEARY, *The use of the L-curve in the regularization of discrete ill-posed problems*, SIAM J. Sci. Comput., v 14 (1993), pp 1487–1503.

[181] P.C. HANSEN, T.K. JENSEN AND G. RODRIGUEZ, *An adaptive pruning algorithm for the discrete L-curve criterion*, J. Comput. Appl. Math., v 198 n 2 (2007), pp 483–492.

[182] R. HAYDOCK, *Comparison of quadrature and termination for estimating the density of states within the recursion method*, J. Phys. C: Solid State Phys., v 17 (1984), pp 4783–4789.

[183] R. HAYDOCK, *Accuracy of the recursion method and basis non-orthogonality*, Comput. Phys. Commun., v 53 (1989), pp 133–139.

[184] R. HAYDOCK AND R.L. TE, *Accuracy of the recursion method*, Phys. Rev. B: Condensed Matter, v 49 (1994), pp 10845–10850.

[185] P. HENRICI, *The quotient-difference algorithm*, NBS Appl. Math. Series, v 49 (1958), pp 23–46.

[186] P. HENRICI, *Quotient-difference algorithms*, in *Mathematical methods for digital computers*, A. Ralston and H.S. Wilf Eds., J. Wiley, New York (1967), pp 37–62.

[187] M.R. HESTENES AND E. STIEFEL, *Methods of conjugate gradients for solving linear systems*, J. Nat. Bur. Standards, v 49 n 6 (1952), pp 409–436.

[188] N. HIGHAM, *Accuracy and stability of numerical algorithms*, second edition, SIAM, Philadelphia (2002).

[189] N. HIGHAM, *Functions of matrices: theory and computation*, SIAM, Philadelphia (2008).

[190] I. HNĚTYNKOVÀ AND Z. STRAKOŠ, *Lanczos tridiagonalization and core problems*, Linear Alg. Appl., v 421 (2007), pp 243–251.

[191] M. HOCHBRUCK AND M. HOCHSTENBACH, *Subspace extraction for matrix functions*, Report Heinrich-Heine-Universität Düsseldorf (2006).

[192] M. HOCHBRUCK AND C. LUBICH, *On Krylov subspace approximations to the matrix exponential operator*, SIAM J. Numer. Anal., v 34 n 5 (1997), pp 1911–1925.

[193] A.S. HOUSEHOLDER, *The theory of matrices in numerical analysis*, Blaisdell, New York (1964). Reprinted by Dover, New York (1975).

[194] M. HUTCHINSON, *Algorithm 642: a fast procedure for calculating minimum cross-validation cubic smoothing spline*, ACM Trans. Math. Soft., v 12 (1986), pp 150–153.

[195] M. HUTCHINSON, *A stochastic estimator of the trace of the influence matrix for Laplacian smoothing splines*, Comm. Statist. Simul., v 18 (1989), pp 1059–1076.

[196] M. HUTCHINSON AND F.R. DE HOOG, *Smoothing noisy data with spline functions*, Numer. Math., v 47 (1985), pp 99–106.

[197] I.C.F. IPSEN AND D.J. LEE, *Determinant approximation*, submitted to Numer. Lin. Alg. Appl. (2005).

[198] J. KAUTSKY, *Matrices related to interpolatory quadrature*, Numer. Math., v 36 (1981), pp 309–318.

[199] J. KAUTSKY, *Gauss quadratures: An inverse problem*, SIAM J. Matrix Anal. Appl., v 13 n 1 (1992), pp 402–417.

[200] J. KAUTSKY AND S. ELHAY, *Calculation of the weights of interpolatory quadratures*, Numer. Math., v 40 (1982), pp 407–422.

[201] J. KAUTSKY AND S. ELHAY, *Gauss quadratures and Jacobi matrices for weight functions not of one sign*, Math. Comput., v 43 (1984), pp 543–550.

[202] J. KAUTSKY AND G.H. GOLUB, *On the calculation of Jacobi matrices*, Linear Alg. Appl., v 52/53 (1983), pp 439–455.

[203] M. KILMER AND D.P. O'LEARY, *Choosing regularization parameter in iterative methods for ill-posed problems*, SIAM J. Matrix Anal. Appl., v 22 (2001), pp 1204–1221.

[204] M. KILMER AND G.W. STEWART, *Iterative regularization and MINRES*, SIAM J. Matrix Anal. Appl., v 21 (1999), pp 613–628.

[205] L. KNIZHNERMAN, *Calculation of functions of unsymmetric matrices using Arnoldi's method*, Comput. Maths. Math. Phys., v 31 (1991), pp 1–9.

[206] L. KNIZHNERMAN, *The simple Lanczos procedure: estimates of the error of the Gauss quadrature formula and their applications*, Comput. Maths. Math. Phys., v 36 n 11 (1996), pp 1481–1492.

[207] L. KNIZHNERMAN, *Gauss-Arnoldi quadrature for $\langle (zI - A)^{-1}\varphi, \varphi \rangle$ and rational Padé-type approximation for Markov-type functions*, Sb. Math., v 199 (2008), pp 185–206.

[208] M.G. KREIN, *Fundamental aspects of the representation theory of Hermitian operators with deficiency index (m, m)*, Ukrain. Mat. Z., v 1 (1949), pp 3–66; Amer. Math. Soc. Transl., v 2 n 97 (1970), pp 75–143.

[209] A.S. KRONROD, *Integration with control of accuracy*, Dokl. Akad. Nauk SSSR, v 154 (1964), pp 283–286.

[210] A.S. KRONROD, *Nodes and weights of quadrature formulas: sixteen-place tables*, Consultants Bureau, New York (1965).

[211] J.V. LAMBERS, *Krylov subspace spectral methods for variable-coefficient initial-boundary value problems*, Elec. Trans. Numer. Anal., v 20 (2005), pp 212–234.

[212] J.V. LAMBERS, *Practical implementation of Krylov subspace spectral methods*, J. Sci. Comput., v 32 (2007), pp 449–476.

[213] J.V. LAMBERS, *Derivation of high-order spectral methods for time-dependent PDE using modified moments*, Elec. Trans. Numer. Anal., v 28 (2008), pp 114–135.

[214] J.V. LAMBERS, *Implicitly defined high-order operator splittings for parabolic and hyperbolic variable-coefficient PDE using modified moments*, Int. J. Comput. Science, v 2 (2008), pp 376–401.

[215] J.V. LAMBERS, *Enhancement of Krylov subspace spectral methods by block Lanczos iteration*, Elec. Trans. Numer. Anal., v 31 (2008), pp 86–109.

[216] C. LANCZOS, *An iteration method for the solution of the eigenvalue problem of linear differential and integral operators*, J. Res. Nat. Bur. Standards, v 45 (1950), pp 255–282.

[217] C. LANCZOS, *Solution of systems of linear equations by minimized iterations*, J. Res. Nat. Bur. Standards, v 49 (1952), pp 33–53.

[218] D.P. LAURIE, *Anti-Gaussian quadrature formulas*, Math. Comput., v 65 n 214 (1996), pp 739–747.

[219] D.P. LAURIE, *Calculation of Gauss–Kronrod quadrature rules*, Math. Comput., v 66 (1997), pp 1133–1145.

[220] D.P. LAURIE, *Accurate recovery of recursion coefficients from Gaussian quadrature formulas*, J. Comput. Appl. Math., v 112 (1999), pp 165–180.

[221] D.P. LAURIE, *Questions related to Gaussian quadrature formulas and two-term recursions*, in *Applications and computation of orthogonal polynomials*, W. Gautschi, G.H. Golub and G. Opfer Eds., Birkhauser, Basel (1999), pp 133–144.

[222] D.P. LAURIE, *Computation of Gauss-type quadrature formulas*, J. Comput. Appl. Math., v 127 (2001), pp 201–217.

[223] C.L. LAWSON AND R.J. HANSON, *Solving least squares problems*, SIAM, Philadelphia (1995).

[224] A.S. LEONOV, *On the choice of regularization parameters by means of the quasi–optimality and ratio criteria*, Soviet Math. Dokl., v 19 (1978), pp 537–540.

[225] K.C. LI, *From Stein's unbiased risk estimates to the method of generalized cross-validation*, Annals Statist., v 13 (1985), pp 1352-1377.

[226] REN-CANG LI, *Solving secular equations stably and efficiently*, Report UCB CSD-94-851, University of California, Berkeley (1994).

[227] K-F. LIU, *A noisy Monte Carlo algorithm with fermion determinant*, Chinese J. Phys., v 38 (2000), pp 605–614.

[228] R. LOBATTO, *Lessen over de differentiaal- en integraal-rekening*, De Gebroeders Van Cleef, 's Gravenhage and Amsterdam, (1851).

[229] H. LU AND F.H. MATHIS, *Surface approximation by spline smoothing and generalized cross-validation*, Math. Comput. Simul., v 34 (1992), pp 541–549.

[230] F. MARCELLÁN AND R. ALVAREZ–NODARSE, *On the "Favard theorem" and its extensions*, J. Comput. Appl. Math., v 17 (2001), pp 231–254.

[231] F. MARCELLÁN AND H.O. YAKHLEF, *Recent trends on analytic properties of matrix orthonormal polynomials*, Elec. Trans. Numer. Anal., v 14 (2002), pp 127–141.

[232] A. MELMAN, *A unifying convergence analysis of second-order methods for secular equation*, Math. Comput., v 66 n 217 (1997), pp 333–344.

[233] A. MELMAN, *A numerical comparison of methods for solving secular equations*, J. Comput. Appl. Math., v 86 (1997), pp 237–249.

[234] G. MEURANT, *A review of the inverse of tridiagonal and block tridiagonal matrices*, SIAM J. Matrix Anal. Appl., v 13 n 3 (1992), pp 707–728.

[235] G. MEURANT, *The computation of bounds for the norm of the error in the conjugate gradient algorithm*, Numer. Algo., v 16 (1997), pp 77–87.

[236] G. MEURANT, *Numerical experiments in computing bounds for the norm of the error in the preconditioned conjugate gradient algorithm*, Numer. Algo., v 22 (1999), pp 353–365.

[237] G. MEURANT, *Computer solution of large linear systems*, North-Holland, Amsterdam (1999).

[238] G. MEURANT, *Estimates of the l_2 norm of the error in the conjugate gradient algorithm*, Numer. Algo., v 40 n 2 (2005), pp 157–169.

[239] G. MEURANT, *The Lanczos and conjugate gradient algorithms, from theory to finite precision computations*, SIAM, Philadelphia (2006).

[240] G. MEURANT, *Gene H. Golub 1932-2007*, Numer. Algo., v 51 n 1 (2009), pp 1–4.

[241] G. MEURANT, *Estimates of the trace of the inverse of a symmetric matrix using the modified Chebyshev algorithm*, Numer. Algo., v 51 n 3 (2009), pp 309–318.

[242] G. MEURANT AND Z. STRAKOŠ, *The Lanczos and conjugate gradient algorithms in finite precision arithmetic*, Acta Numerica, v 15 (2006), pp 471–542.

[243] G. MONEGATO, *An overview of the computational aspects of Kronrod quadrature rules*, Numer. Algo., v 26 (2001), pp 173–196.

[244] V.A. MOROZOV, *Methods for solving incorrectly posed problems*, Springer, Berlin (1984).

[245] M. MÜHLICH AND R. MESTER, *The role of total least squares in motion analysis*, in *Proceedings of the european conference on computer vision*, H. Burkhardt Ed., Springer, Berlin (1998), pp 305–321.

[246] J.G. NAGY, K. PALMER AND L. PERRONE, *Iterative methods for image deblurring: A Matlab object oriented approach*, Numer. Algo., v 36 (2004), pp 73–93.

[247] C.M. NEX, *Estimation of integrals with respect to a density of states*, J. Phys. A, v 11 n 4 (1978), pp 653–663.

[248] C.M. NEX, *The block Lanczos algorithm and the calculation of matrix resolvents*, Comput. Phys. Commun., v 53 (1989), pp 141–146.

[249] D.P. O'LEARY, *The block conjugate gradient algorithm and related methods*, Linear Alg. Appl., v 29 (1980), pp 293–322.

[250] D.P. O'LEARY, *Near-optimal parameters for Tikhonov and other regularization methods*, SIAM J. Sci. Comput., v 23 n 4 (2001), pp 1161–1171.

[251] D.P. O'LEARY AND J.A. SIMMONS, *A bidiagonalization-regularization procedure for large scale discretizations of ill-posed problems*, SIAM J. Sci. Stat. Comput., v 2 (1981), pp 474–489.

[252] D.P. O'LEARY, Z. STRAKOŠ AND P. TICHÝ, *On sensitivity of Gauss–Christoffel quadrature*, Numer. Math., v 107 n 1 (2007), pp 147–174.

[253] B. ORTNER, *On the selection of measurement directions in second-rank tensor (e.g. elastic strain) determination of single crystals*, J. Appl. Cryst., v 22 (1989), pp 216–221.

[254] B. ORTNER AND A.R. KRÄUTER, *Lower bounds for the determinant and the trace of a class of Hermitian matrix*, Linear Alg. Appl., v 236 (1996), pp 147–180.

[255] C.C. PAIGE, *The computation of eigenvalues and eigenvectors of very large sparse matrices*, Ph.D. thesis, University of London (1971).

[256] C.C. PAIGE AND M.A. SAUNDERS, *LSQR: An algorithm for sparse linear equations and sparse least squares*, ACM Trans. Math. Soft., v 8 (1982), pp 43–71.

[257] C.C. PAIGE AND M.A. SAUNDERS, *Algorithm 583, LSQR: Sparse linear equations and least squares problems*, ACM Trans. Math. Soft., v 8 (1982), pp 195–209.

[258] C.C. PAIGE AND Z. STRAKOŠ, *Bounds for the least squares residual using scaled total least squares*, in *Proceedings of the third international workshop on TLS and error-in-variables modelling*, S. Van Huffel and P. Lemmerling Eds., Kluwer, Dordrecht (2001), pp 25–34.

[259] C.C. PAIGE AND Z. STRAKOŠ, *Unifying least squares, total least squares and data least squares*, in *Proceedings of the third international workshop on TLS and error-in-variables modelling*, S. Van Huffel and P. Lemmerling Eds., Kluwer, Dordrecht (2001), pp 35–44.

[260] C.C. PAIGE AND Z. STRAKOŠ, *Bounds for the least squares distance using scaled total least squares problems*, Numer. Math., v 91 (2002), pp 93–115.

[261] C.C. PAIGE AND Z. STRAKOŠ, *Scaled total least squares fundamentals*, Numer. Math., v 91 (2002), pp 117–146.

[262] C.C. PAIGE AND Z. STRAKOŠ, *Core problems in linear algebraic systems*, SIAM J. Matrix Anal. Appl., v 27 n 3 (2006), pp 861–874.

[263] C.C. PAIGE, B.N. PARLETT AND H. VAN DER VORST, *Approximate solutions and eigenvalue bounds from Krylov subspaces*, Numer. Linear Alg. Appl., v 2 (1995), pp 115–133.

[264] B.N. PARLETT, *The development and use of methods of LR type*, SIAM Rev., v 6 n 3 (1964), pp 275–295.

[265] B.N. PARLETT, *The new qd algorithms*, Acta Numerica, v 15 (1995), pp 459-491.

[266] B.N. PARLETT, *The symmetric eigenvalue problem*, Prentice-Hall, Englewood Cliffs (1980), second edition SIAM, Philadelphia (1998).

[267] B.N. PARLETT, *For tridiagonals T replace T with LDL^T*, J. Comput. Appl. Math., v 123 (2000), pp 117–130.

[268] B.N. PARLETT AND I.S. DHILLON, *Fernando's solution to Wilkinson's problem: an application of double factorization*, Linear Alg. Appl., v 267 (1997), pp 247–279.

[269] B.N. PARLETT AND O.A. MARQUES, *An implementation of the dqds algorithm (positive case)*, Linear Alg. Appl., v 309 (2000), pp 217–259.

[270] B.N. PARLETT AND D.S. SCOTT, *The Lanczos algorithm with selective orthogonalization*, Math. Comput., v 33 n 145 (1979), pp 217–238.

[271] S. PASZKOWSKI, *Sur des transformations d'une fonction de poids*, in *Polynômes orthogonaux et applications*, C. Brezinski, A. Draux, A.P. Magnus, P. Maroni and A. Ronveaux Eds., LNM v 1171, Springer-Verlag, Berlin (1985), pp 239–246.

[272] T.N.L. PATTERSON, *The optimal addition of points to quadrature formulae*, Math. Comput., v 22 (1968), pp 847–856.

[273] T.N.L. PATTERSON, *Modified optimal quadrature extensions*, Numer. Math., v 64 (1993), pp 511–520.

[274] T.N.L. PATTERSON, *Stratified nested and related quadrature rules*, J. Comput. Appl. Math., v 112 (1999), pp 243–251.

[275] R. PINTELON, P. GUILLAUME, G. VANDERSTEEN AND Y. ROLAIN, *Analyses, development, and applications of TLS algorithms in frequency domain system identification*, SIAM J. Matrix Anal. Appl., v 19 n 4 (1998), pp 983–1004.

[276] R. RADAU, *Etude sur les formules d'approximation qui servent à calculer la valeur numérique d'une intégrale définie*, J. Math. Pures Appl., v 6 (1880), pp 283–336

[277] T. REGIŃSKA, *A regularization parameter in discrete ill-posed problems*, SIAM J. Sci. Comput., v 17 (1996), pp 740–749.

[278] L. REICHEL, *Fast QR decomposition of Vandermonde-like matrices and polynomial least squares approximation*, SIAM J. Matrix Anal. Appl., v 12 n 3 (1991), pp 552–564.

[279] L. REICHEL, *Construction of polynomials that are orthogonal with respect to a discrete bilinear form*, Adv. Comput. Math., v1 (1993), pp 241–258.

[280] L. REICHEL, G. RODRIGUEZ AND S. SEATZU, *Error estimates for large-scale ill-posed problems*, Numer. Algo., v 51 n 3 (2009), pp 341–361.

[281] C. REINSCH, *Smoothing with spline functions*, Numer. Math., v 10 (1967), pp 177–183.

[282] C. REINSCH, *Smoothing with spline functions II*, Numer. Math., v 16 (1971), pp 451–455.

[283] F. RIESZ AND B. SZ-NAGY *Functional analysis*, Dover, New York (1990). The original edition was published in 1952.

[284] J.L. RIGAL AND J. GACHES, *On the compatibility of a given solution with the data of a linear system*, J. ACM, v 14 (1967), pp 543–548.

[285] P.D. ROBINSON AND A. WATHEN, *Variational bounds on the entries of the inverse of a matrix*, IMA J. Numer. Anal., v 12 (1992), pp 463–486.

[286] H. RUTISHAUSER, *Der Quotienten-Differenzen-Algorithmus*, Zeitschrift für Angewandte Mathematik und Physik (ZAMP), v 5 n 3 (1954), pp 233-251.

[287] H. RUTISHAUSER, *Solution of eigenvalue problems with the LR-transformation* , Nat. Bur. Standards Appl. Math. Ser., v 49 (1958), pp 47–81.

[288] H. RUTISHAUSER, *On Jacobi rotation patterns*, in *Proceedings of symposia in applied mathematics, v 15, Experimental arithmetic, high speed computing and mathematics*, American Mathematical Society, Providence RI (1963), pp 219–239.

[289] R.A. SACK AND A. DONOVAN, *An algorithm for Gaussian quadrature given modified moments*, Numer. Math., v 18 n 5 (1972), pp 465–478.

[290] H.E. SALZER, *A recurrence scheme for converting from one orthogonal expansion into another*, Commun. ACM, v 16 n 11 (1973), pp 705–707.

[291] B. SAPOVAL, T. GOBRON AND A. MARGOLINA, *Vibrations of fractal drums*, Phys. Rev. Lett., v 67 (1991), pp 2974–2977.

[292] M.A. SAUNDERS, H.D. SIMON AND E.L. YIP, *Two conjugate-gradient-type methods for unsymmetric linear equations*, SIAM J. Numer. Anal., v 25 no 4 (1988), pp 927–940.

[293] P.E. SAYLOR AND D.C. SMOLARSKI, *Why Gaussian quadrature in the complex plane?*, Numer. Algo., v 26 (2001), pp 251–280.

[294] P.E. SAYLOR AND D.C. SMOLARSKI, *Addendum to: Why Gaussian quadrature in the complex plane?*, Numer. Algo., v 27 (2001), pp 215–217.

[295] L.L. SCHUMAKER AND F.I. UTRERAS, *On generalized cross-validation for tensor smoothing splines*, SIAM J. Sci. Stat. Comput., v 11 (1990), pp 713–731.

[296] J.C. SEXTON AND D.H. WEINGARTEN, *Systematic expansion for full QCD based on the valence approximation*, Report IBM T.J. Watson Research Center (1994).

[297] B. SHAHRARAY AND D.J. ANDERSON, *Optimal estimation of contour properties by cross-validated regularization*, IEEE Trans. Pattern Anal. Machine Intell., v 11 (1989), pp 600–610.

[298] J. SHOHAT, *The relation of classical orthogonal polynomials to the polynomials of Appell*, Amer. J. Math., v 58 (1936), pp 453–464.

[299] J.A. SHOHAT AND J.D. TAMARKIN, *The problem of moments*, Mathematical surveys, v 1, American Mathematical Society, Providence RI (1943).

[300] R.B. SIDJE, K. BURRAGE AND B. PHILIPPE, *An augmented Lanczos algorithm for the efficient computation of a dot-product of a function of a large sparse symmetric matrix*, in *Computational science - ICCS 2003* Lecture notes in computer science, Springer, Berlin (2003), pp 693–704.

[301] D.M. SIMA AND S. VAN HUFFEL, *Using core formulations for ill-posed linear systems*, in *Proceedings of the 76th GAMM Annual Meeting, Luxembourg*, special issue of PAMM, v 5 n 1 (2005), pp 795-796.

[302] D.M. SIMA, S. VAN HUFFEL AND G.H. GOLUB, *Regularized total least squares based on quadratic eigenvalue problem solvers*, BIT Numer. Math., v 44 (2004), pp 793–812.

[303] H.D. SIMON, *The Lanczos algorithm with partial reorthogonalization*, Math. Comput., v 42 n 165 (1984), pp 115–142.

[304] A. SINAP, *Gaussian quadrature for matrix valued functions on the real line*, J. Comput. Appl. Math., v 65 (1995), pp 369–385.

[305] A. SINAP AND W. VAN ASSCHE, *Polynomial interpolation and Gaussian quadrature for matrix-valued functions*, Linear Alg. Appl., v 207 (1994), pp 71–114.

[306] A. SINAP AND W. VAN ASSCHE, *Orthogonal matrix polynomials and applications*, J. Comput. Appl. Math., v 66 (1996), pp 27–52.

[307] M.R. SKRZIPEK, *Orthogonal polynomials for modified weight functions*, J. Comput. Appl. Math., v 41 (1992), pp 331–346.

[308] R. SKUTSCH, *Ueber formelpaare der mechanischen quadratur*, Arch. Math. Phys., v 13 n 2 (1894), pp 78–83.

[309] M.M. SPALEVIĆ, *On generalized averaged Gaussian formulas*, Math. Comput., v 76 n 259 (2007), pp 1483–1492.

[310] G.W. STEWART, *Matrix algorithms, volume I: basic decompositions*, SIAM, Philadelphia (1998).

[311] E. STIEFEL, *Kernel polynomials in linear algebra and their numerical applications*, Nat. Bur. Standards Appl. Math. Ser., v 49 (1958), pp 1–22.

[312] T.J. STIELTJES, *Quelques recherches sur la théorie des quadratures dites mécaniques*, Ann. Sci. Ecole Norm. Sup. Paris, Sér. 3 tome 1 (1884), pp 409–426. Also in Oeuvres complètes, v 1, Noordhoff, Groningen (1914), pp 377–396.

[313] T.J. STIELTJES, *Recherches sur les fractions continues*, Ann. Fac. Sci. Toulouse, v 8, (1894), J1–122, v 9 (1895), A1–10. Also in Oeuvres complètes, v 2, Noordhoff, Groningen (1918), pp 402–566.

[314] J. STOER AND R. BULIRSCH, *Introduction to numerical analysis*, second edition, Springer-Verlag, Berlin (1983).

[315] Z. STRAKOŠ, *On the real convergence rate of the conjugate gradient method*, Linear Alg. Appl., v 154–156 (1991), pp 535–549.

[316] Z. STRAKOŠ, *Model reduction using the Vorobyev moment problem*, Numer. Algo., v 51 n 3 (2008), pp 363–379.

[317] Z. STRAKOŠ AND P. TICHÝ, *On error estimates in the conjugate gradient method and why it works in finite precision computations*, Elec. Trans. Numer. Anal., v 13 (2002), pp 56–80.

[318] Z. STRAKOŠ AND P. TICHÝ, *Error estimation in preconditioned conjugate gradients*, BIT Numer. Math., v 45 (2005), pp 789–817.

[319] Z. STRAKOŠ AND P. TICHÝ, *Estimation of* $c^* A^{-1} b$ *via matching moments*, submitted to SIAM J. Sci. Comput., (2008).

[320] A.H. STROUD AND D. SECREST, *Gaussian quadrature formulas*, Prentice-Hall, Englewoods Cliffs, NJ (1966).

[321] G.W. STRUBLE, *Orthogonal polynomials: variable-signed weight functions*, Numer. Math., v 5 (1963), pp 88–94.

[322] ZHENG SU, *Computational methods for least squares problems and clinical trials*, Ph.D. thesis, Stanford University (2005).

[323] G. SZEGÖ, *Orthogonal polynomials*, Third Edition, American Mathematical Society, Providence, RI (1974).

[324] A.N. TIKHONOV ED., *Ill-posed problems in natural sciences, Proceedings of the international conference Moscow 1991*, TVP Science publishers, Moscow (1992).

[325] A.N. TIKHONOV AND V.Y. ARSENIN, *Solutions of ill-posed problems*, Wiley, New York (1977).

[326] A.M. THOMPSON, J.C. BROWN, J.W. KAY AND D.M. TITTERINGTON, *A study of methods of choosing the smoothing parameter in image restoration by regularization*, IEEE Trans. Pattern Anal. Machine Intell., v 13 (1991), pp 326–339.

[327] L.N. TREFETHEN, *Is Gauss quadrature better than Clenshaw–Curtis?*, SIAM Rev., v 50 n 1, (2008).

[328] E. TYRTYSHNIKOV, *How bad are Hankel matrices?*, Numer. Math., v 67 (1994), pp 261–269.

[329] F.I. UTRERAS, *Sur le choix du paramètre d'ajustement dans le lissage par fonctions spline*, Numer. Math., v 34 (1980), pp 15–28.

[330] F.I. UTRERAS, *Optimal smoothing of noisy data using spline functions*, SIAM J. Sci. Stat. Comput., v 2 (1981), pp 349–362.

[331] V.B. UVAROV, *Relation between polynomials orthogonal with different weights*, Dokl. Akad. Nauk SSSR, v 126 (1959), pp 33–36. (in Russian),

[332] V.B. UVAROV, *The connection between systems of polynomials that are orthogonal with respect to different distribution functions*, USSR Comput. Math. Phys., v 9 (1969), pp 25–36.

[333] S. VAN HUFFEL AND J. VANDEWALLE, *The total least squares problem: computational aspects and analysis*, SIAM, Philadelphia (1991).

[334] J.M. VARAH, *Pitfalls in the numerical solution of linear ill-posed problems*, SIAM J. Sci. Stat. Comput., v 4 n 2 (1983), pp 164–176.

[335] R. Vio, J. Nagy, L. Tenorio and W. Wamsteker, *Multiple image deblurring with spatially variant PSFs*, Astron. Astrophys., v 434 (2005), pp 795-800.

[336] C.R. Vogel, *Non-convergence of the L-curve regularization parameter selection method*, Inverse Problems, v 12 (1996), pp 535–547.

[337] U. von Matt, *The orthogonal qd-algorithm*, SIAM J. Sci. Comput., v 18, n 4 (1997), pp 1163–1186.

[338] Yu.V. Vorobyev, *Methods of moments in applied mathematics*, Gordon and Breach Science Publishers, New York (1965).

[339] G. Wahba, *Spline bases, regularization and generalized cross-validation for solving approximation problems with large quantities of noisy data*, in *Approximation theory III*, E.W. Cheney Ed., Academic Press, New York (1980), pp 905–912.

[340] G. Wahba, *Spline models for observational data*, CBMS-NSF regional conference series in applied mathematics, v 59, SIAM, Philadelphia (1990).

[341] G. Wahba and S. Wold, *A completely automatic french curve*, Commun. Statist., v 4 (1975), pp 1-17.

[342] G. Wahba, D.R. Johnson, F. Gao and J. Gong, *Adaptive tuning of numerical weather prediction models; randomized GCV in three and four-dimensional data assimilation*, Monthly Weather Review, v 123 (1995), pp 3358–3369.

[343] B. Waldén, R. Karlson and J.G. Sun, *Optimal backward perturbation bounds for the linear least squares problem*, Numer. Lin. Alg. Appl., v 2 n 3 (1995), pp 271–286.

[344] K.F. Warnick, *Continued fraction error bound for conjugate gradient method*, ECEN Department report TR-L100-98.3, Brigham Young University (1997).

[345] K.F. Warnick, *Gaussian quadrature and scattering amplitude*, Report Department of electrical and computer engineering, University of Illinois (2000).

[346] K.F. Warnick, *Nonincreasing error bound for the biconjugate gradient method*, Report Department of electrical and computer engineering, University of Illinois (2000).

[347] J.A.C. Weideman and L.N. Trefethen, *The kink phenomenon in Fejér and Clenshaw–Curtis quadrature*, Numer. Math., v 107 (2007), pp 707–727.

[348] J.C. WHEELER, *Modified moments and Gaussian quadrature*, in *Proceedings of the international conference on Padé approximants, continued fractions and related topics*, Boulder, CO, Rocky Mtn. J. Math., v 4 n 2 (1974), pp 287–296.

[349] H.S. WILF, *Mathematics for the physical sciences*, Wiley, New York (1962).

[350] A. WILLIAMS AND K. BURRAGE, *Surface fitting using GCV smoothing splines on supercomputers*, in *Conference on high performance networking and computing, Proceedings of the 1995 ACM/IEEE conference on supercomputing*, ACM, New York (1995).

[351] S.Y. WU, J.A. COCKS AND C.S. JAYANTHI, *An accelerated inversion algorithm using the resolvent matrix method*, Comput. Phys. Commun., v 71 (1992), pp 15–133.

[352] M. XIA, E. SABER, G. SHARMA AND A. MURAT TEKALP, *End-to-end color printer calibration by total least squares regression*, IEEE Trans. Image Process., v 8 n 5 (1999), pp 700–716.

[353] SHU-FANG XU, *On the Jacobi matrix inverse eigenvalue problem with mixed given data*, SIAM J. Matrix Anal. Appl., v 17 n 3 (1996), pp 632–639.

Index